Green Stormwater Infrastructure for Sustainable Urban and Rural Development

Green Stormwater Infrastructure for Sustainable Urban and Rural Development

Editors

Luis A. Sañudo-Fontaneda
William F. Hunt

MDPI • Basel • Beijing • Wuhan • Barcelona • Belgrade • Manchester • Tokyo • Cluj • Tianjin

Editors
Luis A. Sañudo-Fontaneda
University of Oviedo
Spain

William F. Hunt
North Carolina State University
USA

Editorial Office
MDPI
St. Alban-Anlage 66
4052 Basel, Switzerland

This is a reprint of articles from the Special Issue published online in the open access journal *Sustainability* (ISSN 2071-1050) (available at: https://www.mdpi.com/journal/sustainability/special_issues/Green_Stormwater).

For citation purposes, cite each article independently as indicated on the article page online and as indicated below:

LastName, A.A.; LastName, B.B.; LastName, C.C. Article Title. *Journal Name* **Year**, *Volume Number*, Page Range.

ISBN 978-3-0365-0610-4 (Hbk)
ISBN 978-3-0365-0611-1 (PDF)

© 2021 by the authors. Articles in this book are Open Access and distributed under the Creative Commons Attribution (CC BY) license, which allows users to download, copy and build upon published articles, as long as the author and publisher are properly credited, which ensures maximum dissemination and a wider impact of our publications.

The book as a whole is distributed by MDPI under the terms and conditions of the Creative Commons license CC BY-NC-ND.

Contents

About the Editors vii

Preface to "Green Stormwater Infrastructure for Sustainable Urban and Rural Development" ix

Andrew J. Erickson, Vinicius J. Taguchi and John S. Gulliver
The Challenge of Maintaining Stormwater Control Measures: A Synthesis of Recent Research and Practitioner Experience
Reprinted from: *Sustainability* **2018**, *10*, 3666, doi:10.3390/su10103666 1

Craig Lashford, Matteo Rubinato, Yanpeng Cai, Jingming Hou, Soroush Abolfathi, Stephen Coupe, Susanne Charlesworth and Simon Tait
SuDS & Sponge Cities: A Comparative Analysis of the Implementation of Pluvial Flood Management in the UK and China
Reprinted from: *Sustainability* **2019**, *11*, 213, doi:10.3390/su11010213 17

Elisa Lähde, Ambika Khadka, Outi Tahvonen and Teemu Kokkonen
Can We Really Have It All?—Designing Multifunctionality with Sustainable Urban Drainage System Elements
Reprinted from: *Sustainability* **2019**, *11*, 1854, doi:10.3390/su11071854 31

Sara Lucía Jiménez Ariza, José Alejandro Martínez, Andrés Felipe Muñoz, Juan Pablo Quijano, Juan Pablo Rodríguez, Luis Alejandro Camacho and Mario Díaz-Granados
A Multicriteria Planning Framework to Locate and Select Sustainable Urban Drainage Systems (SUDS) in Consolidated Urban Areas
Reprinted from: *Sustainability* .**2019**, *11*, 2312, doi:10.3390/su11082312 51

Junyu Zhang, Dafang Fu, Christian Urich and Rajendra Prasad Singh
Accelerated Exploration for Long-Term Urban Water Infrastructure Planning through Machine Learning
Reprinted from: *Sustainability* **2018**, *10*, 4600, doi:10.3390/su10124600 85

Outi Tahvonen
Scalable Green Infrastructure—The Case of Domestic Private Gardens in Vuores, Finland
Reprinted from: *Sustainability* **2018**, *10*, 4571, doi:10.3390/su10124571 101

Xiaofan Xu, Dylan S. P. Schreiber, Qing Lu and Qiong Zhang
A GIS-Based Framework Creating Green Stormwater Infrastructure Inventory Relevant to Surface Transportation Planning
Reprinted from: *Sustainability* **2018**, *10*, 4710, doi:10.3390/su10124710 117

Ignacio Andrés-Doménech, Sara Perales-Momparler, Adrián Morales-Torres and Ignacio Escuder-Bueno
Hydrological Performance of Green Roofs at Building and City Scales under Mediterranean Conditions
Reprinted from: *Sustainability* **2018**, *10*, 3105, doi:10.3390/su10093105 131

Bailee N. Young, Jon M. Hathaway, Whitney A. Lisenbee and Qiang He
Assessing the Runoff Reduction Potential of Highway Swales and WinSLAMM as a Predictive Tool
Reprinted from: *Sustainability* **2018**, *10*, 2871, doi:10.3390/su10082871 147

Jeffrey P. Johnson and William F. Hunt
A Retrospective Comparison of Water Quality Treatment in a Bioretention Cell 16 Years Following Initial Analysis
Reprinted from: *Sustainability* **2019**, *11*, 1945, doi:10.3390/su11071945 . . . . **159**

Rajendra Prasad Singh, Fei Zhao, Qian Ji, Jothivel Saravanan and Dafang Fu
Design and Performance Characterization of Roadside Bioretention Systems
Reprinted from: *Sustainability* **2019**, *11*, 2040, doi:10.3390/su11072040 . . . . **171**

Muhammad Shafique, Reeho Kim and Kwon Kyung-Ho
Rainfall Runoff Mitigation by Retrofitted Permeable Pavement in an Urban Area
Reprinted from: *Sustainability* **2018**, *10*, 1231, doi:10.3390/su10041231 . . . . **185**

Carlos Rey-Mahía, Luis A. Sañudo-Fontaneda, Valerio C. Andrés-Valeri, Felipe Pedro Álvarez-Rabanal, Stephen John Coupe and Jorge Roces-García
Evaluating the Thermal Performance of Wet Swales Housing Ground Source Heat Pump Elements through Laboratory Modelling
Reprinted from: *Sustainability* **2019**, *11*, 3118, doi:10.3390/su11113118 . . . . **195**

About the Editors

Luis A. Sañudo-Fontaneda is an Assistant Professor at the Department of Construction and Manufacturing Engineering, University of Oviedo, Spain, where he teaches urban and territorial planning, highway and bridge engineering and green stormwater infrastructure. He served as the first Academic Director of the M.Eng. studies in Civil Engineering at the University of Oviedo for the 2016/2017–2019/2020 academic period. He has been the Academic Director of the Extension Course in Sustainable Drainage Systems (SuDS) and green infrastructure in the University of Oviedo since 2017, having participated in extension courses in this topic in Chile, Colombia, Spain, the UK and the USA. Luis holds a Ph.D., an M.S. and an M.Eng. from the University of Cantabria, Spain, with a special focus in Research in Civil Engineering and SuDS, where he worked from 2007 to 2014 as a researcher, achieving the FPI Grant. Currently, he is the Lead Researcher in the Civil, Environmental and Geomatic Engineering Area in the Institute of Natural Resources and Territorial Planning (INDUROT) and a Researcher in the GICONSIME Research Group of the Construction Engineering Area, both at the University of Oviedo. He is also an Honorary Research Fellow at the Centre for Agroecology, Water and Resilience (CAWR), Coventry University, UK, where he previously worked as Research Fellow between 2013 and 2017. Luis served as a Member of the Engineering and Physical Sciences Panel for the Newton Fund International Programme of the British Council, UK, and as an International Expert for research project proposals evaluations in the CONICYT, Chile. He has been a Guest Editor in Sustainability and *Clean Technologies* (MDPI) and a Member of the Editorial Board in the latter. He has coauthored over 90 journal and conference papers, as well as books and book chapters and technical guides related to green stormwater infrastructure. Luis has participated in over 40 research projects in the topic of this Special Issue.

William F. Hunt III, Ph.D., P.E., is a William Neal Reynolds Professor of Biological and Agricultural Engineering and Extension Specialist at North Carolina State University (N.C. State). As the leader of the Stormwater Engineering Group at N.C. State, his team has designed and/or monitored more than 200 different stormwater practices, published 130 peer-refereed journals and authored 2 books. Professor Hunt served as a Visiting Scholar at the University of Auckland (New Zealand) in 2010–2011, a CUGE Research Fellow with Singapore National Parks in 2012–2013, and as a Royal Society of Engineers Fellow at Coventry University in the United Kingdom in 2018–2019. Among his active research, teaching and outreach areas are assessing a wide range of ecosystem services provided by stormwater measures, creating simple-to-use computer models to design stormwater devices, recording impacts of maintenance on stormwater system performance and pursuing market-based improvements in stormwater design. Professor Hunt teaches 20–25 workshops, field tours and webinars per year through N.C. State University and has been the chair or cochair of three major conferences in the United States. He received degrees in Civil Engineering, Economics and Biological and Agricultural Engineering at N.C. State. His Ph.D. in Agricultural and Biological Engineering was earned from Pennsylvania State University in 2003. Dr. Hunt's greatest joys in life are people, traveling and (most especially) his five entertaining children.

Preface to "Green Stormwater Infrastructure for Sustainable Urban and Rural Development"

Climate change and noncontrolled urbanization schemes are two of the major threats to the stability and resilience of human settlements, raising the risks for flooding and droughts across the world. As a consequence, urban streams receive flashier flow rates and masses of pollution, which contribute to the degradation of the natural water cycle. Infrastructure is key to the adaptation of both urban and rural communities to unpredictable change and varying climate scenarios. Green stormwater infrastructure (GSI) has been highlighted as one of the main lines of work for the years to come by the UN's 2030 Sustainable Development Goals, as it reduces the effects of flooding and droughts, diffuse pollution and unchecked urbanization. GSI also delivers ecosystem services such as carbon sequestration, heat island mitigation, air quality protection, increasing biodiversity and enhancing the liveability of communities.

This book presents 13 chapters featuring some of the main lines of research around GSI, emphasizing international experiences across several continents. These chapters represent a collection of articles published in the Special Issue entitled "Green Stormwater Infrastructure for Sustainable Urban and Rural Development" published by Sustainability (MDPI) in 2019, a journal indexed in the JCR 2019 Environmental Sciences and Green & Sustainable Science & Technology categories. The Editors of this book would like to acknowledge the excellent guidance and efforts from the Editorial Team at MDPI as well as the quality of the experience and research presented by the 53 authors who have contributed towards the academic and technical success of the abovementioned Special Issue.

This book covers a wide range of GSI practices, highlighting literature reviews of the latest lines of research published internationally, allowing the reader to identify the main knowledge gaps in this field. Moreover, the book presents the main philosophies underpinning the implementation of GSI, such as water-sensitive urban design and sponge cities, the pillars of design for multifunctionality purposes, a multicriteria framework for sustainable drainage systems in urban environments and a GIS-based application for the efficient location of GSI, amongst others. The reader will find studies applied under different climate conditions across the world and different environments, such as consolidated urban areas, periurban zones and rural environments. Finally, transportation infrastructure is also presented as the main link between communities and how GSI could be implemented in it. This book also explores the hydrological and water quality performance of GSI, assessment tools to be used both at the design and operation stages, retrospective analyses, maintenance and operation guidance and an initial laboratory experience of the water–energy nexus using GSI.

We hope that this collection of papers may be of use to academics and practitioners in helping provide resilience to our communities and infrastructure against climate change.

Luis A. Sañudo-Fontaneda, William F. Hunt
Editors

Review

The Challenge of Maintaining Stormwater Control Measures: A Synthesis of Recent Research and Practitioner Experience

Andrew J. Erickson [1,*], Vinicius J. Taguchi [1,2] and John S. Gulliver [1,2]

[1] St. Anthony Falls Laboratory, University of Minnesota, Minneapolis, MN 55414, USA; taguc006@umn.edu (V.J.T.); gulli003@umn.edu (J.S.G.)
[2] Department of Civil, Environmental and Geo-Engineering, University of Minnesota, Minneapolis, MN 55455, USA
* Correspondence: eric0706@umn.edu; Tel.: +1-612-624-4629

Received: 11 August 2018; Accepted: 8 October 2018; Published: 13 October 2018

Abstract: The methods for properly executing inspection and maintenance of stormwater control measures are often ambiguous and inconsistently applied. This paper presents specific guidelines for inspecting and maintaining stormwater practices involving media filtration, infiltration, ponds, and permeable pavements because these tend to be widely implemented and often unsatisfactorily maintained. Guidelines and examples are based on recent scientific research and practitioner experience. Of special note are new assessment and maintenance methods, such as testing enhanced filtration media that targets dissolved constituents, maintaining proper vegetation coverage in infiltration practices, assessing phosphorus release from pond sediments, and the development of compressed impermeable regions in permeable pavements and their implications for runoff. Inspection and maintenance examples provided in this paper are drawn from practical examples in Northern Midwest USA, but most of the maintenance recommendations do not depend on regional characteristics, and guidance from around the world has been reviewed and cited herein.

Keywords: maintenance; stormwater; treatment; assessment; stormwater control measure; sustainable drainage system; best management practice; green infrastructure; filtration; infiltration; retention pond; permeable pavement

1. Introduction

As urbanized areas around the world wrestle with growing pains and shifting ideologies on urban planning, stormwater control measures (SCMs) and green infrastructure are becoming increasingly popular for managing urban hydrology and stormwater. However effective newly-constructed SCMs and newly-installed proprietary devices may be, none can be expected to continue functioning effectively without regular and well-informed maintenance and inspections [1]. These efforts are best conducted by individuals experienced in stormwater management, which requires designating and training a dedicated stormwater work crew or contracting a stormwater engineer for consultations [2]. Even if the need is not immediately obvious (primarily because it is underground or under water), maintenance may still be required and can be identified with timely and thorough inspections. The frequency with which maintenance is needed can only be informed by periodic inspections but should occur at least once per year [1]. Additionally, properly budgeting and assigning responsibility for these activities is paramount for them to occur [3–5], especially considering that the total cost of maintenance for SCMs typically approximates the original construction cost over its designed lifetime [6]. Additional maintenance may be necessary to sustain site-specific performance criteria, such as managing erosion due to landslide concerns, managing vegetation due to wildfire concerns, or managing water quality to protect sensitive fisheries [5].

A regular inspection begins with visual observations and ends with detailed documentation. Excessive sedimentation, bank destabilization and erosion, invasive vegetation, or problematic wildlife could all lead to costly maintenance if left unresolved [1]. Any evidence of illicit discharges should be carefully noted, and other problems beyond the normal loading conditions of the watershed should be documented [3]. Steps should be taken to raise public awareness of stormwater infrastructure and its connections to water bodies rather than to sewage treatment facilities, a common misconception [7,8]. While previous work has set base guidelines for the maintenance of common stormwater control measures, the purpose of this paper is to address new and emerging challenges faced by stormwater professionals. Thus, scientific research is combined with practitioner experience to develop guidelines for the proper maintenance of high-priority SCMs, including media filtration practices, infiltration practices, stormwater wet ponds, and permeable pavements. These four SCM types were selected because they are in widespread use and are often inadequately maintained. While the observations are drawn from practical examples in Northern Midwest USA, most of the maintenance recommendations do not depend on regional characteristics and guidance from around the world has been reviewed and cited herein whenever possible. This information is intended to serve as a supplement to currently-available assessment and maintenance manuals (e.g., [1]) that have been developed globally, including the Pacific Northwest USA [7,9–11], New England USA [8,12–16], Mid-Atlantic USA [17–20], South Central USA [21–23], Southwest USA [5,24–27], Canada [28,29], New Zealand [30], the United Kingdom [31], Australia [32–35], Malaysia [36], Singapore [37,38], and South Korea [39], among others.

2. Media Filtration

Media filtration is the process by which particles suspended in stormwater are removed while water is passing through granular media [1]. The design of media filtration for stormwater treatment is simple and well-defined [40], and the maintenance of these filters has been similarly studied and documented [1]. The greatest need of stormwater professionals maintaining media filtration practices arises from the development of new additives, which are added to filtration media to capture soluble reactive pollutants, such as phosphate [41,42], nitrate [43], metals [44,45], bacteria [46], and others [47]. Primarily, stormwater professionals are unsure of how to determine when to maintain the additives within media filtration practices because there is a lack of visual indicators of when additives are no longer functional. To overcome these challenges, stormwater professionals must adopt more advanced assessment methods and rigorous documentation.

Visual inspection is a simple assessment method that can be used to identify poor performance within an SCM, assess the cause of the poor performance, and determine the necessary maintenance to restore the practice to proper functionality [1]. An example for filtration is slow drainage (poor performance indicator), which is often caused by clogged media as a result of accumulation of stormwater sediment or erosion from misaligned inlet and outlet structures and/or around the exterior of the filtration practice (cause of poor performance). Corrective maintenance involves removing the accumulated sediment and restoring the hydraulic conductivity of the media surface. Visual inspection, however, typically cannot identify poorly functioning additives because the additives are commonly mixed into or installed within the media itself, and thus are not visible from the surface of the filtration practice. To properly assess the performance of media filtration additives, more intense assessment, such as capacity or synthetic runoff testing [1] or monitoring, may be necessary. These methods can be expensive, and thus cost-prohibitive, to deploy throughout a municipality or other jurisdiction with tens, hundreds, or even thousands of SCMs. Annual maintenance cost estimates for media filtration practices range from 1–10% of the original construction cost [1]. However, these assessment methods can be modified to simplify the process, reduce costs, and specifically assess media filtration practices with additives.

Capacity testing is an assessment method that measures the capacity of an SCM to perform its intended function [1]. Typically employed to measure sedimentation or infiltration, capacity testing can be modified to incorporate a batch jar test and directly measure the capacity of media filtration additives to capture their target pollutants. Measuring the sorption capacity of media filtration additives will provide a snapshot of the remaining capacity, which can be used to estimate when additives need to be replaced. Optionally, this procedure can be performed prior to installation of the filtration media and additive(s) to determine a 'baseline' by which subsequent tests can be compared to determine the rate of degradation. A sample protocol of such batch tests is described in the following steps:

Step 1 Collect a representative sample of the filtration media, including additive(s), with a known volume and mass. It is important to know the mass of media, including the mass of sand (if applicable) and each additive individually, to determine the ratio of these masses to the pollutant(s) captured in subsequent steps. In addition, the (bulk) volume of the sample can be used to expand the results to the full-scale media filtration practice.

Step 2 Place the filtration media in a container of clean water with a known concentration of pollutant(s) that the additive is intended to capture. The mass ratio of water to additive should be approximately 100:1, and the mass ratio of pollutant(s) to additives should be approximately equal to the capacity of the additives to capture that pollutant. For example, a 10 g sample is collected of a mixed filtration media comprising sand (8 g) and a commercial adsorbent media (2 g). The capacity of the commercial additive to capture arsenic (As) is reported to be 12 mg As per kg sorbent. Thus, 2 g of additive within the sample can be expected to capture 24 μg of As. Using a mass ratio of water to additive of 100:1, the mass of water should be 200 g, which is approximately 0.2 L. The mass of As (24 μg) in this volume of water yields an As concentration of 120 μg/L.

Step 3 Thoroughly mix the additive in the water for at least a length of time equal to the contact time between the additive and the pollutant in the full-scale SCM, or up to 24 hours. Selecting a shorter mixing time will often result in less pollutant(s) capture and thus a more conservative measure of remaining sorption capacity.

Step 4 Collect samples from the water and measure pollutant concentration. This should be performed at the beginning of the test to verify the initial pollutant concentration, and at the end of Step 3 to confirm performance. This step can be performed throughout the duration of Step 3 to measure the change in concentration as a function of time, which can be used to estimate the relative rate of removal. Pollutant concentration can be measured following Standard Methods [48], other approved laboratory methods, using analytical laboratory services, or by chemical analysis kits that can be purchased online.

Step 5 Determine the pollutant capture ratio as the ratio of captured pollutant mass to additive mass. For example, if the mass of As in solution is reduced by 10 μg, then the remaining capacity of the additive to capture As is 14 μg As per 2 g of additive, or 7 mg per kg. Thus, the capacity has been reduced from 12 mg per kg to 7 mg per kg.

Table 1 provides a list of base guidelines for the maintenance of media filtration practices, including those with additives for enhanced performance.

Table 1. Maintenance recommendations for media filtration practices [1,13,17,21,22,30,32].

Task	Frequency	Notes
Inspection	Annually or after every two-year storm	
Remove trash and debris	Annually	Increase frequency, if needed
Remove obstructions to outlet structures and underdrain systems	As needed	Cleanouts can simplify obstruction removal from underdrain systems and should be included in all filtration designs
Remove vegetation from filter surface, if applicable	Once per year	Increase frequency, if needed

Table 1. *Cont.*

Task	Frequency	Notes
Perform testing to determine filtration rates	Whenever visual inspection identifies the need	
Remove retained sediment, typically the top 5–20 cm of discolored surface media	Variable (once every five to ten years is typical in stable watersheds)	In unstable watersheds (i.e., those with active construction), the frequency is typically once per year
Effluent sampling and analysis of enhanced media	Annually, or when amendment performance is in question as needed	
Capacity testing for pollutant capture by additives	As needed, when effluent samples suggest reduced pollutant capture capacity	

3. Infiltration Practices

Infiltration practices capture stormwater runoff and allow it to flow into the ground rather than into a collection system [1]. Infiltration practices vary in design and appearance and include practices such as infiltration basins, trenches, and rain gardens (bioretention, bioinfiltration), among others. Visual inspection will identify poor performance in a manner similar to filtration practices. In addition, capacity testing of the infiltration rate is often conducted on infiltration practices through measurement with field infiltrometers. It has been found that the infiltration rate (as indicated by saturated hydraulic conductivity) will vary substantially over most infiltration practices, even with engineered soil [49,50]. A representative infiltration rate for the whole practice can be determined with the appropriate mean value of hydraulic conductivity [51].

Many of these infiltration practices rely on vegetation to support infiltration through the soil surface [52], evapotranspiration, pollutant capture [47,53,54], and microbial breakdown of captured pollutants [55]. Thus, managing proper vegetation is one of the greatest challenges for stormwater professionals. The aspects of managing proper vegetation in infiltration practices include maintaining proper coverage and species and also ground cover management because it affects the health and diversity of vegetation. Proper vegetation coverage is important because a lack of vegetation results in open and exposed soils, which are susceptible to erosion and weed germination. In addition, fine sediment removed from the stormwater runoff often clogs the soil surface of an infiltration basin. Healthy vegetation in SCMs can create macropores by which stormwater can pass through a clogged soil surface [52]. Thus, a lack of proper vegetation coverage can reduce infiltration, which subsequently increases the amount of time that water is stored within an infiltration basin. This periodic inundation can further impact vegetation, beginning a cycle of reduced vegetation coverage, reduced infiltration, and increased ponding time until the infiltration practice completely fails.

Vegetation coverage can also be over-abundant, which potentially limits access for inspection and corresponding maintenance. The most common cause of over-abundant vegetation is a lack of vegetation management, often resulting in undesirable vegetation species (e.g., invasive weeds) that can quickly outcompete and dominate native or selected vegetation species. In fact, a major challenge in managing infiltration practices is maintaining the proper vegetation diversity. Native plants are typically better-suited to their environment and will require less fertilizer to become established [19]. Working with local partners can facilitate the selection of appropriate species [33]. Infiltration practices are often designed with between one and ten different vegetation species, ranging from native prairie grasses and sedges to wildflowers and pollinator-supporting plant species in the upper Midwest USA [56], to forbs, rushes, and trees in Australia [57,58], and to succulents and forbs in drier climates [25,26]. Maintaining a plant species palette requires knowledge in plant species identification to ensure that non-design species are removed and design species are healthy and present. Some plants may also require specialized care such as limited pruning to minimize stress and maximize health [23]. In appropriate regions, desert vegetation may require little maintenance [26]. Vegetation management for aesthetics will depend on site characteristics. In some cases, a more natural appearance can be desirable, while a manicured landscape is preferable in others [19]. In applications where longer vegetation is desired, it can be cut back just enough to show that it is being maintained [39]; the appearance of maintenance is important to discourage littering and vandalism. Site-specific safety considerations regarding overgrown vegetation should also be considered (e.g., blocking vehicle lines-of-sight or allowing individuals to hide) [19].

Proper ground cover (e.g., mulch) management can also limit an over-abundance of design vegetation and invasion of undesirable species. Ground cover includes mulch, landscaping stone, rock, and recycled materials, such as shredded tires. These materials provide aesthetic benefits, but when properly selected, designed, and maintained, can also limit erosion, weed germination, and vegetation overabundance. Proper inspection frequency and effectiveness can identify issues related to poor vegetation cover, poor species diversity, and improper ground cover management. Annual maintenance cost estimates for infiltration practices range from 3–5% of the original construction cost [1]. Table 2 provides a list of base guidelines for the maintenance of infiltration practices.

Table 2. Maintenance recommendations for infiltration practices [1,12,17,21,30,31].

Task	Frequency
Remove sediment and oil/grease from pretreatment devices and overflow structures	As Needed
Mow and remove litter and debris	As Needed
Stabilize eroded banks, repair undercut and eroded areas at inflow and outflow structure	As Needed
Inspect pretreatment devices and diversion structures for signs of sediment buildup and structural damage	Semi-Annual Inspection
If dead or dying grass is evident at the bottom or the basin/trench, check to ensure water infiltrates within two days following significant rain events	Semi-Annual Inspection
Disc or otherwise aerate bottom	As Needed
De-thatch basin bottom	Annually
Provide an extended dry period, if bypass capability is available, to regain or increase the infiltration rate in the short term	Five-year Maintenance

4. Ponds

Despite being one of the most abundant SCMs, many stormwater ponds (also known as retention ponds, wet detention ponds, or wet ponds) are seldom maintained [29]. First and foremost, a stormwater pond must be designed with maintenance in mind. This includes everything from having an easily-accessible sedimentation forebay or other pretreatment practice, to budgeting for and scheduling both routine and non-routine maintenance activities. Annual maintenance cost estimates for ponds range from 2–10% of the original construction cost [1]. A standardized inspection schedule may not be appropriate for all ponds because watershed and even pond characteristics vary greatly, and the frequency with which maintenance is needed may change as the watershed becomes more developed [59]. For example, poor upstream erosion control can drastically shorten a pond's lifespan due to increased sediment loads, requiring more frequent maintenance [60]. As physical changes to the pond or watershed occur, or water quality treatment goals intensify, the need may arise to increase the hydraulic residence time by adding screens or flow-lengthening baffles [59].

The design of a pond must be suited to a specific purpose, and performance goals must be appropriate for the given watershed and site constraints. Whether a pond addresses volume control, water quality, ornamental purposes, or a combination of these, other priorities will determine how an optimally functional system should look and how it needs to be maintained. Water features often provide ecosystem services in addition to their hydraulic and hydrologic functions. Aside from bringing open green spaces to urban environments, which can provide socioeconomic benefits [20,29,61], ponds can contribute to carbon sequestration, biodiversity, and cultural services. The first two of these are facilitated by the presence of a littoral shelf, which must be maintained to promote non-invasive, emergent vegetation that provides habitat for various species, including predators of mosquitos [62]; where mosquitos are a particular threat, regular inspections and treatments may be necessary [5,20,38]. The abundance and general variability of stormwater ponds further magnify their potential benefits to biodiversity [63,64]. Cultural services, such as recreation and education, will depend on accessibility, proper landscape management, and the maintenance of trails, infrastructure, and signage [62], in addition to public understanding and aesthetic preferences, which will vary and must be determined locally [65,66]. It is therefore important to educate the public on the function of stormwater ponds and safety concerns related to coming into contact with the water [29,38].

When tasked with performing such a broad range of functions, ponds may need additional improvements and enhancements. Stormwater ponds are effectively sacrificial water bodies aimed at protecting downstream waters; however, residents and other stakeholders will often expect to use ponds for ornamental or recreational purposes, which can cause concern when water clarity decreases and nuisance vegetation or algae begin to take over [59,60,66,67]. Deriving additional benefits from stormwater ponds may therefore necessitate providing additional pretreatment for the ponds in the form of pretreatment sumps (potentially including sediment separation devices). In some cases, direct treatment of the undesired symptoms, such as algae growth, may become necessary by means of mechanical or chemical methods [67]. Because of the complex functions and roles expected of stormwater ponds by the public, it is important that all stakeholders be given a voice regarding large-scale maintenance or construction activities [33]. Residents around stormwater ponds tend to view them as natural water bodies and may even resist maintenance efforts that could be seen as destroying 'natural' habitats [29].

The baseline for any SCM should come from the as-built condition. A thorough assessment following construction can help trace future problems back to issues with the design, construction, operation, and/or maintenance [5]. The sooner deviations from designs are discovered, the easier it will be to have the construction contractor rectify them [3]. Clearly communicating to contractors the intricacies and special considerations involved in constructing SCMs is paramount to minimizing such design deviations [34]. As-built drawings are usually not available for ponds that were constructed by retrofitting existing wetlands with poorly-defined elevations [4]. A follow-up assessment two years after construction or the most recent dredging can help estimate targeted characteristics, such as the sedimentation rate, to approximate when dredging will need to occur (often at 50% sediment accumulation [35]), which is particularly important because it can take a year or more for the excavation to occur once the need for it has been established [68]. Most, if not all, municipal separate storm sewer system (MS4) permits require regular outfall inspections (approximately every five years), at which time the bathymetry of the pond can be recorded to update models and keep track of sediment deltas [60]. Knowing the volume of sediment to be removed also allows the number of trucks necessary to haul the dredged material, and therefore the number of days required, to be estimated [4].

Most stormwater pond maintenance efforts evaluate 'success' as removing particulates to restore volume [2,4], and many stormwater ponds have water quality goals that include phosphorus removal associated with these solids. However, a portion of the phosphorus in the sediments is bound to be redox-sensitive ions, which means that oxygen must be present in the water to keep phosphorus in its particulate form. When dissolved oxygen (DO) drops below 1 mg/L, the pond is considered anoxic and redox-sensitive phosphorus will be released into the water column as soluble reactive phosphorus (also called orthophosphorus, ortho-P, or phosphate, PO_4^{-3}). This is particularly problematic because this is the most bio-available form of phosphorus and can lead to harmful algal blooms of blue-green algae (cyanobacteria) in addition to contributing to eutrophication and other water quality problems. Cyanobacterial growth rates will depend on water temperatures and pond residence time, so it is recommended that residence times in warmer regions be reduced according to the average summer water temperature to minimize harmful algal blooms [35]. Blooms can also be suppressed by applying beneficial bacteria, aerators, or specific chemicals [20,67].

Stormwater ponds that release ortho-P from the sediments will appear to capture less phosphorus overall and could be a net source of phosphorus to the receiving water body. To keep ponds from becoming anoxic, early design recommendations from the National Urban Runoff Program (NURP) called for stormwater ponds to be between 1 and 8 m in depth [69]; current design standards typically specify a depth of approximately 1 to 3 m [70–72]. This depth was assumed to allow for settling of suspended sediments containing particulate phosphorus, while remaining shallow enough to be fully-mixed by wind and storm events and therefore remain oxic [69]. However, periodic and sometimes regular and persistent thermal stratification has been observed during summer months, even in ponds less than 1 or 2 m in depth [72,73]. Thermal stratification can be especially problematic in warmer climates [36]. It is recommended that DO and temperature profiles be measured during regular inspections to evaluate which ponds are experiencing anoxic conditions that may trigger phosphorus release from the sediments [74]. Sheltering from wind (see Figure 1) by trees can prevent destratification, so vegetative growth around stormwater ponds should be controlled when possible [75]. In addition, conductivity profiles should also be measured in colder regions because road salt applications from winter deicing operations can accumulate in stormwater ponds and contribute to stratification [76]. Alum treatment [77] or iron treatment [78] can be used to fix phosphate in the sediments. In certain cases, aeration systems could be used to avoid stratification, although these systems must be run continuously and must aerate across as much of the pond area as possible to be effective.

Figure 1. A well-sheltered stormwater pond (**a**) and a poorly-sheltered stormwater pond (**b**).

The worst-performing ponds usually get maintained first, but a cost-effectiveness approach looking at pounds of phosphorus removed per dollar spent within a given watershed would allow funds to be spent more efficiently on a greater number of ponds [60]. Additionally, there may be times where maintaining one pond over another is necessary because of connectedness to downstream protected water bodies or water bodies of interest [4,5].

With any maintenance strategy, careful coordination and planning is integral to success. Maintenance access along the edge of the pond and through easements must be maintained over time, both to facilitate access and to keep the easements identifiable. It is also important to inform residents of maintenance/access agreements whenever property changes ownership. Any long-term plan must clearly hold specific individuals and entities responsible so that the required tasks occur as intended [3–5]. Poorly maintained easements are occasionally unintentionally annexed by residents who may place permanent structures or plant trees that block the path of larger equipment. Often, these may have special sentimental value to the residents, composing the 'human dimension' of challenges [4]. In such cases, opposition can be circumvented by working with residents to enable the establishment of temporary easements to minimize disruptions to their yards. If trees must be felled, an offer can be made, for example, to replace them at a ratio of 2:1, potentially even allowing residents to select the species and placement of the new trees [3]. However, care must be taken that tree roots are not at risk of destabilizing banks [35] or infiltrating pipes [20].

The timing of maintenance activities can also be optimized. Retrofit projects to meet increasing needs and standards can be used as opportunities to improve performance and increase the time until the next maintenance activity [60]. Different settings will require activities to be conducted at different times of day to minimize traffic and noise disruptions. In climates that have a season where frozen soil is common, full-pond dredging is often done in the winter because soils are hard; this will make heavy equipment movements simpler and minimize undesirable impacts to surrounding soils [60,68]. Ponds can also be more easily dewatered with lower liquid precipitation in winter [68]. Otherwise, special care should be taken to ensure that dewatering operations do not cause erosion downstream of the pond [79]. If dewatering is not necessary and only the sediment delta is being removed from the forebay, a temporary silt screen can be deployed to minimize suspended sediment dispersal and impacts to the rest of the pond. However, this dredging will typically be limited to the sediment that can be reached from the shore by an excavator [3]. Adverse impacts to fish and wildlife should also be considered. Special requirements and permitting may also be required for wetlands that were converted from natural wetlands. In this case, permit applications should specify that only non-natural materials are being removed [3].

Forebay dredging can occur at any time of year. For summer operations, temporary shield plates can be placed over grass and soft soils to minimize impacts. For sediment delta dredging, sediments can be deposited into a vacuum dredge box (a metal trough to which a vacuum hose can be connected, as shown in Figure 2) and be collected by a vacuum truck rather than having to be transferred directly from excavators to dump trucks [2]. This method can minimize 'human dimension' challenges by greatly reducing impacts to yards and lawns via reducing vehicular traffic through the easement and the width of easement that is required [4]. Efforts made to minimize impact and disruptions also simplify and shorten restoration efforts following maintenance operations, which can make up approximately 30–50% of total costs. Communication is also crucial in this phase to make sure that all parties involved understand what is expected of them so that any new sod or replacement trees are adequately watered until roots can become established [2,4,18]. In wetter regions, drainage may be necessary to keep seeds from drowning, and in all cases, plantings should occur at the appropriate time of the year for vegetation to properly establish [35]. The potential for herbivory must also be considered [18].

Figure 2. Dredged sediment being deposited into a vacuum dredge box (**a**) and vacuumed away to a truck on the street (**b**). Photos copyright of the City of Eden Prairie, MN, USA.

During dredging, junk materials from illicit dumping are often discovered and may complicate dredging, depending on the sizes of the materials. It may also be discovered that as-built drawings are inaccurate and not representative of the conditions encountered at the site. An inaccurately-defined pond bottom coupled with unexpected underlying pervious soils can lead to groundwater impacts and the unintended, and perhaps undesirable, conversion of a stormwater pond into an infiltration basin [4]. Placing a hard surface as a reference point (e.g., concrete or rocks at the bottom of the forebay) can facilitate identifying the bottom of the pond during dredging [35].

After dredging, sediments should be dewatered to reduce the mass that must be transported [29,35]. Sediments can be reused or disposed of according to concentrations of various contaminants, such as heavy metals [29,80]. The contaminants of concern in pond sediments are polycyclic aromatic hydrocarbons (PAHs), which are carcinogenic products of incomplete combustion, primarily originating from coal tar sealants and vehicular combustion [81]. Different PAHs vary in carcinogenic risk and bioavailability, but typically must be disposed of in confined disposal facilities due to their perceived danger [80]. This alone can triple the cost of dredging a pond [82]. In some cases, the cost of dredging ponds can become so high that it becomes preferable to reroute stormwater to an entirely new pond and abandon the original pond [4]. When handling potentially hazardous materials, appropriate personal protective equipment (PPE) should be worn. In this case, soils should not be handled or disturbed until laboratory results have been received. If soils are determined to be hazardous, only professionals trained to safely and properly handle the soils should do so [19].

Ultimately, the most effective pond maintenance technique is proactive load reduction. Vocal residents may tend to call stormwater pond managers with questions and concerns regarding a pond's appearance. These are opportunities for energetic residents to be activated to raise awareness about how stormwater ponds function and promote watershed management for nutrient load reduction [60]. Table 3 provides a list of base guidelines for maintenance of stormwater ponds.

Table 3. Maintenance recommendations for ponds [1,17].

Task	Frequency	Notes
Inspection	Annually or after every two-year storm	
Monitor sediment depth in forebay and deep pools	Once per year	Can be performed with capacity testing
Measure pond bathymetry	After construction/dredging and then every five years	Calculate sedimentation rates to estimate dredging timeline
Inspect outlet structures	Annually or after every 2-year storm	Follow visual inspection guidelines
Remove trash and debris	Annually	Increase frequency, if needed
Remove vegetation from dam top and faces, if applicable	Once per year	Increase frequency, if needed
Mow wet pond perimeter	As needed	
Remove burrowing animals and beavers, if present	As needed	Destroy burrow holes whenever present; contact a professional trapper to remove beavers; nuisance animals may return after removal
Measure dissolved oxygen, temperature, and conductivity profiles	As frequently as possible	Frequency can be increased or decreased once trends are observed
Collect total phosphorus surface water samples	As frequently as possible	Frequency can be increased or decreased once trends are observed
Remove all sediment from forebay and deep pool (dredging)	Variable (Once every five to ten years is typical in stable watersheds)	In unstable watersheds (i.e., those with active construction), the frequency is typically once per year
Treat phosphorus release with alum or iron filings	As needed	Harmful algal blooms resulting from high phosphorus may have to be treated directly with beneficial bacteria, aeration, or chemicals
Maintain easements accessible	Annually	Maintaining a regular presence can discourage homeowners from obstructing passage

5. Permeable Pavements

Permeable pavements are an alternative to conventional asphalt or concrete pavement material where the porosity of the pavement is increased to allow transport of water from the surface through the pavement to the materials below. Permeable pavements include asphalt, concrete, and modular permeable block systems, where the water passes either through or between the blocks. Permeable pavements are often designed with up to 90 cm of large gravel below the pavement to temporarily store water that infiltrates through the permeable pavement.

A major challenge in the maintenance of permeable pavements is the development of depressed areas where vehicle tires commonly impact the pavement surface. In some cases, the cause of this depression is poor pavement strength because of poor design or construction, resulting in pore space collapse and reduced infiltration capacity. In addition, particles from vehicle tires and wheel wells tend to be deposited within these depressed areas, and water preferentially accumulates and infiltrates into these depressed areas, causing an accumulation of particles that can clog the permeable pavement surface. In permeable pavements that do not develop depressed areas, sediment from vehicles can still clog the pavement surface preferentially in the areas in which tires impact the pavement surface. As a result of collapsed pore space and/or accumulated sediment, the infiltration capacity can be substantially reduced. Because depressed areas are lower in elevation than the surrounding permeable pavement, these linear channels can become surface conveyances and create runoff from an area intended for infiltration.

Simple methods have been developed to determine whether collapsed pores or sediment accumulation have reduced infiltration rates through permeable pavements [83]. Maintenance activities for permeable pavements have been shown to restore up to 90% of the original infiltration capacity [84]. In a comparison of mechanical street sweeping, regenerative-air street sweeping, vacuum street sweeping, hand-held vacuuming, high-pressure washing, and milling of porous asphalt, the most successful methods were milling 2.5 cm from the surface and vacuum street sweeping [84]. In some areas with high debris loading, multiple passes with a vacuum street sweeper were needed to increase surface infiltration rates above acceptable thresholds [84]. While vacuum street sweeping can remove sediment, none of the surface cleaning maintenance methods can restore infiltration capacity in collapsed pores. If the collapsed pores are only near the surface, milling may be the only maintenance activity that will restore infiltration capacity. Milling as a maintenance activity on permeable pavement requires some additional research, though, to determine how clean pavement can be added to the surface or whether pavement sections can be designed such that milled pavement can be removed without replacement. Table 4 provides a list of base guidelines for the maintenance of permeable pavements.

Table 4. Maintenance recommendations for permeable pavements [1,17,84].

Task	Frequency	Notes
Inspection	Annually or after every two-year storm	
Vacuum street sweeping	Variable (three to four times per year recommended)	More frequent cleanings may be required in watersheds with large debris loads
Measure surface infiltration rate	As needed, when inspections indicate reduced infiltration rate (i.e., surface ponding)	
Milling the top 1–2.5 cm	As needed, when vacuum sweeping does not restore infiltration capacity	
Where areas of paving settle, lift blocks, re-level bedding material, and lay blocks at new level	As needed	
Do not sand or salt during the winter	Annually	
Maintain landscaped areas that may run-on to pavement; reseed bare areas	As needed; inspect annually	

6. Future Research

As the above review suggests, recent research has found new methods for improving how engineers maintain SCMs. New research is continually expanding the types of SCMs available to engineers and improving the performance of existing SCM designs. As new mechanisms are added to existing practices and new practices are developed, still more research is needed to determine the best maintenance methods and the frequency, effort, and costs associated with the maintenance. In addition, more research is needed to better understand the relationship between the performance of a practice (e.g., runoff volume reduction, pollutant capture) and maintenance activities. While this has been done for a select few practices and maintenance activities (e.g., [84]), more research like this for more practices and more maintenance activities is needed to better understand the cost-effectiveness of maintenance throughout the life-cycle of an SCM.

7. Summary and Conclusions

The function of a stormwater control measure (SCM) needs to be maintained and should not be ignored in determining life-cycle costs. A rule-of-thumb is that the maintenance of a SCM throughout its life will cost as much in current currency as the construction cost of the practice. As the treatment of stormwater becomes more complex, new concerns for SCM assessment emerge, such as the capacity of media filtration additives targeting specific dissolved pollutants and the implications of permeable pavement compression for runoff. There are also older stormwater practices that have developed new problems, such as retention ponds that are sheltered by large trees and can therefore stratify and develop low dissolved oxygen concentrations at the bottom, which can in turn lead to phosphate release from the sediments that can flow into receiving water bodies. The maintenance of an SCM is therefore a continuous adaptation to changes in the practices and condition of the practices.

Author Contributions: A.J.E. drafted the abstract and Sections 2 and 3 and Sections 5–7. V.J.T. drafted Sections 1 and 4. All authors revised all sections, responded to comments from anonymous reviewers, and prepared the final manuscript.

Funding: A portion of this research was managed by David Fairbairn of the Minnesota Pollution Control Agency through funding from the Minnesota Clean Water Council. Funding for the first author was provided by the National Science Foundation (grant number 00039202).

Acknowledgments: The authors wish to thank John Carlon, Shahram Missaghi and Ross Bintner, who agreed to be interviewed for this article. The authors also wish to thank the three anonymous reviewers that provided feedback and suggestions on the manuscript.

Conflicts of Interest: The authors declare no conflict of interest.

References

1. Erickson, A.J.; Weiss, P.T.; Gulliver, J.S. *Optimizing Stormwater Treatment Practices: A Handbook of Assessment and Maintenance*; Springer-Verlag: New York, NY, USA, 2013; ISBN 978-1-4614-4624-8.
2. Carlon, J. (Public Works Department, City of Eden Prairie, Minnesota, USA); Taguchi, V. (University of Minnesota, Minneapolis, Minnesota, USA). Personal communication, 2018.
3. University of Minnesota Stormwater and Erosion Control Certification Program; University of Minnesota Extension. *Stormwater Practice Maintenance Recertification Workshop*; University of Minnesota Stormwater: Minneapolis, MN, USA, 2018.
4. Missaghi, S. (University of Minnesota, St. Paul, Minnesota, USA); Taguchi, V. (University of Minnesota, Minneapolis, Minnesota, USA). Personal communication, 2018.
5. Caltrans. Statewide Stormwater Management Plan. 2016. Available online: https://www.waterboards.ca.gov/water_issues/programs/stormwater/docs/caltrans/swmp/swmp_approved.pdf (accessed on 24 September 2018).
6. Erickson, A.J.; Gulliver, J.S.; Kang, J.H.; Weiss, P.T.; Wilson, C.B. Maintenance for Stormwater Treatment Practices. *J. Contemp. Water Res. Educ.* **2010**, *146*, 75–82. [CrossRef]
7. City of Vancouver Department of Public Works. General Storm Water Construction Notes. 2017. Available online: https://www.cityofvancouver.us/sites/default/files/fileattachments/public_works/page/11891/surfacewaterstormwaterdetails.pdf (accessed on 23 September 2018).
8. Commonwealth of Massachusetts. Massachusetts Stormwater Handbook. 2018. Available online: https://www.mass.gov/guides/massachusetts-stormwater-handbook-and-stormwater-standards (accessed on 23 September 2018).
9. Herrera, Inc. Guidance Document Western Washington Low Impact Development (LID) Operation and Maintenance (O&M). Prepared for Washington State Department of Ecology Water Quality Program. 2013. Available online: https://ecology.wa.gov/Regulations-Permits/Guidance-technical-assistance/Stormwater-permittee-guidance-resources/Low-Impact-Development-guidance (accessed on 24 September 2018).

10. City of Portland. Stormwater Management Manual. 2016. Available online: https://www.portlandoregon.gov/bes/64040 (accessed on 23 September 2018).
11. Seattle Public Utilities. Green Stormwater Operations and Maintenance Manual. 2009. Available online: http://www.seattle.gov/util/cs/groups/public/@spu/@usm/documents/webcontent/spu02_020023.pdf (accessed on 23 September 2018).
12. Philadelphia Water Department. Post-Construction and Operations and Maintenance Guidance. Stormwater Plan Review. 2018. Available online: https://www.pwdplanreview.org/manual/chapter-6 (accessed on 24 September 2018).
13. Geosyntec Consultants. *Stormwater Best Management Practices: Guidance Document*; Boston Water and Sewer Commission: Boston, MA, USA, 2018.
14. University of New Hampshire Stormwater Center. Regular Inspection and Maintenance Guidance for Porous Pavements. 2011. Available online: https://www.unh.edu/unhsc/sites/unh.edu.unhsc/files/UNHSC%20Porous%20Pavement%20Routine%20Maintenance%20Guidance%20and%20Checklist%202-11.pdf (accessed on 23 September 2018).
15. University of New Hampshire Stormwater Center. Regular Inspection and Maintenance Guidance for Bioretention Systems/Tree Filters. 2011. Available online: https://www.unh.edu/unhsc/sites/unh.edu.unhsc/files/UNHSC%20Biofilter%20Maintenance%20Guidance%20and%20Checklist%201-11_0.pdf (accessed on 23 September 2018).
16. Maine Department of Environmental Protection. Maine Stormwater Best Management Practices Manual: Technical Design Manual. 2016. Available online: https://www.maine.gov/dep/land/stormwater/stormwaterbmps/vol3/volume%20III%20May%202016.pdf (accessed on 24 September 2018).
17. Wossink, A.; Hunt, B. *The Economics of Structural Stormwater BMPs in North Carolina*; Report 2003–344; University of North Carolina Water Resources Research Institute: Chapel Hill, NC, USA, 2003.
18. Northern Virginia Regional Commission. Maintaining Stormwater Systems: A Guidebook for Private Owners and Operators in Northern Virginia. 2007. Available online: http://www.novaregion.org/DocumentCenter/View/1675/MaintainingYourStormwaterSystem-2007?bidId (accessed on 24 September 2018).
19. Tetra Tech, Inc. Operation and Maintenance of Green Infrastructure Receiving Runoff from Roads and Parking Lots, United Stated Environmental Protection Agency. 2016. Available online: https://www.epa.gov/sites/production/files/2016-11/documents/final_gi_maintenance_508.pdf (accessed on 23 September 2018).
20. Tetra Tech, Inc. Stormwater Wet Pond and Wetland Management Guidebook, United Stated Environmental Protection Agency. 2009. Available online: https://www.epa.gov/sites/production/files/2015-11/documents/pondmgmtguide.pdf (accessed on 23 September 2018).
21. Landphair, H.C.; McFalls, J.A.; Thompson, D. *Design Methods, Selection, and Cost-Effectiveness of Stormwater Quality Structures*; Report 1837–1; Texas Department of Transportation: Austin, TX, USA, 2000.
22. Bayouland RC&D Council. *Stormwater BMP Guidance Tool: A Stormwater Best Management Practices Guide for Orleans and Jefferson Parishes*; Louisiana Department of Environmental Quality: Baton Rouge, LA, USA, 2010.
23. City of Austin. Green Stormwater Infrastructure Maintenance Manual. 2018. Available online: http://www.austintexas.gov/sites/default/files/files/Watershed/stormwater/GSI-Maintenance-Manual.pdf (accessed on 23 September 2018).
24. Hester, B.J. Stormwater Management Program (SWMP), City of Tucson Department of Transportation. 2014. Available online: https://www.tucsonaz.gov/files/transportation/SWMP_2014.pdf (accessed on 24 September 2018).
25. Phillips, A.A. City of Tucson Water Harvesting Guidance Manual. City of Tucson Department of Transportation, Stormwater Management Section. 2005. Available online: https://www.tucsonaz.gov/files/transportation/2006WaterHarvesting.pdf (accessed on 23 September 2018).
26. Holcomb, S.; Romero, J.; Huddleson, S.; Smith, N. Green Infrastructure Implementation in New Mexico, New Mexico Environment Department. 2017. Available online: http://www.ose.state.nm.us/PIO/News/2017/Green%20Infrastructure%20FAQs_May%202017_Final.pdf (accessed on 23 September 2018).

27. City of Aurora. City of Aurora Storm Drainage Design and Technical Criteria. 2010. Available online: https://auroragov.org/UserFiles/Servers/Server_1881137/File/Business%20Services/Development%20Center/Code%20&%20Rules/Design%20Standard/Engineering%20Design%20Standard/001861.pdf (accessed on 23 September 2018).
28. Ontario Ministry of the Environment. *Stormwater Management Planning and Design Manual*; Queen's Printer for Ontario; Ontario Ministry of the Environment: Vancouver, ON, USA, 2003; Available online: https://www.ontario.ca/document/stormwater-management-planning-and-design-manual-0 (accessed on 23 September 2018).
29. Drake, J.; Guo, Y. Maintenance of Wet Stormwater Ponds in Ontario. *Can. Water Resour. J.* **2008**, *34*, 351–368. [CrossRef]
30. Auckland Council. Stormwater Forms and Guides. Auckland Council Operation & Maintanance [sic] Guide. Stormwater Device Information Series; 2018. Available online: https://www.aucklandcouncil.govt.nz/environment/stormwater/Pages/stormwater-forms-and-guides.aspx (accessed on 23 September 2018).
31. Woods Ballard, B.; Udale-Clarke, H.; Illman, S.; Scott, T.; Ashley, R.; Kellagher, R. *The SuDS Manual*; CIRIA: London, UK, 2015; ISBN 978-0-86017-760-9.
32. City of Melbourne. Water Sensitive Urban Design Guidelines. 2006. Available online: https://www.melbourne.vic.gov.au/building-and-development/sustainable-building/Pages/water-sensitive-urban-design.aspx (accessed on 23 September 2018).
33. Monk, E.; Chalmers, L. Stormwater Management Manual for Western Australia, Government of Western Australia Department of Water. 2007. Available online: http://www.water.wa.gov.au/__data/assets/pdf_file/0019/1774/84954.pdf (accessed on 23 September 2018).
34. Victoria Stormwater Committee. *Urban Stormwater: Best Practice Environmental Management Guidelines*; CSIRO Publishing: Collingwood, Australia, 1999; Available online: http://www.publish.csiro.au/ebook/download/pdf/2190 (accessed on 24 September 2018).
35. Melbourne Water. WSUD Engineering Procedures–Stormwater. 2005. Available online: https://app.knovel.com/hotlink/toc/id:kpWSUDEPS1/wsud-engineering-procedures/wsud-engineering-procedures (accessed on 24 September 2018).
36. Government of Malaysia Department of Irrigation and Drainage. Urban Stormwater Management Manual for Malaysia. 2012. Available online: https://www.water.gov.my/jps/resources/PDF/MSMA2ndEdition_august_2012.pdf (accessed on 23 September 2018).
37. Singapore Public Utilities Board. Active Beautiful Clean Waters Design Guidelines. 2014. Available online: https://www.pub.gov.sg/Documents/ABC_DG_2014.pdf (accessed on 23 September 2018).
38. Singapore Public Utilities Board. Managing Urban Runoff–Drainage Handbook. 2013. Available online: https://www.pub.gov.sg/Documents/managingUrbanRunoff.pdf (accessed on 23 September 2018).
39. D'Arcy, B.; Kim, K.H.; Maniquiz-Redillas, M. (Eds.) *Wealth Creation without Pollution*; IWA Publishing: London, UK, 2018.
40. Claytor, R.A.; Schueler, T.R. *Design of Stormwater Filtering System*; Center for Watershed Protection for Chesapeake Research Consortium and U.S. EPA: Washington, DC, USA, 1996; Available online: https://owl.cwp.org/?mdocs-file=4553 (accessed on 24 September 2018).
41. Erickson, A.J.; Gulliver, J.S.; Weiss, P.T. Capturing phosphates with iron enhanced sand filtration. *Water Res.* **2012**, *46*, 3032–3042. [CrossRef] [PubMed]
42. Erickson, A.J.; Gulliver, J.S.; Weiss, P.T. Phosphate removal from agricultural tile drainage with iron enhanced sand. *Water* **2017**, *9*, 672. [CrossRef]
43. Erickson, A.J.; Gulliver, J.S.; Arnold, W.A.; Brekke, C.; Bredal, M. Abiotic capture of stormwater nitrates with granular activated carbon. *Environ. Eng. Sci.* **2016**, *33*, 354–363. [CrossRef]
44. Paus, K.H.; Morgan, J.; Gulliver, J.S.; Hozalski, R.M. Effects of bioretention media compost volume fraction on toxic metals removal, hydraulic conductivity, and phosphorous release. *J. Environ. Eng.* **2014**, *140*. [CrossRef]
45. Paus, K.H.; Morgan, J.; Gulliver, J.S.; Leiknes, T.; Hozalski, R.M. Effects of temperature and NaCl on toxic metal retention in bioretention media. *J. Environ. Eng.* **2014**, *140*. [CrossRef]
46. Clary, J.; Jones, J.; Urbonas, B.; Quigley, M.; Strecker, E.; Wagner, T. Can Stormwater BMPs Remove Bacteria? New Findings from the International Stormwater BMP Database. Stormwater. 2008. Available online: http://citeseerx.ist.psu.edu/viewdoc/summary?doi=10.1.1.506.7468 (accessed on 24 September 2018).

47. LeFevre, G.H.; Paus, K.H.; Natarajan, P.; Gulliver, J.S.; Novak, P.J.; Hozalski, R.M. Review of dissolved pollutants in urban storm water and their removal and fate in bioretention cells. *J. Environ. Eng.* **2015**, *141*. [CrossRef]
48. American Public Health Association (APHA). 4500-p phosphorus. In *Standard Methods for the Examination of Water and Wastewater*, 20th ed.; American Public Health Association (APHA): Washington, DC, USA, 1998.
49. Asleson, B.C.; Nestingen, R.S.; Gulliver, J.S.; Hozalski, R.M.; Nieber, J.L. Performance assessment of rain gardens. *J. Am. Water Resour. Assoc.* **2009**, *45*, 1019–1031. [CrossRef]
50. Ahmed, F.; Gulliver, J.S.; Nieber, J.L. Field infiltration measurements in grassed roadside drainage ditches: Spatial and temporal variability. *J. Hydrol.* **2015**, *530*, 604–611. [CrossRef]
51. Weiss, P.T.; Gulliver, J.S. Effective saturated hydraulic conductivity of an infiltration-based stormwater control measure. *J. Sustain. Water Built Environ.* **2015**, *1*, 5. [CrossRef]
52. Le Coustumer, S.; Fletcher, T.D.; Deletic, A.; Barraud, S.; Poelsma, P. The influence of design parameters on clogging of stormwater biofilters: A large-scale column study. *Water Res* **2012**, *46*, 6743–6752. [CrossRef] [PubMed]
53. Paus, K.H.; Morgan, J.; Gulliver, J.S.; Leiknes, T.; Hozalski, R.M. Assessment of the hydraulic and toxic metal removal capacities of bioretention cells after 2 to 8 years of service. *Water Air Soil Pollut.* **2014**, *225*, 1803. [CrossRef]
54. Lucas, W.C.; Greenway, M. Nutrient retention in vegetated and nonvegetated bioretention mesocosms. *J. Irrig. Drain. Eng.* **2008**, *134*, 613–623. [CrossRef]
55. LeFevre, G.H.; Novak, P.J.; Hozalski, R.M. Fate of naphthalene in laboratory-scale bioretention cells: Implications for sustainable stormwater management. *Environ. Sci. Technol.* **2012**, *46*, 995–1002. [CrossRef] [PubMed]
56. Shaw, D.; Schmidt, R. *Plants for Stormwater Design: Species Selection for the Upper Midwest*; Minnesota Pollution Control Agency: St. Paul, MN, USA, 2003. Available online: https://www.pca.state.mn.us/water/plants-stormwater-design (accessed on 21 September 2018).
57. Kazemi, F.; Beecham, S.; Gibbs, J. Streetscape biodiversity and the role of bioretention swales in an Australian urban environment. *Landsc. Urban Plan.* **2011**, *101*, 139–148. [CrossRef]
58. Winfrey, B.K.; Hatt, B.E.; Ambrose, R.F. Biodiversity and functional diversity of Australian stormwater biofilter plant communities. *Landsc. Urban Plan.* **2018**, *170*, 112–137. [CrossRef]
59. Anderson, B.; Watt, W.; Marsalek, J. Critical issues for stormwater ponds: Learning from a decade of research. *Water Sci. Technol.* **2002**, *45*, 277–283. [CrossRef] [PubMed]
60. Bintner, R.; City of Edina, Edina, Minnesota, USA; Taguchi, V.; University of Minnesota, Minneapolis, Minnesota, USA. Personal communication, 2018.
61. Benedict, M.A.; McMahon, E.T. *Green Infrastructure*; Sprawl Watch Clearinghouse: Washington, DC, USA, 2006.
62. Moore, T.L.; Hunt, W.F. Ecosystem service provision by stormwater wetlands and ponds–a means for evaluation? *Water Res.* **2012**, *46*, 6811–6823. [CrossRef] [PubMed]
63. Clifford, C.C.; Heffernan, J.B. Artificial Aquatic Ecosystems. *Water* **2018**, *10*, 1096. [CrossRef]
64. Blicharska, M.; Andersson, J.; Bergsten, J.; Bjelke, U.; Hilding-Rydevik, T.; Johansson, F. Effects of management intensity, function and vegetation on the biodiversity in urban ponds. *Urban For. Urban Green.* **2016**, *20*, 103–112. [CrossRef]
65. Dobbie, M.F. Public austhetic preferences to inform sustainable wetland management in Victoria, Australia. *Landsc. Urban Plan.* **2013**, *120*, 178–189. [CrossRef]
66. Monaghan, P.; Hu, S.; Hansen, G.; Ott, E.; Nealis, C.; Morera, M. Balancing the Ecological Function of Residential Stormwater Ponds with Homeowner Landscaping Practices. *Environ. Manag.* **2016**, *58*, 843–856. [CrossRef] [PubMed]
67. Christenson, M.; Lokke, A.; Rickbeil, D.; Taguchi, V.; Weis, R. *The Draw: Algal Removal Feasibility Study*; University of Minnesota: Minneapolis, MN, USA, 2017; Available online: http://hdl.handle.net/11299/193498 (accessed on 23 September 2018).
68. Hafner, J.; Panzer, M. Case study #11: Stormwater retention ponds: Maintenance vs. Efficiency. In *Stormwater Treatment: Assessment and Maintenance*; Gulliver, J.S., Erickson, A.J., Weiss, P.T., Eds.; University of Minnesota, St. Anthony Falls Laboratory: Minneapolis, MN, USA, 2010. Available online: http://stormwaterbook.safl.umn.edu/case-studies/case-study-11-stormwater-retention-ponds-maintenance-vs-efficiency (accessed on 23 September 2018).

69. Walker, W.W., Jr. Phosphorus removal by urban runoff detention basins. *Lake Reserv. Manag.* **1987**, *3*, 314–326. [CrossRef]
70. Minnesota Pollution Control Agency. Design Criteria for Stormwater Ponds—Minnesota Stormwater Manual. Available online: https://stormwater.pca.state.mn.us/index.php?title=Design_criteria_for_stormwater_ponds (accessed on 30 July 2018).
71. North Carolina Department of Environmental Quality. Wet Pond. Stormwater Design Manual. Available online: https://files.nc.gov/ncdeq/Energy%20Mineral%20and%20Land%20Resources/Stormwater/BMP%20Manual/C-3%20%20Wet%20Pond%2004-17-17.pdf (accessed on 30 July 2018).
72. Lake Simcoe Region Conservation Authority. Stormwater Pond Maintenance and Anoxic Conditions Investigation Final Report. 2011. Available online: https://www.lsrca.on.ca/Shared%20Documents/reports/stormwater_maintenance.pdf (accessed on 23 September 2018).
73. Song, K.; Xenopoulos, M.A.; Buttle, J.M.; Marsalek, J.; Wagner, N.D.; Pick, F.R.; Frost, P.C. Thermal stratification patterns in urban ponds and their relationships with vertical nutrient gradients. *J. Manag.* **2013**, *127*, 317–323. [CrossRef] [PubMed]
74. Nürnberg, G.K. Assessing internal phosphorus load–problems to be solved. *Lake Reserv. Manag.* **2009**, *25*, 419–432. [CrossRef]
75. McEnroe, N.; Buttle, J.; Marsalek, J.; Pick, F.; Xenopoulos, M.; Frost, P. Thermal and chemical stratification of urban ponds: Are they 'completely mixed reactors'? *Urban Ecosyst.* **2013**, *16*, 327–339. [CrossRef]
76. Marsalek, J. Road salts in urban stormwater: An emerging issue in stormwater management in cold climates. *Water Sci. Technol.* **2003**, *48*, 61–70. [CrossRef] [PubMed]
77. Minnesota Pollution Control Agency. Tanners Lake–Alum Injection for Phosphorus Removal–Minnesota Stormwater Manual. Available online: https://stormwater.pca.state.mn.us/index.php?title=Tanners_Lake_-_alum_injection_for_phosphorus_removal (accessed on 30 July 2018).
78. Natarajan, P.; Gulliver, J.S.; Arnold, W.A. *Internal Phosphorus Load Reduction with Iron Filings*; St. Anthony Falls Laboratory P.R. 582; University of Minnesota: Minneapolis, MN, USA, 2017. Available online: https://conservancy.umn.edu/handle/11299/195677 (accessed on 30 July 2018).
79. Caltrans. Construction Site Best Management Practices Manual: Dewatering Operations. 2004. Available online: http://www.dot.ca.gov/hq/construc/stormwater/NS02Update.pdf (accessed on 23 September 2018).
80. Kyser, S.; Hozalski, R.; Gulliver, J.S. *Use of Compost to Biodegrade SEDIMENTS Contaminated with Polycyclic Aromatic Hydrocarbons*; St. Anthony Falls Laboratory P.R. 582; University of Minnesota: Minneapolis, MN, USA, 2010; Available online: https://conservancy.umn.edu/handle/11299/196261 (accessed on 23 September 2018).
81. Crane, J.L. Source apportionment and distribution of polycyclic aromatic hydrocarbons, risk considerations, and management implications for urban stormwater pond sediments in Minnesota, USA. *Arch. Environ. Contam. Toxicol.* **2014**, *66*, 176–200. [CrossRef] [PubMed]
82. Kyser, S. The Fate of Polycyclic Aromatic Hydrocarbons Bound to Stormwater Pond Sediment During Composting. Master's Thesis, University of Minnesota, Minneapolis, MN, USA, 2010. Available online: http://hdl.handle.net/11299/104204 (accessed on 10 July 2018).
83. Winston, R.J.; Al-Rubaei, A.M.; Blecken, G.T.; Hunt, W.F. A simple infiltration test for determination of permeable pavement maintenance need. *J. Environ. Eng.* **2016**, *142*, 06016005. [CrossRef]
84. Winston, R.J.; Al-Rubaei, A.M.; Blecken, G.T.; Viklander, M.; Hunt, W.F. Maintenance measures for preservation and recovery of permeable pavement surface infiltration rate–the effects of street sweeping, vacuum cleaning, high pressure washing, and milling. *J. Environ. Eng.* **2016**, *169*, 132–144. [CrossRef] [PubMed]

© 2018 by the authors. Licensee MDPI, Basel, Switzerland. This article is an open access article distributed under the terms and conditions of the Creative Commons Attribution (CC BY) license (http://creativecommons.org/licenses/by/4.0/).

Review

SuDS & Sponge Cities: A Comparative Analysis of the Implementation of Pluvial Flood Management in the UK and China

Craig Lashford [1,*], Matteo Rubinato [2], Yanpeng Cai [3], Jingming Hou [4], Soroush Abolfathi [1], Stephen Coupe [1], Susanne Charlesworth [1] and Simon Tait [2]

1. Centre for Agroecology, Water & Resilience, Coventry University, Priory Street, Coventry CV1 5FB, West Midlands, UK; soroush.abolfathi@coventry.ac.uk (S.A.); steve.coupe@coventry.ac.uk (S.C.); s.charlesworth@coventry.ac.uk (S.C.)
2. Civil and Structural Engineering Department, University of Sheffield, Sir Frederick Mappin Building, Mappin Street, Sheffield S1 3JD, South Yorkshire, UK; m.rubinato@sheffield.ac.uk (M.R.); s.tait@sheffield.ac.uk (S.T.)
3. School of Environment, Beijing Normal University, 19 Xinjiekou Outer St, Bei Tai Ping Zhuang, Haidian Qu, Beijing 100875, China; yanpeng.cai@bnu.edu.cn
4. School of Environmental and Municipal Engineering, Xi'an University of Architecture & Technology, Yanta IT Shangquan, Beilin, Xi'an 710055, China; jingming.hou@xaut.edu.cn
* Correspondence: craig.lashford@coventry.ac.uk or ab0874@coventry.ac.uk

Received: 31 October 2018; Accepted: 27 December 2018; Published: 4 January 2019

Abstract: In recent decades, rapid urbanization has resulted in a growing urban population, transformed into regions of exceptional socio-economic value. By removing vegetation and soil, grading the land surface and saturating soil air content, urban developments are more likely to be flooded, which will be further exacerbated by an anticipated increase in the number of intense rainfall events, due to climate change. To date, data collected show that urban pluvial flood events are on the rise for both the UK and China. This paper presents a critical review of existing sustainable approaches to urban flood management, by comparing UK practice with that in China and critically assessing whether lessons can be learnt from the Sponge City initiative. The authors have identified a strategic research plan to ensure that the sponge city initiative can successfully respond to extreme climatic events and tackle pluvial flooding. Hence, this review suggests that future research should focus on (1) the development of a more localized rainfall model for the Chinese climate; (2) the role of retrofit SuDS (Sustainable Drainage Systems) in challenging water environments; (3) the development of a robust SuDS selection tool, ensuring that the most effective devices are installed, based on local factors; and (4) dissemination of current information, and increased understanding of maintenance and whole life-costing, alongside monitoring the success of sponge cities to increase the confidence of decision makers (5) the community engagement and education about sponge cities.

Keywords: flood management; urban flooding; Sustainable Drainage Systems; sponge cities; lessons to be learnt; future opportunities

1. Introduction

Flooding impacted approximately 78 million people globally in 2016 [1]. In China, between June and July an estimated 32 million people were affected by flooding [2], whilst flooding in the UK caused damage in excess of £1.6 billion over the winter of 2015–2016 [3]. Climate change projections show that even for a moderate climate change scenario, an increase in the intensity and frequency of global flood events is likely [4]. With a worldwide anticipated increase in the urban population from 55%

to 68% by 2050 [5], and the impact of climate change, a sustainable solution to flood management is essential for the socio-economic growth of nations.

Urban developments across the globe are regularly built in close proximity to rivers, and near or on floodplains, with natural drainage replaced by hard engineering solutions, such as piped sub-surface drainage [6], maintaining the protection from rivers via engineered flood protection systems [7]. This, alongside catchments that change characteristics from permeable to impermeable surfaces, results in a reduction in infiltration and an increased hydrological response, ultimately increasing flood risk for even low-intensity events [8–10].

Traditional pipe-based drainage has largely been implemented globally, particularly in urban areas. However, with the UK population increasing by nearly 14% over the last 20 years [11], and a shift to more people living in urban areas, the existing drainage systems are not sufficient, and require enhancement. Consequently, a number of major UK cities have been exposed to pluvial flooding since 2011 (e.g., Birmingham in 2016, London in 2012 and Edinburgh in 2011) [12,13]. In China, Zheng et al. [14] show that the increased rate of urbanization since the economic reform in 1978 has coincided with a steady increase in large flood events. The 2016 floods impacted 26 southern provinces in China, with estimated losses in excess of USD 500 billion [15]. In July 2012, the Fangshan District of Beijing experienced 460 mm of rainfall in 24 hours, three times the daily average, which caused over USD 1.86 billion of damage and impacted 1.6 million people [16].

In order to address these impacts, China has adopted a top-down policy whereby cities are directed to become "sponges" and manage 70% of incident rainfall using sustainable drainage techniques. They are funded to do so, but if they are not successful, funding is withdrawn. In contrast, in the UK, implementation of SuDS is not supported by legislation, and is a piecemeal, bottom-up approach essentially relying on local "SuDS Champions" to support the concept.

Due to an increasing flood risk, climate change, urbanization and the change in flooding patterns in the UK and China, a critical review of sustainable approaches to flood management is necessary to improve existing flood management practice and tools to deliver new solutions [17]. The purpose of this paper is to present a review of sustainable flood management in the UK and the move to create "sponge cities" in China, determining the lessons that can be learnt from both approaches.

2. Methodology

A systematic review of literature linked to sustainable flood management in China provided the basis for this review (see Figure 1). An initial database search was completed using SCOPUS, to identify suitable publications, using the search terms "sponge cities" and "China" in the title, abstract or keywords. To recognize existing challenges in China, journal articles were considered from 2014, to coincide with the implementation of sponge cities [18]. Only journal articles were considered, which had to be either *in Press* or published at the time of the search (October 2018) and written in either English or Chinese. The literature was initially screened for their suitability, based on title and abstract. Those excluded either repeated points already raised by previous articles, or were not suitable for this review. Articles were then reviewed in their entirety, with 14 subsequently excluded if they simply described specific individual SuDS methods adopted in China, without examining the wider sponge city process, or if it failed to identify any issues or challenges with any sponge city plan.

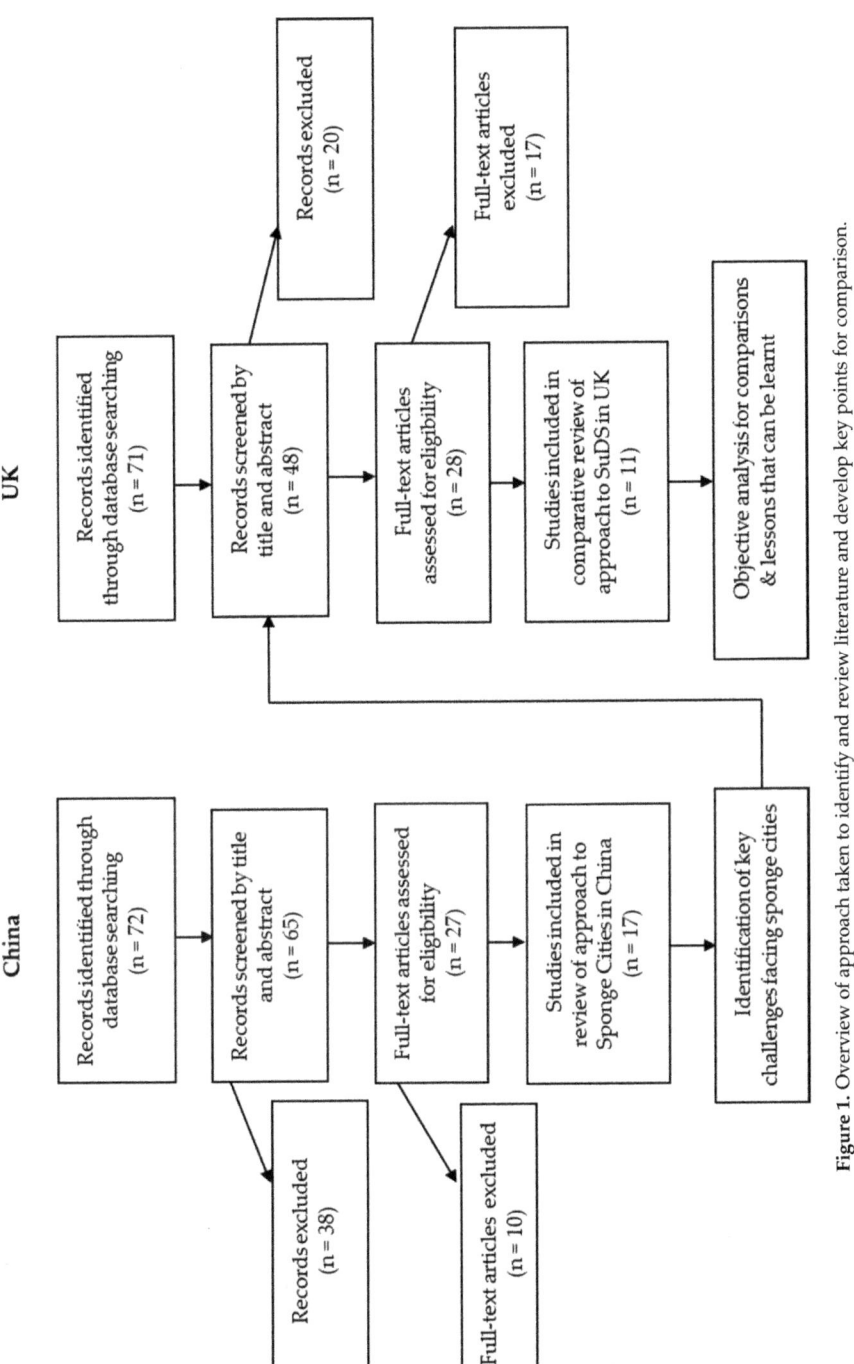

Figure 1. Overview of approach taken to identify and review literature and develop key points for comparison.

The literature analysis identified a number of key underlying themes, which were consistently highlighted as challenges for sponge cities (Section 4.1). With this in mind, UK literature was sifted to identify future research projects that would reduce the anticipated challenges. SCOPUS was again used to search for literature, using the terms "SuDS" and "UK" in the title, abstract or keywords. As was the case with the literature for China, only journal articles were used, which had to be either *in Press* or published at the time of the search. The literature had to be published post-2010, when schedule 3 of the UK Flood and Water Management Act 2010 [19] highlighted the need to incorporate more SuDS into design in the UK. Both screening phases were based on the suitability of the literature to provide answers for the challenges presented in the literature on sponge cities. This therefore enabled the identification of key similarities and differences between the Chinese and UK approaches to flood management and mitigation, the challenges that are faced in China, and approaches to the challenges faced in the UK.

3. Historical Pluvial Flood Management

Post-Industrial Revolution, flood management in the UK has taken the approach of efficiently moving water from an urban area to a downstream location, typically a nearby watercourse, using a network of pipes [20]. The London main drainage project was built in the mid-19th Century to manage both sewage and runoff, using piped methods [21]. Similarly in China, conventional pluvial flood management is achieved using pipes and sewage treatment plants. Due to rapid socioeconomic development, capacities of existing drainage systems proved inefficient, and the conventional mode of flood management has become insufficient [22]. However, with limited expansion space, many cities around the world and particularly China, rely on old drainage pipe networks [23], leading to frequent large-scale pluvial flooding and considerable loss of property and life.

In the context of a changing climate however, it is unlikely that existing conventional drainage will manage events within their designed capacity [24,25]. For this reason, more sustainable approaches are beginning to be used to manage pluvial flooding.

4. Flood Management in China & UK

4.1. Sponge Cities: Pluvial Flood Management in China

Due to pluvial flooding in high rainfall season and the lack of water in the dry season, the sponge city concept was proposed as an alternative solution for better urban water management in China [26]. The term "sponge cities" was first proposed in the early 2000s, however, it was not widely adopted with reference to an integrated approach to urban water management in China until 2013, with technical guidance published in 2014 [27–29]. A top down approach to the implementation of sponge cities is largely applied by the Ministry of Housing and Rural-Urban Development (MHURD), Ministry of Finance (MOF) and the Ministry of Water Resources (MWR), who created the Sponge City Construction national guidelines in 2014 [30]. The initiatives are implemented at the city-scale, with some test-cities creating local guidance which is heavily informed by guidelines from the USA, and rarely consider variability in local climate, soil or topography, and often have a preference for grey infrastructure [30,31]. Following the development of guidelines and the desire of the government to implement sustainable urban flood management infrastructure, the aim is that 20% of Chinese cities will use modern drainage techniques, integrating green infrastructure, by 2020, and 80% by 2030, indicating a reliance on retrofit [32].

A sponge city refers to an approach of sustainably managing water, and is based on the "six-word" principle; infiltrate, detain, store, cleanse, use, and drain [30,33,34]. The sponge city model draws on influences from SuDS. Figure 2 illustrates the underlying principles of sponge city and compares it with conventional flood management.

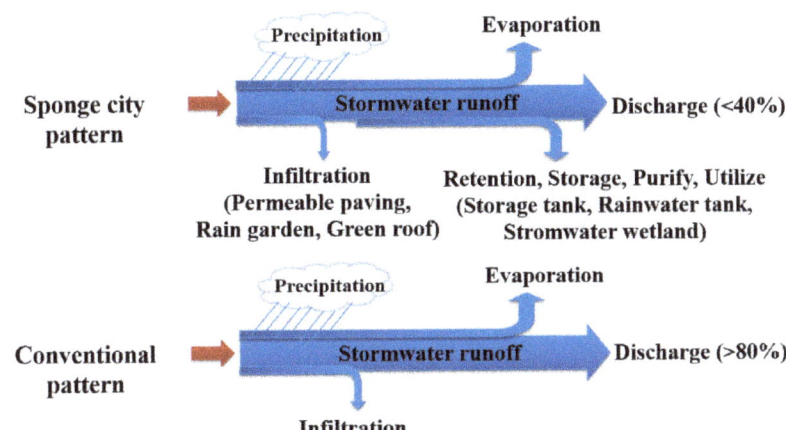

Figure 2. Overview of conventional urban pluvial flood management pattern and sponge city pattern.

Thirty cities, including Beijing and Shanghai, were designated sponge cities across two selection periods in 2014 and 2016 [18]. Most pilot cities are located in central and southern parts of China, with annual precipitation varying from 410 to 1830 mm and annual average temperatures from 4.6 to 25.5 °C. However, regardless of the spatial climatic variability of China, a national approach to standards are taken for sponge cities, with more general guidelines typically outweighing local needs. Li et al. [30] studied two sponge cities, each with a differing climate. Baicheng City, Jilin Province, suffers from water shortages, as annual evaporation outweighs annual rainfall, whereas Shenzen, a low-lying coastal city impacted by seasonal tropical storms, utilized largely consistent SuDS designs and devices, independent of their location. A similar approach has previously been adopted in China for conventional drainage, with cities configured to manage a rainstorm of 187 mm/24 h; a 100% Annual Exceedance Probability (AEP) scenario [35].

To ensure the initiative has the best possibility to succeed, designated sponge city sites should cover more than 20% of the city. The sponge City guideline document stipulates that provision has to be made to drain runoff from up to 3% AEP 24-h rainfall, as opposed to traditional drainage, which is designed to manage runoff from the 100% AEP 24-h event [34].

The development of both retrofit and new build infrastructure is driven by initial central government funding, alongside public-private-partnership (PPP) funds and local subsidies [30]. The amount of funds received is entirely dependent on the administrative levels of candidate cities, for example USD 88 million is given to those guided by State Council, with USD 73 million to provincial capitals, and USD 59 million to other cities [36]. However, due to the scale and need for development as part of the sponge city plan, it is estimated that governmental funding account for just 33% of the total costs, which are expected to be at least USD 22 millions of investment per square kilometer [36]. Additional PPP funds are therefore required to ensure continued growth and maintenance, particularly post the three-year initial funding [30]. Nonetheless, the arrangements for the adoption of SuDS, and ultimately continued maintenance upon completion is unclear.

Beijing was selected in the second phase in 2016, with the primary aim of reducing the impacts of large pluvial flood events, such as the 2012 floods [37]. The sponge city construction in Beijing is expected to control 85% of annual runoff, and manage flooding up to the 2% AEP scenario, through green and grey infrastructure, such as permeable pavements, bio-retention ditches and rain gardens [30]. The plan includes 55 projects over 19.36 km^2, however as of 2017, only eight had been completed, with a further seven under way [18]. Zhang et al. [38] suggest that more needs to be done before the plan can be considered a success, with more emphasis needed on increasing water scarcity issues in Beijing. There are therefore a number of challenges that have arisen as part of the sponge city process which are mainly associated with the differences in natural conditions, financial uncertainty, complexity of the legislation and regulations as well as the degree of public acceptance [27,30].

4.2. Sustainable Drainage Systems: Pluvial Flooding and Management in the UK

Returning drainage to natural processes is one of the chief aims of sustainable drainage (SuDS), increasing infiltration by reducing the amount of impermeable surface. The concept of SuDS arose during the late-1980s due to a philosophy shift favoring sustainable management over hard engineered solutions [39], with Butler and Parkinson [40] questioning the sustainability of traditional piped drainage in urban environments, highlighting the need for an alternative approach. Whilst the main purpose of SuDS is to provide a nature-based drainage system capable of managing large volumes of runoff, they also have wider benefits; improving water quality, enhancing amenity, aesthetics and biodiversity [41].

SuDS installation in the UK has typically centered on single, standalone disconnected devices. However, combining devices to make a "management train", provides a cumulative approach to runoff management, focusing on swales as opposed to pipes for conveyance, increasing opportunities for infiltration [42]. Hamilton in Leicester, UK, (Figure 3) is a 26 ha new development site built in 2003, with 1500 houses built on land previously used for farming, and located in the Environment Agency 'Flood Zone 1'; 0.1% AEP. Pipes are used at the site to transport runoff from the impermeable surfaces to a network of swales, vegetated ponds, filter strips and detention basins, offering a greener, more natural approach to drainage.

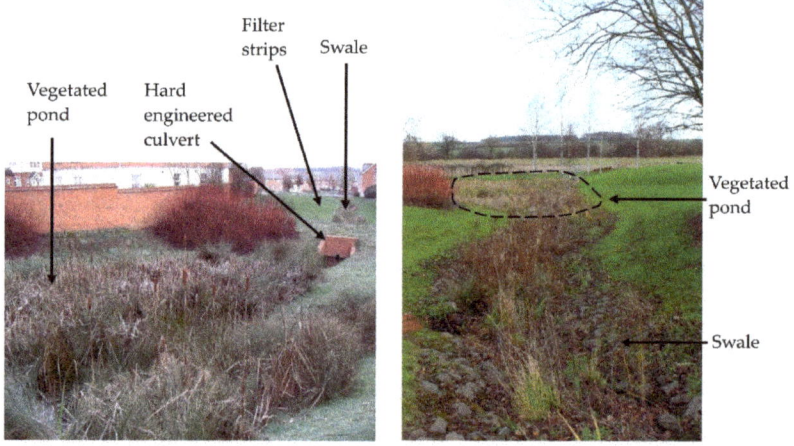

Figure 3. The SuDs management train in Hamilton, Leicester, UK—runoff is conveyed between a series of linked vegetated ponds by small swales, to the nearby watercourse.

SuDS installation is often focused on new build sites, however, only 1% of all buildings in the UK are new builds [43]. The focus therefore should be shifted to retrofitting SuDS; disconnecting stormwater from the existing conventional drainage network into a SuDS device [44]. Consequently, a combined strategy for dealing with both new and old builds is essential to effectively manage pluvial flooding. There are however limited examples of SuDS retrofit across the UK. Lamond et al. [45] highlight this in the UK, attributing it to high initial costs, demands on space in urban environments, the disturbance associated with disconnection from the conventional system and access for maintenance. For this reason, the current focus is on creating an integrated SuDS and conventional drainage approach. Although evidence supports the benefits of SuDS [24,31,44,46,47], there remains an issue in the UK regarding its ownership and ultimately who should be maintaining the systems [48,49].

SuDS require regular maintenance to ensure their continued success, whether that be trimming vegetation, or removing pollutants and de-clogging [50]. Failure to regularly maintain measures, reduces the impact of SuDS and increases the risk of flooding. From a technical perspective, successful maintenance is feasible for all UK SuDS, as the operation of single SuDS devices is sufficiently well understood [41]. In the UK, the barriers to successful SuDS maintenance, including possible retrofit SuDS, are due to difficulties with ownership for the long-term responsibilities and costs of maintenance activity [48,49]. Where ownership of liabilities is uncertain, this inhibits the production of an acceptable site drainage plan, and may well prevent a SuDS scheme from being installed. Incentives in place to produce a plan that facilitates SuDS, include involvement of internal drainage boards (IDB) that charge a commuted sum to help underpin longer-term operations [51]. This can assist in meeting the upfront costs, as the developer contributes to construction, and the asset may be eligible for local authority adoption. Water and Sewerage companies can adopt SuDS assets from a developer, taking on the responsibility of long-term operation and maintenance; both of these options are considered to be low risk as IDB have wide powers locally to intervene on drainage matters and receive reliable local authority and private funding [41,52].

In England and Wales, by default, landowners are the responsible party for maintaining SuDS, but there are options that may be taken to pass responsibility for SuDS assets to a third party. These include adoption by a local authority, a water company or another private company [53]. In Scotland, maintenance is significantly different from the rest of the UK, in that the Scottish Environmental Protection Agency explicitly prefer Scottish Water to adopt the SuDS assets where they are not part of a privately-owned development. In addition, Section 7 of the Sewerage (Scotland) Act 1968 and the 2010 SuDS for Roads guidance are part of a strategy to integrate roads, sewers and surface drainage infrastructure, including SuDS, more effectively [54].

Policy developments in the UK have attempted to highlight the importance of SuDS, particularly in response to recent large flood events. The Non-statutory SuDS standards [55] suggest that all new developments should manage runoff for events up to and including the 1% AEP scenario, with infiltration being the priority destination for runoff, followed by disposal to a nearby watercourse. SuDS design is typically advised by the Construction Industry Research and Information Association (CIRIA), through design guidance [41], the National Planning Policy Framework [56] and opportunity maps which support decision making, created by the British Geological Society [57]. These guidance documents, alongside the Environment Agency and Lead Local Flood Authorities (LLFAs) minimize the impact of new developments on flood risk. However, due to the non-statutory nature of SuDS standards and guidance, many developers continue with conventional piped drainage methods. To ensure the local climate is considered during the site development stage, the Flood Estimation Handbook (FEH) is used to manage the spatial variation of rainfall [58] and the design of rainfall events. The FEH uses depth-duration-frequency models to predict rainfall at the catchment scale for a given storm duration and return period [58].

5. Objective Analysis for Comparisons

Using the approach outlined in the methodology and the information above, there are evident themes and challenges arising regarding urban flood management in China. Although the UK and China are contrasting examples, the UK has attempted to integrate SuDS since the late 1980s, whereas the sponge city initiative in China is a more recent development, beginning in 2014. A number of issues have arisen in the UK regarding the wider integration of SuDS with existing drainage, and implementation as a method of sustainable flood management, therefore a number of comparisons can be drawn, and ultimately lessons that can be learnt and shared. Developing a coherent research strategy to share knowledge is critical, to ensure mistakes are not replicated and new plans can be developed.

A crucial difference between China and the UK is in terms of climate, since in China it is extremely varied, with annual rainfall ranging from as low as 100 mm in the north-west to upwards of 7000 mm in the south-east [18,30]. This is in comparison to a range of 450 mm to 3000 mm in the UK. The Tibetan Plateau to the southwest of China reults in alpine conditions, with sub-arctic conditions possible in the far north, and warmer cities to the south-east, with average temperatures ranging from 4–25 °C [30], compared to 5.5–13 °C average annual temperature range in the UK. This creates a series of challenges across China when attempting to ensure that standards are in place to manage runoff locally under diverse climate conditions. However, although the UK has a much less-varied climate, it has adopted a design methodology to ensure that flood management infrastructure is fit for local purpose; the Flood Estimation Handbook [58]. Comparing this approach for understanding local climate conditions is key to guaranteeing the long-term sustainability of urban pluvial flood management projects in China.

The lack of UK regulations for SuDS has resulted in limited SuDS retrofit examples [45]. Integrating retrofit SuDS is a key aspect of sponge cities, therefore a comparison with the UK can ensure that process and regulations are stringent enough to guarantee the same mistakes are not replicated in China. A strict maintenance and adoption regime is also necessary to ensure the long-term success of different techniques. As highlighted in Section 4.2, current UK practice has resulted in less ownership and maintenance plans of SuDS, reducing their effectiveness, particularly regarding flood management [49]. It is crucial that this is also provided in the context of sponge cities, for the initiative to be successful, and devices to provide the best possible solution to urban pluvial flooding.

As highlighted previously (Section 4.1), consistent funding is necessary to ensure the requirements of flood management of sponge cities are met. However, initial funding is only due to last for three years [59], with a requirement for further funding to be sourced from PPP funds. Comparing this to how SuDS are funded in the UK, and methods for increasing the awareness of SuDS, to generate funding, will ensure that money is available to continue the initiative post-Government funding.

6. Lessons to be Learnt

6.1. Climate

The design process for sponge cities fundamentally relies on design criteria based on nationwide standards, regardless of the local climate [30]. This can result in either an over, or under-engineered solution, depending on the location, however more importantly can further exacerbate issues regarding water availability and/or flooding.

Research is therefore needed to better understand the spatial distribution of rainfall across China, which can inform future sponge city design. An increased network of rain gauges, particularly in major cities, will assist depth-duration-frequency modelling to analyze how the design standard storms vary between regions, similar to the FEH process used in the UK. Implementing such strategies at the city scale will ensure that SuDS are designed to manage local needs, whether that is water scarcity or water excess. Integrating local site and climatic characteristics, such as geology and evaporation rates, alongside rainfall data, will further ensure that the most appropriate SuDS are installed, therefore providing the best possible solution for pluvial flood management.

However, due to the size of China, adopting a similar method to the FEH used in the UK may prove unfeasible in the short term, as the process analyses all catchments across the UK. Nevertheless, a more robust city-scale method of defining rainfall return periods is required to ensure a more transparent process across the country. This can then be adopted by the MWR, and implemented more widely as part of the regulatory process to ensure that local needs are met.

6.2. Regulations

Due to the existing urban infrastructure in China, the high population density and the desire to achieve 80% disconnection from the existing sewer network by 2030, there is a reliance on retrofit systems as part of the sponge city initiative [32]. Retrofit SuDS are still a necessity in the UK, but a more effective framework is required to offer more opportunities for sustainable pluvial flood management. As part of the regulation process in China, more feasibility and opportunity mapping would assist in highlighting locations where retrofit SuDS would have the greatest impact, as has been undertaken in the UK by the British Geological Society [57]. Opportunity mapping ensures that the most suitable sites are developed, and therefore have the maximum impact on runoff reduction. Although mapping has not yet resulted in the widespread implementation of retrofit SuDS in the UK, if it were to be enforced through regulations in China, which are more rigorously applied than in the UK, it is likely to be more readily adopted as part of the wider sponge city initiative.

Nonetheless, many sponge city applications are on a case by case site basis, with a limited view of the impacts of the wider drainage system [30]. Whilst Mei et al. [60] underlined the extent to which different SuDS can work in Liangshuihe, south of Beijing, more research is needed at the city scale by demonstrating the role of disconnection from the central drainage through retrofit SuDS. Disconnection from the existing stormwater system is becoming a more common new build process in the UK, by integrating management trains. Although SuDS management trains typically require large open space, combining SuDS in sponge cities with the existing stormwater drainage network has the potential to reduce total flows by detaining and storing runoff, therefore possibly achieving the desired 80% disconnections discussed above. This approach would offer wider pluvial flood management, as opposed to a series of non-linked or disconnected devices, which could have a limited impact on runoff volumes. However, existing codes of practice do not indicate how SuDS methods can be integrated into existing drainage design [35].

In principle, UK retrofit SuDS should not should be more problematic than new-build schemes from a planning and financial feasibility view. As discussed above, many of the incentives that facilitate SuDS installation, such as commuted sums and underwriting by local authorities and sewage or water companies, are equally applicable to retrofit. In practice, the association between retrofit and high value urbanized locations, particularly in the case of sponge city type initiatives, may make the upfront cost of retrofit prohibitive. It is also well established that urban environments exclude certain types of SuDS such as ponds, wetlands and extensive green infrastructure [41]. A possible option for successful urban SuDS in the UK and China, particularly with retrofit, may be to encourage more disconnection of individual properties from the drainage network by using inexpensive rainwater harvesting systems, raingardens and permeable pavements. This would avoid high upfront costs, be achievable during refurbishment and be a feasible proposition for a householder to maintain.

As is also necessary for the UK, research regarding maintenance is required to directly inform guidance and regulation for SuDS. To better understand what the necessary design requirements are for retrofit SuDS as part of the sponge city process, a robust site-selection tool is required, accounting for localized factors. A series of vulnerability assessments are also essential to identify those areas susceptible to pluvial flooding. This will ensure that the most suitable SuDS are installed for different environments, ensuring the best value for money, and the future success of devices.

6.3. Funding for Development

The large initial outlay for funding for sponge cities in China is at odds with the UK approach, where developers are expected to adhere to non-statutory standards with small financial incentives provided if sustainable flood management is integrated into development plans. However, the Chinese central funding plan is only for three years, with sponge cities expected to raise further funding through PPP, requiring greater community engagement to develop necessary links. To do this, similar to the UK, there needs to be more community incentives to drive projects ensuring small, local scale installation, and engagement in the sponge city process to ensure the long-term success and viability of projects [61]. However, as was identified by Wang et al. [61], only 61% of a sample population were aware of the sponge city program. Consequently, although expenditure is high, public engagement in the projects remains low. Research is therefore needed to develop plans for engagement and education of the sponge city process to better engage the public with sustainable pluvial flood management techniques, but importantly, also engage potential future funders through the PPP approach.

As the Chinese investment model for sponge cities relies heavily on external funding sources, funders expect a return on their investment, but this is unlikely [36]. For this reason, it is possible that sponge city construction will slow down, and achieving the government target of 80% of all cities to utilize more modern, green infrastructure drainage techniques by 2030 will be challenging. The UK has had similar problems, particularly in the context of retrofit SuDS, with a lack of ownership, resulting in reduced maintenance of devices, and an overall lack of desire to integrate SuDS into drainage schemes [31,62]. Undertaking an assessment of the whole life cost of sponge city developments is therefore essential in order to ensure that PPP funding can support the initiatives, post-government funding support.

Funding support can be further reinforced by educating key stakeholders on the benefits of SuDS in urbanized environments, regarding flood risk management. To do this, field monitoring of before and after construction of SuDS in sponge cities will provide evidence of the likely reduction in runoff and ultimately, reduction in urban flood extents. Assessments of the success of sponge cities are often calculated through modelling the after impacts of SuDS [63–65], with monitoring practices undertaken sporadically [32,64,66] and no formal approach to monitoring available [18]. Creating a robust monitoring approach will provide evidence of the success of devices, therefore, understanding the impact of SuDS in sponge cities will increase the confidence of key decision makers in the initiative, and their likely engagement in the process.

7. Summary: Identifying Opportunities for Future Research

This review paper has outlined how both the UK and China manage pluvial flooding, with a view to examining the strengths and weaknesses of SuDS and the sponge cities approach. The paper presents novel research topics that are required to ensure the long-term success of the sponge city project in China, based on lessons learnt from the UK. The Chinese Government have spent approximately USD 25 billion on the 30 sponge city projects, therefore they need to be sustainable, whilst also working efficiently and effectively [30]. Sponge cities are not solely focused on managing flooding, but must also achieve all facets of sustainable drainage; infiltrate, detain, store, cleanse, use, and drain. With this in mind, the following areas have been identified, based on the challenges posed, as priority research topics to generate future research:

1. Develop a more localized rainfall model for China, to ensure that local climate characteristics are accounted for in the design of sponge cities and therefore the most appropriate SuDS are integrated dependent on the population needs.
2. Understand the role and cost benefits of retrofit SuDS in challenging water environments at the city scale.
3. Mapping vulnerability, undertaking feasibility assessments and the potential of disconnections to provide sustainable pluvial flood management, and create a robust SuDS selection tool, ensuring that the most effective devices are installed, based on local factors.
4. Bring maintenance, whole life costing approaches and before-after implementation monitoring to SuDS and sponge City developments to increase and disseminate current information and increase the confidence of decision makers when choosing unfamiliar drainage solutions.
5. Assess how community engagement and education of sponge cities can be better developed to foster potential funding and develop more local partnerships.

The review concludes that each of these five research recommendations are crucial for ensuring the future success of the sponge City programme. Furthermore, the underlying research will better inform global practice by developing retrofit pluvial flood management schemes in urban environments, in the context of a changing climate.

Author Contributions: Conceptualization, C.L., M.R. and S.T.; Methodology, C.L.; Investigation, C.L., M.R., Y.C., J.H., S.A., S.C. (Stephen Coupe), S.C. (Susanne Charlesworth); Formal Analysis C.L., M.R., Y.C., J.H., S.C. (Stephen Coupe), S.C. (Susanne Charlesworth), S.T.; Data Curation, C.L., M.R., Y.C., J.H., S.A., S.C. (Stephen Coupe); Writing—Original Draft, C.L., M.R., Y.C., J.H.; Writing—Reviewing and Editing M.R., S.A., S.C. (Stephen Coupe), S.C. (Susanne Charlesworth); Visualization, C.L., Y.C., J.H. and S.A.; Supervision, S.C. (Susanne Charlesworth), S.T.; Funding Acquisition, M.R., S.T.

Funding: This work was completed under the EPSRC grant—EP/K040405/1.

Acknowledgments: Particular thanks to: Dragan Savić, Guangtao Fu, Albert Chen, Fanlin Meng (University of Exeter), Yuntao Guan and Haifeng Jia (Tsinghua University) and Huapeng Qin (Peking University), for organizing the workshop "Urban Flooding and Sponge cities", held on 3–5 July 2017, at Shenzhen, China, facilitating discussion between the participants and enabling new collaboration between early career researchers; David Butler, Guangtao Fu, Fanlin Meng (University of Exeter), Xiaochang Wang (Xi'an University of Architecture and Technology) Nanqi Ren (Harbin Institute of Technology) and Qiang He (Chongqing University) for organizing the workshop "Water-Wise Cities and Smart Water Systems", held on 11–13 September 2018, at Xi'an, China, facilitating discussion between the participants and enabling new collaboration between early career researchers.

Conflicts of Interest: The authors declare no conflict of interest. The funders had no role in the design of the study; in the collection, analyses, or interpretation of data; in the writing of the manuscript, or in the decision to publish the results.

References

1. Guha-Sapir, D.; Hoyois, P.; Wallemacq, P.; Below, R. Annual Disaster Statistical Review 2016: The Numbers and Trends. 2017. Available online: https://reliefweb.int/sites/reliefweb.int/files/resources/adsr_2016.pdf (accessed on 18 August 2018).
2. Tang, G.; Zeng, Z.; Ma, M.; Liu, R.; Wen, Y.; Hong, Y. Can near-real-time satellite precipitation products capture rainstorms and guide flood warning for the 2016 summer in South China? *IEEE Geosci. Remote Sens. Lett.* **2017**, *14*, 1208–1212. [CrossRef]
3. Environment Agency. Estimating the Economic Costs of the Winter Floods 2015 to 2016. 2018. Available online: https://assets.publishing.service.gov.uk/government/uploads/system/uploads/attachment_data/file/672087/Estimating_the_economic_costs_of_the_winter_floods_2015_to_2016.pdf (accessed on 18 August 2018).
4. Intergovernmental Panel on Climate Change. Climate Change 2013: Physical Science Basis (AR5). 2013. Available online: https://www.ipcc.ch/report/ar5/wg1/ (accessed on 18 August 2018).
5. United Nations. World Urbanization Prospects: The 2018 Revision. 2018. Available online: https://population.un.org/wup/Publications/Files/WUP2018-KeyFacts.pdf (accessed on 18 August 2018).
6. Miller, J.D.; Hutchins, M. The impacts of urbanisation and climate change on urban flooding and urban water quality: A review of the evidence concerning the United Kingdom. *J. Hydrol. Reg. Stud.* **2017**, *12*, 345–362. [CrossRef]
7. Kuriqi, A.; Ardiçlioglu, M.; Muceku, Y. Investigation of seepage effect on river dike's stability under steady state and transient conditions. *Pollack Period.* **2016**, *11*, 87–104. [CrossRef]
8. Leopold, L.B. Hydrology for urban planning—A guidebook on the hydrological effects of urban land use. *Geol. Surv. Circ.* **1968**, *554*, 1–18.
9. Miller, J.D.; Kim, H.; Kjeldsen, T.R.; Packman, J.; Grebby, S.; Dearden, R. Assessing the impact of urbanization on storm runoff in a peri-urban catchment using historical change in impervious cover. *J. Hydrol.* **2014**, *515*, 59–70. [CrossRef]
10. Ahilan, S.; Guan, M.; Sleigh, A.; Wright, N.; Chang, H. The influence of floodplain restoration on flow and sediment dynamics in an urban river. *J. Flood Risk Manag.* **2018**, *11*, 986–1001. [CrossRef]
11. Office for National Statistics. Population estimates for the UK, England and Wales, Scotland and Northern Ireland: Mid 2017. 2018. Available online: https://www.ons.gov.uk/peoplepopulationandcommunity/populationandmigration/populationestimates/bulletins/annualmidyearpopulationestimates/mid2017/pdf (accessed on 8 September 2018).
12. Golding, B.; Roberts, N.; Leoncini, G.; Mylne, K.; Swinbank, R. MOGREPS-UK Convection-Permitting Ensemble Products for Surface Water Flood Forecasting: Rationale and First Results. *J. Hydrometeorol.* **2016**, *17*, 1383–1406. [CrossRef]

13. Soetanto, R.; Mullins, A.; Achour, N. The perceptions of social responsibility for community resilience to flooding: The impact of past experience, age, gender and ethnicity. *Nat. Hazards* **2017**, *86*, 1105–1126. [CrossRef]
14. Zheng, Z.; Qi, S.; Xu, Y. Questionable frequent occurrence of urban flood hazards in modern cities of China. *Nat. Hazards* **2013**, *65*, 1009–1010. [CrossRef]
15. Lyu, H.-M.; Xu, Y.-S.; Cheng, W.-C.; Arulrajah, A. Flooding Hazards across Southern China and Prospective Sustainability Measures. *Sustainability* **2018**, *10*, 1682. [CrossRef]
16. Jiang, X.; Yuan, H.; Xue, M.; Xue, M.; Chen, X.; Tan, X. Analysis of a heavy rainfall event over Beijing during 21–22 July 2012 based on high resolution model analyses and forecasts. *J. Meteorol. Res.* **2014**, *28*, 199–212. [CrossRef]
17. García-Feal, O.; González-Cao, J.; Gómez-Gesteira, M.; Cea, L.; Domínguez, J.; Formella, A. An Accelerated Tool for Flood Modelling Based on Iber. *Water* **2018**, *10*, 1459. [CrossRef]
18. Jiang, Y.; Zevenbergen, C.; Ma, Y. Urban pluvial flooding and stormwater management: A contemporary review of China's challenges and "sponge cities" strategy. *Environ. Sci. Policy* **2018**, *80*, 132–143. [CrossRef]
19. Flood and Water Management Act. Chapter 29. 2010. Available online: http://www.legislation.gov.uk/ukpga/2010/29/contents (accessed on 8 September 2018).
20. Wheater, H.; Evans, E. Land use, water management and future flood risk. *Land Use Policy* **2009**, *26*, S251–S264. [CrossRef]
21. Hughes, M. The Victorian London sanitation projects and the sanitation of projects. *Int. J. Proj. Manag.* **2013**, *31*, 682–691. [CrossRef]
22. Qiu, B.X. The connotation, approach and perspective of Sponge city and LID. *Water Wastewater Eng.* **2015**, *41*, 1–7. (In Chinese)
23. Zhou, Z.; Qu, L.; Zou, T. Quantitative Analysis of Urban Pluvial Flood Alleviation by Open Surface Water Systems in New Towns: Comparing Almere and Tianjin Eco-City. *Sustainability* **2015**, *7*, 13378–13398. [CrossRef]
24. Semadeni-Davies, A.; Hernebring, C.; Svensson, G.; Gustafsson, L.-G. The impacts of climate change and urbanisation on drainage in Helsingborg, Sweden: Suburban stormwater. *J. Hydrol.* **2008**, *350*, 114–125. [CrossRef]
25. Bell, V.A.; Kay, A.L.; Cole, S.J.; Jones, R.G.; Moore, R.J.; Reynard, N.S. How might climate change affect river flows across the Thames Basin? An area-wide analysis using the UKCP09 Regional Climate Model ensemble. *J. Hydrol.* **2012**, *442–443*, 89–104. [CrossRef]
26. Zhang, P.; Cai, Y.; Wang, J. A simulation-based real-time control system for reducing urban runoff pollution through a stormwater storage tank. *J. Clean. Prod.* **2018**, *183*, 641–652. [CrossRef]
27. Wang, H.; Chao, M.; Liu, J.H.; Shao, W.W. A new strategy for integrated urban water management in China: Sponge city. *Sci. China Technol. Sci.* **2018**, *3*, 1–13. [CrossRef]
28. Rooijen, D.J.V.; Turral, H.; Biggs, T.W. Sponge city: Water balance of mega-city water use and wastewater use in Hyderabad, India. *Irrig. Drain.* **2010**, *54*, S81–S91. [CrossRef]
29. Zhang, J.Y.; Wang, Y.T.; Hu, Q.F. Discussion and views on some issues of the sponge city construction in China. *Adv. Water Sci.* **2016**, *27*, 793–799. (In Chinese)
30. Li, H.; Ding, L.; Ren, M.; Li, C.; Wang, H. Sponge city construction in China: A survey of the challenges and opportunities. *Water* **2017**, *9*, 594. [CrossRef]
31. Hoang, L.; Fenner, R.A. System interactions of stormwater management using sustainable urban drainage systems and green infrastructure. *Urban Water J.* **2015**, 1–20. [CrossRef]
32. Nguyen, T.T.; Ngo, H.H.; Guo, W.; Wang, X.C.; Ren, N.; Li, G.; Ding, J.; Liang, H. Implementation of a specific urban water management—Sponge City. *Sci. Total Environ.* **2019**, *652*, 147–162. [CrossRef] [PubMed]
33. Hu, C.W. Reconstruction of urban water ecology by "Sponge city". *Ecol. Econ.* **2015**, *31*, 10–13. (In Chinese)
34. Ministry of Housing and Urban-Rural Development. Technical Guide for Sponge Cities-Water System Construction of Low Impact Development. 2014. Available online: http://www.mohurd.gov.cn/zcfg/jsbwj_0/jsbwjcsjs/201411/W020141102041225.pdf (accessed on 16 September 2018).
35. Chan, F.K.S.; Griffiths, J.A.; Higgitt, D.; Xu, S.; Zhu, F.; Tang, Y.T.; Xu, Y.; Thorne, C.R. "Sponge City" in China-A breakthrough of planning and flood risk management in the urban context. *Land Use Policy* **2018**. [CrossRef]

36. Dai, L.; van Rijswick, H.F.M.W.; Driessen, P.P.J.; Keessen, A.M. Governance of the Sponge City Programme in China with Wuhan as a case study. *Int. J. Water Resour. Dev.* **2017**, *34*, 1–19. [CrossRef]
37. Li, N.; Qin, C.; Du, P. Optimization of China Sponge City Design: The Case of Lincang Technology Innovation Park. *Water* **2018**, *10*, 1189. [CrossRef]
38. Zhang, S.; Yongkun, L.; Ma, M.; Song, T.; Ruining, S. Storm Water Management and Flood Control in Sponge City Construction of Beijing. *Water* **2018**, *10*, 1040. [CrossRef]
39. Pompêo, C. Development of a state policy for sustainable urban drainage. *Urban Water* **1999**, *1*, 155–160. [CrossRef]
40. Butler, D.; Parkinson, J. Towards sustainable urban drainage. *Water Sci. Technol.* **1997**, *35*, 53–63. [CrossRef]
41. Woods Ballard, B.; Wilson, S.; Udale-Clarke, H.; Illman, S.; Scott, T.; Ashley, R.; Kellagher, R. *The SuDS Manual (C753)*; Construction Industry Research and Information Association (CIRIA): London, UK, 2015.
42. Lashford, C.; Charlesworth, S.; Warwick, F.; Blackett, M. Deconstructing the Sustainable Drainage Management Train in Terms of Water Quantity—Preliminary Results for Coventry, UK. *Clean* **2014**, *42*, 187–192. [CrossRef]
43. Committee on Climate Change. Climate change—Is the UK Preparing for Flooding and Water Scarcity? 2012. Available online: https://www.theccc.org.uk/archive/aws/ASC/CCC_ASC_2012_bookmarked_2.pdf (accessed on 6 October 2018).
44. Stovin, V. The potential of green roofs to manage Urban Stormwater. *Water Environ. J.* **2010**, *24*, 192–199. [CrossRef]
45. Lamond, J.E.; Rose, C.; Booth, C.A. Evidence for improved urban flood resilience by sustainable drainage retrofit. *Proc. Inst. Civ. Eng. Urban Des. Plan.* **2015**, *168*, 101–111. [CrossRef]
46. Ellis, J.B. Sustainable surface water management and green infrastructure in UK urban catchment planning. *J. Environ. Plan. Manag.* **2016**, *56*, 24–41. [CrossRef]
47. Ellis, J.B.; Viavattene, C. Sustainable urban drainage system modeling for managing urban surface water flood risk. *Clean* **2014**, *42*, 153–159. [CrossRef]
48. Everett, G.; Lamond, J.; Morzillo, A.; Chan, F.K.S.; Matsler, A.M. Sustainable drainage systems: Helping people live with water. *Proc. Inst. Civ. Eng.* **2016**, *169*, 94–104. [CrossRef]
49. Melville-Shreeve, P.; Cotterill, S.; Grant, L.; Arahuetes, A.; Stovin, V.; Farmani, R.; Butler, D. State of SuDS delivery in the United Kingdom. *Water Environ. J.* **2018**, *32*, 9–16. [CrossRef]
50. Scholz, M. Case study: Design, operation, maintenance and water quality management of sustainable storm water ponds for roof runoff. *Bioresour. Technol.* **2004**, *95*, 269–279. [CrossRef] [PubMed]
51. Ellis, J.B.; Lundy, L. Implementing sustainable drainage systems for urban surface water management within the regulatory framework in England and Wales. *J. Environ. Manag.* **2016**, *183*, 630–636. [CrossRef] [PubMed]
52. Association of Drainage Authorities. An introduction to Internal Drainage Boards. 2017. Available online: https://www.ada.org.uk/wp-content/uploads/2017/12/IDBs_An_Introduction_A5_2017_web.pdf (accessed on 24 November 2018).
53. Susdrain. Sustainable Drainage Systems (SuDS) Maintenance and Adoption Options (England). 2015. Available online: https://www.susdrain.org/files/resources/fact_sheets/09_15_fact_sheet_suds_maintenance_and_adoption_options_england_.pdf (accessed on 24 November 2018).
54. Susdrain. SuDS Adoption in Scotland. Available online: https://www.susdrain.org/delivering-suds/using-suds/adoption-and-maintenance-of-suds/adoption/SuDS-adoption-in-Scotland.html (accessed on 24 November 2018).
55. Department for Environment, Food and Rural Affairs. Non-Statutory Technical Standards for Sustainable Drainage Systems. 2015. Available online: https://www.gov.uk/government/uploads/system/uploads/attachment_data/file/415773/sustainable-drainage-technical-standards.pdf (accessed on 24 November 2018).
56. Department for Communities and Local Government. National Planning Policy Framework. 2012. Available online: https://www.gov.uk/government/uploads/system/uploads/attachment_data/file/6077/2116950.pdf (accessed on 24 November 2018).
57. Dearden, R.; Marchant, A.; Royse, K. Development of a suitability map for infiltration sustainable drainage systems (SuDS). *Environ. Earth Sci.* **2013**, *70*, 2587–2602. [CrossRef]
58. Institute of Hydrology. *Flood Estimation Handbook*; Institute of Hydrology: London, UK, 1999.

59. Liang, X. Integrated Economic and Financial Analysis of China's Sponge City Program for Water-resilient Urban Development. *Sustainability* **2018**, *10*, 669. [CrossRef]
60. Mei, C.; Liu, J.; Wang, H.; Yang, Z.; Ding, X.; Shao, W. Integrated assessments of green infrastructure for flood mitigation to support robust decision-making for sponge city construction in an urbanized watershed. *Sci. Total Environ.* **2018**, *639*, 1394–1407. [CrossRef] [PubMed]
61. Wang, Y.; Sun, M.; Song, B. Public perceptions of and willingness to pay for sponge city initiatives in China. *Resour. Conserv. Recycl.* **2017**, *122*, 11–20. [CrossRef]
62. Wright, G.B.; Jack, L.B. Property-level stormwater drainage systems: Integrated flow simulation and whole-life costs. *Build. Res. Inf.* **2013**, *41*, 223–236. [CrossRef]
63. Luan, Q.; Fu, X.; Song, C.; Wang, H.; Liu, J.; Wang, Y. Runoff Effect Evaluation of LID through SWMM in Typical Mountainous, Low-Lying Urban Areas: A Case Study in China. *Water* **2017**, *9*, 439. [CrossRef]
64. Yuan, Y.; Xu, Y.-S.; Arulrajah, A. Sustainable Measures for Mitigation of Flooding Hazards: A Case Study in Shanghai, China. *Water* **2017**, *9*, 310. [CrossRef]
65. Li, Q.; Wang, F.; Yu, Y.; Huang, Z.; Li, M.; Guan, Y. Comprehensive performance evaluation of LID practices for the sponge city construction: A case study in Guangxi, China. *J. Environ. Manag.* **2019**, *231*, 10–20. [CrossRef]
66. Jiang, Y.; Zevenbergen, C.; Fu, D. Understanding the challenges for the governance of China's "sponge cities" initiative to sustainably manage urban stormwater and flooding. *Nat. Hazards* **2017**, *89*, 521–529. [CrossRef]

© 2019 by the authors. Licensee MDPI, Basel, Switzerland. This article is an open access article distributed under the terms and conditions of the Creative Commons Attribution (CC BY) license (http://creativecommons.org/licenses/by/4.0/).

Article

Can We Really Have It All?—Designing Multifunctionality with Sustainable Urban Drainage System Elements

Elisa Lähde [1],*, Ambika Khadka [2], Outi Tahvonen [1] and Teemu Kokkonen [2]

[1] Department of Architecture, School of Arts, Design and Architecture, Aalto University, 00076 Espoo, Finland; outi.tahvonen@aalto.fi
[2] Department of Built Environment, School of Engineering, Aalto University, 00076 Espoo, Finland; ambika.khadka@aalto.fi (A.K.); teemu.kokkonen@aalto.fi (T.K.)
* Correspondence: elisa.lahde@aalto.fi

Received: 10 February 2019; Accepted: 21 March 2019; Published: 28 March 2019

Abstract: Multifunctionality is seen as one of the key benefits delivered by sustainable urban drainage systems (SUDS). It has been promoted by both scientific research and practical guidelines. However, interrelations between different benefits are vaguely defined, thus highlighting a lack of knowledge on ways they could be promoted in the actual design process. In this research, multifunctionality has been studied with the help of scenario analysis. Three stormwater scenarios involving different range of SUDS elements have been designed for the case area of Kirstinpuisto in the city of Turku, Finland. Thereafter, the alternative design scenarios have been assessed with four criteria related to multifunctionality (water quantity, water quality, amenity, and biodiversity). The results showed that multifunctionality could be analyzed in the design phase itself, and thus provided knowingly. However, assessing amenity and biodiversity values is more complex and in addition, we still lack proper methods. As the four criteria have mutual interconnections, multifunctionality should be considered during the landscape architectural design, or else we could likely lose some benefits related to multifunctionality. This reinforces emerging understanding that an interdisciplinary approach is needed to combine ecological comprehension together with the system thinking into SUDS design, locating them not as individual elements or as a part of the treatment train, but in connection with wider social ecological framework of urban landscape.

Keywords: stormwater management; multifunctionality; landscape design; water sensitive urban design (WSUD)

---

1. Introduction

During the last decade, with the emergence of the concept of green infrastructure (GI) and its recognition as a network of natural and semi-natural areas delivering multiple benefits [1] into urban landscape planning, multifunctionality has subsequently crystallized as a defining criterion for ascertaining the quality of this urban landscape [2–6]. As it has become desirable for the capacity of the urban landscape to expand to provide multiple benefits, multifunctionality has emerged as an aspect of great importance. This has been further enhanced by the compact city ideology promoted by agencies, such as the UN's New Urban Agenda [7]. Indeed, this compact city structure reduces opportunities for urban greenspaces and inevitably requires them to be multifunctional [8].

As the GI approach becomes adopted, there is an on-going and simultaneous transformation towards water sensitive urban design (WSUD), due to climate adaptation and water quality issues [9]. WSUD offers an alternative to sewer based urban drainage systems and covers a series of ecosystem service based approaches to urban stormwater management. Furthermore, it encourages the use of above-ground solutions, such as rain gardens, swales, green roofs, and wetlands (i.e., technologies called sustainable urban drainage systems (SUDS)); in fact, the delivery of multiple benefits is an essential part of the approach [10]. The role of SUDS is to harvest, infiltrate, slow, store, convey, and treat runoff on site [11] to sustain the existing local hydrology.

In addition, along with direct water-related benefits, SUDS possess the potential to create synergies with other functions in urban areas. An increased amount of vegetation combined with visible water provides several ecosystem services, such as habitat provision, erosion control, microclimate regulation, recreation, and aesthetical experiences [12,13]. However, there is no precise understanding of the ways multifunctionality and a combination of benefits can be promoted with SUDS in an urban landscape, due to related research concentrating mainly on the evaluation of individual benefits [14,15]. Moreover, studies that simultaneously touch on hydrological and ecological benefits do not consider the design process, but mainly evaluate existing structures [10,15].

This paper examines opportunities to design multifunctional urban greenspaces by integrating SUDS elements into the urban landscape. The aims are to shed light on the preconditions required for the provision of different SUDS related benefits, and further discuss the ways they can be addressed in the landscape architectural design of urban greenspaces. This paper answers the question of how the multifunctionality of SUDS can be estimated during the landscape architectural design process. The results are discussed to additionally understand the relations between different criteria of multifunctional SUDS, as well as ways of consequently incorporating this understanding during the design phase.

A scenario analysis is the method chosen with the research being conducted in three phases. First, three scenarios have been created representing three different strategies of stormwater management: (1) substituting part of the pipe network with open swales, (2) adding SUDS elements that allow water detention, and (3) maximizing the amount of SUDS elements on the site. This is based on an approach that combines SUDS elements into differing treatment trains allowing the formation of a portfolio of options, which contribute to a variety of benefits [16]. The scenarios have been designed based on a case study area of the site of Kirstinpuisto in Turku, Finland; each of them is composed of a varying combination of SUDS elements to create three different treatment trains.

In the second phase, methods to measure the potentially provided benefits (stormwater quantity and quality management, amenity, and biodiversity) are studied and tested with the three scenarios. Finally, possible synergies or conflicts among different benefits are scrutinized and discussed, including the potential of the landscape architectural design process to provide multifunctional greenspaces through stormwater management.

2. Multifunctional SUDS

Multifunctionality is defined as "an integration and interaction between functions" [17] (p. 655) or as an ability of GI to "perform several functions and provide several benefits on the same spatial area" [3]. Multifunctionality is also described as the capacity of GI to provide multiple ecosystem services (ESS) [18]. In the ESS approach, benefits are commonly divided further into provisioning, regulating, and cultural ecosystem services, according to the Common International Classification for Ecosystem Services, with the understanding being that by simultaneously providing these, it could help achieve several environmental, social, and economic urban policy aims [19].

The ESS approach is closely linked to the cascade model of ecosystem services [20] stating that without correct biological structures, processes, and functions, the provision of ecosystem services is incomplete. Furthermore, the provision of services leads to human well-being and valuation of the provided services (e.g., monetary value). Hansen and Pauleit [4] have underlined that in GI approaches, the term "functions" can be confusingly used to mean the same as "services," whereas in the ESS concept, "functions" are understood as an intermediary step of the biophysical structures and processes needed to provide ESS. In this paper, "functions" and "services" are understood in line with the ESS cascade model, highlighting our dependency on well-functioning urban green elements. Such elements should be planned, designed, and managed in a way that is "sensitive to, and includes provision for, natural features and systems" [3].

Although multifunctionality is regarded as being essential and its connection to biological structures and functions is commonly recognized, the conflicts or synergies between different benefits have not been adequately studied. Meerow and Newell [21] have argued that most green infrastructure related research and planning focus only on a handful of benefits, despite a major demand for the use of GI to mediate between different and potentially conflicting demands [19]. If multifunctionality is seen as the main feature of GI, which delivers solutions to urban environmental challenges and maintains the quality of life [6], it is essential that research is the framework through which we understand the potential synergies or conflicts among its assigned benefits as well as the limitations of providing them through landscape architectural design.

More specifically, urban planning and design outline facilities for urban multifunctionality. In the context of a green infrastructure, it means the integration of systems supporting vegetation growth, such as water, vegetation, or carbon cycles. However, the operationalization of multifunctionality in planning [4,22] and practical examples are still lacking in GI planning and design.

CIRIA, a well-known and respected British forum for water sector industry improvement, has defined the multifunctionality of SUDS. In its guidelines [11], CIRIA has provided four criteria for the design of SUDS—water quantity, water quality, amenity, and biodiversity (Figure 1). Despite these guidelines, the design, implementation, and maintenance of SUDS often emphasize drainage functions over its additional benefits [16,23]. Moreover, when measuring SUDS multifunctionality, a mostly natural sciences approach has been utilized to explore and enumerate the provision of quantity and quality management; in addition, amenity and biodiversity provision have been less well researched [12,14,15,24]. Thus, the authentication of multifunctionality with SUDS in landscape architectural design of urban greenspaces still lacks precise indicators. In this study, the design criteria provided by the aforementioned C753 SUDS Manual [11] are utilized as a framework to define the multifunctionality of SUDS solutions.

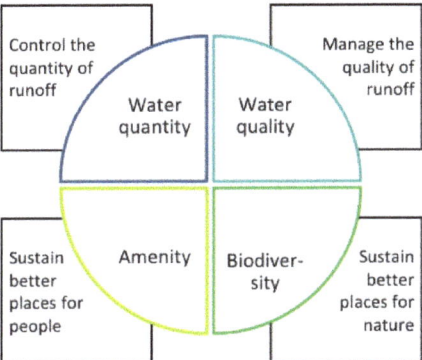

Figure 1. According to CIRIA [11] multifunctionality of sustainable urban drainage systems (SUDS) based on the simultaneous existence of four criteria; quality and quantity control, biodiversity, and amenity. However, any mutual interconnections are not presented (figure adapted from CIRIA [11]).

SUDS are inherently multifunctional structures if the criteria are considered sufficiently early on and are fully integrated into the urban design [11]. In the following section, each of the criteria is shortly introduced together with an understanding of the ways they can be promoted through design. Additionally, the four criteria provided by SUDS are not independent of each other [14,15,25]; thus, mutual interconnections are also clarified.

Being part of the drainage network, the primary function of SUDS is to control water quantity [13] to prevent both flooding on-site and in downstream areas. Additionally, on-site water quantity management helps to preserve the natural hydrological functions of a catchment. We are aware that different SUDS elements possess a varying effectiveness to perform run-off regulation [26]; for example, bioretention cells infiltrate water and slow down surface flow together with vegetation that additionally intercepts and evaporates water [27]. In the design process, varying SUDS elements can be chosen and combined depending on the qualities of the site; for example, if there is an abundance of space available, aboveground elements can be used, but if the urban structure is very dense, green roofs might be needed. Furthermore, both the location in the watershed and runoff coefficient affect the amount of stormwater, which then specify dimensioning of elements and choice of vegetation.

On-site water quality management safeguards water quality in the receiving surface waters and ground waters. This impacts the living conditions of a variety of water-related flora and fauna as well as the wellbeing of local residents. The overall impact of a site on water quality is dependent on types of pollutants, the peak flow pollutant concentrations, and the total pollutant load in the runoff [11].

SUDS elements provide water quality improvements by reducing sediment and contaminants from runoff either through settlement or biological breakdown of pollutants. Multiple plant-related mechanisms, such as phytoextraction and phytodegradation [27], are important for biological treatment and pollutant removal. Again, different SUDS have different impacts; i.e., bioretention cells are effective in filtration, sedimentation, adsorption, and plant uptake [28], while extensive green roofs have a varying ability to retain pollutants depending on the season, substrate type, event size, and rainfall regime [29,30]. If the functions of different SUDS are known, it is possible to match the right SUDS elements to meet local stormwater quality management needs in the design process.

Amenity is related to the attractiveness of the site and the provision of recreation and leisure services [12]. Echols and Pennypacker [31] have listed amenity goals as being education, recreation, safety, public relations, and aesthetics. Furthermore, visible water and SUDS increase the amenity of urban green areas [32]. The amenity values experienced in existing urban greenspaces can be measured by scoring systems [13] or by investigating public perception (i.e., with questionnaires, such as those conducted by Bastien et al. [33]). During the design process, amenity values are challenging to measure, but opportunities for recreation, education, and human contact with nature, can be maximized by enhancing ease of public access and social interaction.

In addition, increased biodiversity affects perceived amenity in positive ways [12,34]; hence, SUDS with vegetation potentially adds amenity values. These values are increased by using above-ground SUDS and linking stormwater management to other functions in urban landscape [32]. Thus, already in the design phase, the proximity of SUDS elements to other structures, such as pathways, urban squares, and residential buildings allowing interplay with water, can actualize amenity values.

Biodiversity supports human wellbeing in various direct and indirect ways as biophysical structures, including functions related to biodiversity, are essential for ecosystem service provision [35]. Urban biodiversity relies on urban greenspaces in which human activities affect ecological processes [36]. In urban conditions, the land use changes, and the transformation of technical and social infrastructures as well as management practices can cause a loss of biodiversity [37].

Furthermore, biodiversity is based on ecological processes including decomposition, nutrient cycling, and fluxes of nutrients and energy [38], in which the hydrological cycle and water availability are essential features. Thus, SUDS contribute positively to local biodiversity [14,39], but for vegetation, it is a risk to consider SUDS only as a part of urban drainage systems. SUDS with vegetation, as with any biophysical structure, require physical inputs of nutrients and water to provide ecological functions [16]. Habitat heterogeneity, biomass production, and biodiversity benefit from the storing and infiltration of rainwater into the soil, instead of turning it into surface flow [15,38,40,41].

Similar to amenity values, there are ways of measuring the biodiversity of existing greenspaces [32,33,42]. In the landscape architectural design process, conditions for biodiversity are created through the vegetation and microbiology of soils; in this way, the implemented design later provides a platform for animal diversity. However, in the design phase, it is difficult to measure future level of biodiversity as it depends on factors, such as the level of maintenance and scale of ecological succession once the design has been realized [43]. Nevertheless, there are some factors that support development of local biodiversity and could be enhanced in design. Structural habitat heterogeneity that is created by abiotic and biotic components of SUDS solution is associated with a high degree of biological diversity, and can already be used as a proxy for biodiversity [39] in the design phase. Furthermore, biodiversity correlates with the size of the habitat, edge effect and connectivity of habitats [2,42]. When emphasizing the biodiversity aspect of SUDS elements, it is important to relate them to neighboring habitats and the larger ecological network.

3. Case Study and Methodology

This section introduces the case site of Kirstinpuisto and three stormwater management scenarios, as well as presents the methodology used to assess water quantity and quality by modeling. It is followed by the presentation and testing of two new assessment methods for amenity and biodiversity values. The results are shared in Section 4.

3.1. Kirstinpuisto Site and the Scenarios

In order to assess the multifunctionality of different treatment trains combined from SUDS elements, three scenarios were created. Each of the scenarios includes a different composition of the SUDS elements, designed together in the context of the Kirstinpuisto site. The site is part of a large brownfield area close to the harbor of Turku that will be gradually transformed into a highly dense residential site. A detailed plan is underway (Figure 2).

The planning principles of Kirstinpuisto, 14 ha, are to create a lively neighborhood with good cycling and pedestrian connections to the city center. Most of the existing land uses will be transformed except for some land uses in the southern corner of the site. A thirty-five meters wide park forms the central axis through the site and four to six storied residential buildings will be built adjacent to the park. Traffic moves along two main streets, which intersect in the middle of the site. The main urban square is located by this intersection. On the streets, the pedestrian traffic is separated from the cars by green strips. The northwest corner of the site is left for parking and recreation.

Figure 2. Detail plan draft of Kirstinpuisto site (figure adapted from the City of Turku).

The site has an existing drainage network, which will remain to be used in future, thus including it as part of the scenarios studied. The existing drainage network has had stormwater flooding issues in the past primarily due to the shortage of the existing drainage capacity. The aim of the scenarios is to create an alternative hybrid model utilizing the SUDS approach to substitute for the existing and malfunctioning drainage network.

The soil type on the site is clay, potentially rendering infiltration an ineffective stormwater management strategy; nevertheless, storing water would allow for some infiltration into the soil.

The site is ideal for the study, because the general aim is to turn former brownfield sites from industrial use into residential areas; therefore, some new urban greenspaces need to be created in this conversion for residential use.

To increase the knowledge base concerning the green infrastructure solutions among local authorities as well as to gather understanding of local interest towards the site, an ESS workshop was held in August 2016 with city planners. The aim of the workshop was to familiarize participants with the concept of ESS and discern local demand. As a result of the workshop, five aspects rose to the fore: (1) the creation of a recreational and restorative living environment is important for future residents; (2) stormwater quality and quantity management are both essential on the site; (3) innovative green infrastructure solutions can help to create new identity to former brownfield area; (4) a diverse urban green will safeguard important regulating services, such as microclimate regulation, habitat provision, and pollution control; and (5) all previous goals can be achieved with a multifunctional and connected green structure. Based on these five points, three scenarios were designed to supplement the plan of Kirstinpuisto, which indicates the location of building masses and street network.

The scenarios have been designed to be realistic concerning the planned urban functions and Finnish building regulations. However, the space requirements and design of the SUDS elements have retained a rather simple and formal level for modeling purposes. The three scenarios (presented in Tables 1–3) have been entitled *RUN* (supplementing the existing pipe network on streets and in the central park with open swales), *NORM* (adding SUDS elements that allow water detention especially on residential yards), and *MAX* (maximizing the amount of SUDS elements everywhere—in the central park, residential yards, parking areas, and close to business premises).

Scenarios have been designed on top of each other, thus retaining the main features from the previous one(s). Available space and building regulations concerning features, such as emergency services access, have set the boundary conditions for the location and dimensioning of SUDS elements. Left over space outside SUDS elements is assumed to be asphalt or other hard surface expect in the park, in which it is assumed to be lawn with random singular trees. In order to estimate the fulfilment of the four criteria of multifunctionality in the scenarios, each of them were estimated in four different ways presented in the following sub-section.

Table 1. Description of RUN (supplementing the existing pipe network on streets and in the central park with open swales) scenario.

Scenario	Intent	Range of SUDS	Area (ha)
RUN	Selection of the SUDS elementes is based on the main objectiv: to delay and conduct water away from the site through above-ground vegetated structures and a supplementing pipe drainage network. Additionally, there are rain gardens to promote on-site treatment.	Vegetated swales	0.6
		Rain gardens	0.6
		SUDS Total	1.2

Table 2. Description of NORM (adding SUDS elements that allow water detention especially on residential yards) scenario.

Scenario	Intent	Range of SUDS	Area (ha)
NORM	Scenario is an upgrade of RUN. It utilises a multiple SUDS approach and additional SUDS are selected based on their ability for local detention, without compromising other urban functions, such as traffic connections and recreation. Bioretention cells are constructed in the yards for stormwater treatment and paved parking lots are replaced with permeable pavement. Use of SUDS is limited to prevailing conventions of the city of Turku; for example, green roofs are only integrated into one-storey buildings.	Vegetated swales	0.6
		Rain gardens	0.9
		Green roofs	0.4
		Bioretention cell	0.1
		Permable pavements	1.3
		SUDS Total	3.3

Table 3. Description of MAX (maximizing the amount of SUDS elements everywhere—in the central park, residential yards, parking areas, and close to business premises) scenario.

Scenario	Intent	Range of SUDS	Area (ha)
MAX	Scenario is an ambitious upgrade of RUN. The amount of SUDS elements have been maximised and selected based on their ability to store and infiltrate stormwater: all roofs are green and all yards and parking lots are covered with permeable surfaces or extended rain gardens. The internal park area is fully utilised for stomrwater management.	Vegetated swales	0.6
		Rain gardens	1.8
		Green roofs	3.3
		Permable pavements	3.3
		SUDS Total	9.0

3.2. Water Quantity and Quality Assessment through Modeling

This study models the current state and the three designed SUDS scenarios using the stormwater management model (SWMM) (EPA, Washington, DC, USA [44]) to assess the impact of SUDS on water quantity and quality. SWMM [44–49] is a widely used tool for single event and long-term simulations of different water balance components, such as surface runoff, flood volume, discharge, and losses in urban areas. Losses refer to water lost from the system in the form of evaporation and infiltration. The SWMM model was first parameterized for the case study area in its current state, with the model subsequently being calibrated against two rainfall-runoff events (SC1 and SC2) and validated against one rainfall-runoff event (SV1) measured on-site between October 2017 and January 2018 [50]. The performance of the SWMM model was evaluated using the Nash-Sutcliffe efficiency (NSE) [51]. The calibrated model was then applied to the three SUDS scenarios presented in Tables 4 and 5 using SUDS parameters adopted from studies conducted in Finland [50].

The effects of SUDS scenarios on water quantity were studied for a seven-month period (E1) consisting of an extreme event during summer (E2) and an intense event after summer (E3). Rainfall data for E1, E2, and E3 are available from a station operated by the City of Turku (Table 4). The station is located about 5 km away from the case study area.

An adaptive neuro-fuzzy inference system (ANFIS) is a fuzzy inference system formulated with a learning algorithm [52]. Proposed by [53], ANFIS is based on the first-order Sugeno fuzzy model. In this study, the five water quality input variables (Table 5) were first clustered by the fuzzy c-means clustering algorithm to place them into different classes. The fuzzy c-means clustering allows a set of data to belong to one or two classes. ANFIS was utilized by defining the Sugeno reasoning and a number of rules to develop a prediction model for turbidity by using these classes. The Sugeno model utilizes "if then" rules to produce an output for each rule. ANFIS uses the input and output variables to construct a FIS whose membership function (generalized bell) parameters are tuned using a back propagation algorithm [52]. Thus, the FIS can learn from the training data (AT1). The measured four input variables and one output variable were used to train (AT1, Table 4) and test (AT2, Table 4) the ANFIS model. The ANFIS model consists of five blocks [52]:

1. A rule base containing a number of if-then rules.
2. A database which defines the membership function.
3. A decision-making interface that operates the given rules.
4. A fuzzification interface that converts the crisp inputs into "degree of match" with the linguistic values, such as high or low.
5. A defuzzification interface that reconverts to a crisp output.

The input variables for the ANFIS model were the 10-minutely rainfall, discharge, temperature, and electrical conductivity with the output variable being turbidity measured continuously on-site from November 2017 to January 2018 by Luode Consulting (Table 5). The rainfall was measured with a Vaisala Rain gauge, discharge was measured with an acoustic StarFlow sensor, and water quality variables measured continuously with an YSI multiparameter sensor placed in the same manhole with the flow sensor. In addition, 16 grab samples from the study site and surrounding areas representing different land uses including forest, railway station, and brownfield areas were collected. From the samples turbidity, total suspended solids (TSS) and metals, chromium (Cr), copper (Cu), lead (Pb), zinc (Zn) were analyzed in the laboratory. The performance of the ANFIS model was evaluated using the coefficient of determination (R^2) and the Nash–Sutcliffe efficiency (NSE). The rainfall data available for event AT1 was used to simulate the discharge output for the current and three SUDS scenarios with the calibrated SWMM model [50]. Subsequently, the trained and tested ANFIS model was used to predict turbidity for the three SUDS scenarios for event AT1.

Table 4. Rainfall events used in the stormwater management model (SWMM) and adaptive neuro-fuzzy inference system (ANFIS) model simulations.

Events	Rainfall Depth (mm)	Start Date Time	Duration	Peak Intensity (mm/10min)	Return Period	Model
SC1	35	11.11.2017 11:00	7:00	2.0	-	SWMM calibration
SC2	26	26.12.2017 20:10	8:50	1.2	-	SWMM calibration
SV1	18	04.01.2018 20:10	6:04	0.6	-	SWMM validation
E1	450	May 2012	7 months	-	-	SWMM scenarios
E2	71.0	27.08.2012 00:00	6:04	18	95 years	SWMM scenarios
E3	42.0	04.10.2012 00:00	12:00	9	30 years	SWMM scenarios
AT1	46.8	13.12.2017 23:40	24 days	0.7	-	ANFIS training and ANFIS scenarios
AT2	19.6	15.12.2017 19:00	10 days	0.7	-	ANFIS testing

Table 5. Basic statistics of the measured water quality input and output variables.

Variables	Min *	Max **	Mean	SD ***	Median	Type
Rainfall depth (mm)	0.4	27.7	2.4	2.6	1.3	Input
Discharge (l/s)	0.0	0.700	0.017	0.058	0.0	Input
Temperature (°C)	1.7	12.4	6.6	1.3	7.2	Input
Electrical conductivity (µS/cm)	33.0	701.0	497.7	152.0	557.0	Input
Turbidity (NTU)	0.1	560.3	27.4	60.4	2.4	Output

* Min, minimum; ** Max, maximum; *** SD, standard deviation.

The effects on water quantity are quantified as changes in peak flows, total flow, and flood volume in the three SUDS scenarios as compared to the current state for E1, E2, and E3 along with losses for E1. For the long-term period (E1), the empirical cumulative distribution of flow rate is analyzed. The simulated flow rate below 0.025 l/s is considered zero.

Similar to water quantity, the effect of SUDS on water quality has been assessed using the ANFIS model for the current state and for the three SUDS scenarios. This study used turbidity as a proxy indicator for water quality after establishing significant correlations between turbidity and total suspended solids (TSS) and concentrations of chromium (Cr) and copper (Cu). The linear regressions for the 16 grab samples are shown in Figure S2. Turbidity is a measure of water clarity and the extent to which the material (e.g., soil, pollution, metals, and solids) suspended in water decreases the passage of light through the water. Memon et al. [54] showed a high correlation between turbidity and suspended solids in the stormwater runoff specifically in a construction site. They suggest turbidity be used as a substitute for total suspended solids (TSS) due to the ease of continuous measurement as compared to laboratory measurement for TSS. Likewise, Nasrabadi et al. [55] used continuous turbidity as a proxy for evaluation of metal transport in river water after establishing meaningful correlation between turbidity and TSS.

3.3. Assessment of Amenity and Biodiversity Values

Amenity and biodiversity values are inherently different from water quantity and quality management as the former two are much more related to the surroundings of SUDS elements: functions, materials, and environment impact amenity and biodiversity values as described in Section 2. The amenity values are assessed based on their links with mental health benefits provided by urban green and blue structures. Green and blue structures affect mental health through various mechanisms [56,57]—viewing and observing green and blue areas yield a restorative impact, environmental health (clean air, less noise) affects residential health and opportunities to perform physical activities, and social interaction also impacts health.

The provided health benefits of each scenario were assessed by applying two parameters (Figure 3). The first parameter involved measuring the total area of SUDS elements with vegetation easily visible from residential windows or from yards, streets, or other public spaces. Green roofs on top of one story buildings were included, but not from multistory houses. Permeable pavement was not counted, as there is no vegetation to observe.

The second parameter involved measuring the total area of surfaces in which people can perform activities or interact together close to SUDS elements with vegetation. Residential yards were included, if SUDS elements were present and in the immediate proximity of the user of the yard. The lawn areas allowing sports and leisure activities were included. The second parameter indicates the extent to which SUDS elements overwhelm other functions in yards or public open areas. If water management structures are too extensive, play areas, pathways, and squares enabling physical exercise and social interaction can be hard to fit in.

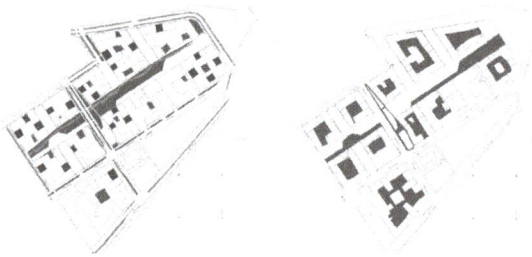

Figure 3. Diagram of NORM scenario presenting two parameters of amenity assessment: area of visible SUDS elements (left) and active spaces with vegetated SUDS elements close by (right).

Similar to amenity, two parameters were utilized to assess biodiversity values of SUDS scenarios (Figure 4). The first parameter utilized the structural heterogeneity index score developed by Monberg et al. [39]. Their study developed an index score for different types of SUDS reflecting the structural heterogeneity potential to "assess potential ecological benefits of SUDS during the design phase". The index scores are based on an expert analysis and reflect the capacity of SUDS elements to host abiotic and biotic components that increase structural heterogeneity. Thus, the same index scores are utilized in the study to evaluate the ability of treatment trains to enhance biodiversity by measuring their potential to enable structural heterogeneity. The approximate value for biodiversity is calculated by multiplying index scores with the surface area of each SUDS structure, thus reflecting the importance of size of habitat.

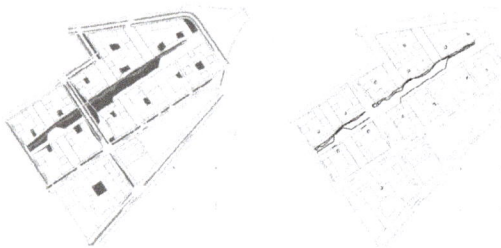

Figure 4. Diagram of NORM scenario presenting two parameters of biodiversity assessment: SUDS elements with structural heterogeneity index value (left) and edge lines of two vegetated surfaces (right).

Monberg et al. [39] provided an index score for six different types of SUDS including swales (Index score 1.8) and rain gardens (Index score 1.0). The bioretention cell has been embraced as a dry basin (Index score 2.2), which is described to be "depressions ... with straight edges designed to delay water and drain slowly until dry" [39] (p. 5). Green roofs were not included in Monberg's study, and permeable pavements do not host any vegetation, thus, receiving an index score of 0.

The second parameter is derived from connectivity and edge effect as these factors also enhance biodiversity. The edge line of each SUDS element uniting with other vegetated surface (other SUDS element or lawn) was measured reflecting a connection to other green structures as well as the ability to create conditions for edge effect, that is, changes in species structure at the boundary of two habitats. Edge lines to non-vegetated surfaces were not measured, as they do not create ecological network connectivity.

4. Results

4.1. Water Quantity

The SWMM model for current state revealed a consistent performance in reproducing a measured discharge with the Nash–Sutcliffe efficiency of 0.69 and 0.82 for the calibration events (SC1 and SC2), and 0.86 for the validation event (SV1). Modeling showed that all scenarios had an impact on water quantity. Table 6 displays the changes in peak flows, total runoff and flood volumes for SUDS scenarios as compared to the current state for a seven-month period (E1), a short-extreme rain event (E2), and a short-intense rain event (E3). The RUN scenario is efficient at conveying stormwater aboveground in a vegetated channel in a controlled manner resulting in the reduction of 65–91% in flood volume. Thus, the RUN scenario is a good conveyance system, which also helped to reduce peak flows (18–24%) for all simulated events. However, for NORM and MAX scenarios, both peak flow and total flow volume of stormwater are reduced. The MAX scenario is the most efficient in reducing both peak flow rates and total volumes in the drainage network, even for the short-extreme event (E2). Furthermore, it produces negligible flooding for both simulated events.

Table 6. Changes in peak flow, total runoff, and flood volume for SUDS scenarios compared to the current state. Increase in losses also shown for the seven-month period, E1.

Events	Scenarios	Peakflow Rate with SUDS [l/s]	Current State Peak Flow (l/s)	Decrease in Peak Flow (%)	Reduction in Total Volume (%)	Reduction in Flooding Volume (%)	Increase in Losses (%)
E1	RUN	1493	1876	20.5	2.0	66.0	1.2
	NORM	989	1876	47.3	39.9	81.1	30.9
	MAX	458	1876	75.6	81.0	98.7	58.9
E2	RUN	1493	1834	18.6	1.4	65.0	–
	NORM	957	1834	47.8	25.6	81.8	–
	MAX	442	1834	75.9	67.8	98.9	–
E3	RUN	360	474	24.2	-8.8	91.1	–
	NORM	249	474	47.6	33.8	98.5	–
	MAX	94	474	80.3	82.0	100.0	–

For the seven-month period, E1, all SUDS scenarios showed a decrease in peak flow as well as a reduction in total and flood volumes as compared to the current state. The reduction of volume can be seen as an increase in losses, which comprise the total evaporation and infiltration. Losses are dominated by infiltration in NORM scenario and evaporation in MAX scenario (Table 6). For the short-extreme event, E2, the total runoff volume is reduced for all scenarios; this is mainly due to the temporary storage of stormwater in the SUDS as contribution by losses is negligible. The temporary storage provided by SUDS also helped reduce peak flow and volume for E2. The increase in the runoff volume in RUN scenario was due to the increased imperviousness from 63 to 80% from the current state. Despite the increased imperviousness due to the planned development, the RUN scenario still diminished the peak flows as a result of the stormwater retention and delayed conveyance in the vegetated channel. Thus, the SUDS in studied scenarios has helped manage water quantity on site through controlled conveyance in the RUN scenario as well as temporary storage and losses from the system in the NORM and MAX scenarios.

Figure 5 shows the cumulative distribution of SWMM model simulated flow rate for the current state and the three SUDS scenarios for the longer simulation (E1). From Figure 5a, it can be seen that the share of zero flows clearly increased for scenarios NORM and MAX, whereas only scenario MAX seems to be effective in decreasing high flow rates (Figure 5b).

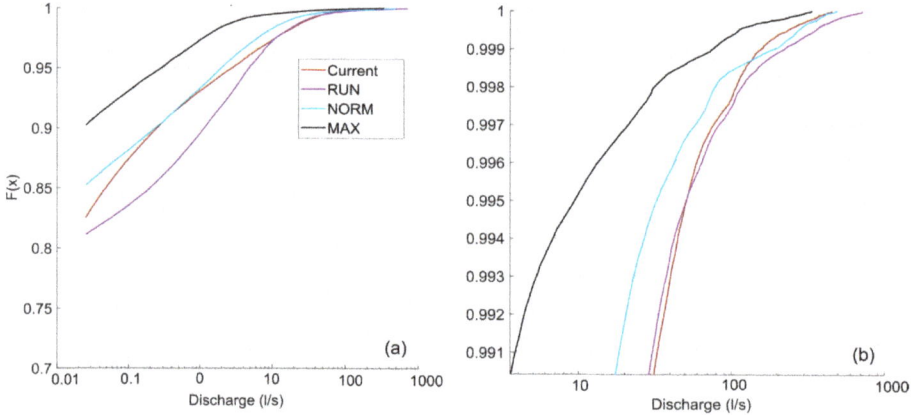

Figure 5. The cumulative distribution of the flow rate for the current state and the three SUDS scenarios for the longer simulation E1 (**a**) and zoomed-in cumulative distribution of the high flow rates (**b**).

4.2. Water Quality

The 95% confidence intervals of slope for TSS, Cr, and Cu excluded zero (Table 7), indicating that there is a significant relationship between turbidity and each of the TSS, Cr, and Cu.

The ability of SUDS scenarios to affect the flow volumes (Figure 5) indicates their ability to manage water quality on-site as turbidity reaches high values with high flow volumes. The trained ANFIS model has the coefficient of determination (R^2) and the Nash–Sutcliffe efficiency (NSE) of 0.86 and 0.78, respectively. The statistics for the tested ANFIS model are 0.74 and 0.59 for R^2 and NSE values, respectively. The comparison of measured and predicted turbidity for calibration and validation periods is shown in Figure S1.

Table 7. Reduction in mean turbidity, and concentrations of total suspended solids, chromium, and copper for SUDS scenarios compared to the current state.

	Unit	RUN	NORM	MAX	Linear Relationship	Coefficient of Determination (R^2)	a* (95% CI ** of a)
Turbidity (T)	NTU	−1.6%	11.6%	46.5%		–	–
Total suspended solids (TSS)	mg/l	−0.4%	3.0%	12.2%	$TSS = aT + 46.763$	0.89	0.404 (0.332, 0.477)
Chromium (Cr)	μg/l	−2.6%	18.3%	73.5%	$Cr = aT - 4.10$	0.95	0.067 (0.061, 0.073)
Copper (Cu)	μg/l	−0.2%	1.7%	6.8%	$Cu = aT + 15.411$	0.83	0.064 (0.049, 0.079)

* a, slope of regression line; ** CI, confidence interval.

The results show that the model performed consistently for both calibration and validation datasets. The correlation between turbidity and total suspended solids is 0.89. The correlation between turbidity and concentration of chromium and copper is 0.95 and 0.83, respectively (Figure S2). The MAX scenario is able to reduce 46.5% of the mean value of turbidity with a corresponding reduction in mean concentrations of TSS and metals of 7–73% when compared to the current state of the catchment. The corresponding reduction in mean value of turbidity for the NORM scenario is 11.6%. The reduction in water quality indicators is the highest for MAX followed by NORM. However, for RUN scenario, the turbidity value increased by 1.6%, and the concentrations of TSS, Cr, and Cu increased by 0.2–2.6% (Table 7). This is likely to be a result of increased imperviousness leading to larger flow volumes, and the positive relationship between flow volume and turbidity identified in the ANFIS water quality model.

4.3. Amenity

Amenity values consist of two parameters: the surface area of visible SUDS elements and the surface area of active spaces close to SUDS elements. For each scenario, both parameters are presented in Table 8 together with the total score. The MAX scenario delivers the highest amenity value through visible green and blue structures, but the other scenarios deliver more opportunities for physical activity and social interactions close to SUDS elements. Thus, the NORM scenario promises to deliver the highest amenity values as it contains both abundant visual interest and space for active interaction with one's surroundings.

Table 8. Total scores of amenity values.

	Elements	RUN	NORM	MAX
Visible SUDS elements	Swales	0.6	0.6	0.6
	Rain gardens	0.6	0.9	1.8
	Bioretention cell		0.1	
	Visible green roofs		0.4	0.4
		1.2	2	2.8
Active Spaces Close SUDS elements	Lawns	1.7	0.7	
	Urban Square	0.3	0.3	0.1
	Yards		1.9	1.8
		2	2.9	1.9
Total Score	(ha)	3.2	4.9	4.7

4.4. Biodiversity

Biodiversity values also consist of two parameters. The potential structural heterogeneity of the scenarios is calculated by multiplying the index score of each SUDS type with their surface area with the results being presented below in Table 9.

Table 9. Total scores of structural heterogeneity (left) and edge line (right).

Elements	RUN	NORM	MAX	Elements	RUN	NORM	MAX
Swales (18)	11	11	11	Between two SUDS el.		945	875
Rain gardens (10)	6	9	18	Between SUDS el. and lawn	875	410	
Bioretention cell (22)		2		Total score	875	1355	875
Total score	17	22	29				

The RUN scenario has only two different types of SUDS elements (swales and rain gardens) with the total structural heterogeneity reaching 17. Swales have a high index value of 18, indicating good opportunities for habitat enhancement by increasing abiotic and biotic components through design, but as the surface area is low, the end score remains moderate. In the NORM and MAX scenarios, the total score is higher as surface areas as well as the range of adapted SUDS elements in NORM are higher.

The values of the other biodiversity parameter, namely the edge lines of two vegetated surfaces, are presented in Table 9. The length of the edge line is equal for RUN and MAX, with the difference being that in RUN, the edge is between the swales and lawn, while in MAX, it is between the swales and rain garden. The edge line length is considerably longer in NORM, which also consists of different types of edges, hence providing better preconditions for connectivity and edge effect, as well as onwards for biodiversity.

5. Discussion

The aim of the research was to study means of assessing multifunctionality during the landscape architectural design process. A widely used SWMM model was parameterized for assessing the impacts of SUDS scenarios with respect to the water quantity criterion [44–49]. Likewise, data-driven ANFIS model was used for assessing the impacts of SUDS scenarios with consideration of the water quality criterion [52]. Amenity and biodiversity values of different types of existing SUDS structures have been assessed in earlier studies [13,32,33], but analyses of landscape architectural designs are rare. In this study, a biphasic assessment was created for both values.

One major consideration is that the amenity and biodiversity values delivered are dependent on the surroundings of SUDS elements. Therefore, the results reflect the qualities of the detail plan draft—the residential blocks are in a row next to the central park and all the adapted SUDS elements on the streets or in the park are easily visible from the apartments. Nevertheless, inner yards are mainly visually closed from the park and if there are no SUDS elements in the yards, neither amenity values related to green and blue structures are delivered. The same feature also hinders opportunities of creating a connected network of green and blue structures that would deliver high biodiversity values.

Moreover, the results are to some extent theoretical, especially concerning biodiversity values. The greatest weakness of the study is poor recognition of the benefits deliverable by green roofs. As there was no index value of structural heterogeneity available for green roofs [39] and they were not directly connected to other vegetated structures, green roofs were not taken into account in the biodiversity assessment. Nevertheless, we know that green roofs have a good potential to enhance local biodiversity [58,59].

Based on the results, the MAX scenario is the most multifunctional option. It works well with water quantity and quality management and delivers high biodiversity values and almost as high amenity values as the NORM scenario. This leads to a discussion of the interrelations of the different criteria. Although the ability of SUDS to provide multifunctionality is continuously enhanced by both the research literature and practical guidelines and links, the interrelations and possible synergies between the four criteria are seldom discussed [19,21,22]. The individual results of the four criteria do not directly indicate a mutual interrelationship between them. However, some processes in SUDS clearly overlap concerning the criteria; for instance, evapotranspiration serves for stormwater quantity control like in MAX scenario, but occurs through vegetation whilst simultaneously supporting microclimatic control for the needs of people. Therefore, it is important to study the ways in which the criteria are interrelated in order to provide a more holistic understanding concerning the provision of multifunctionality in the landscape architectural design process.

The results show that NORM and MAX scenarios that combine several SUDS with different features provide better quantity and quality management in conjunction with higher biodiversity and amenity values. This confirms the relationship between different criteria presented in literature [15]; the ability of SUDS to store and ensure the availability of water for vegetation enhances biodiversity through ecological processes. In turn, biodiversity and the amount of vegetation in SUDS enhance evaporation and infiltration, subsequently affecting water quality. Additionally, increased biodiversity positively affects perceived amenity, but an increased amount of water in urban greenspaces simultaneously requires higher design skills to provide amenity values [32].

Understanding these mutual interconnections and relations presented in Figure 6 will help to design and implement simultaneous functions of the four criteria. Based on the results above, three principles can be outlined for promoting multifunctionality. First, designing SUDS requires a thorough understanding of the hydrological process in order to create high amenity values in urban greenspaces.

The results indicate that SUDS elements with a high capacity for run-off regulation and water detention should be implemented to enhance water quality management. However, such SUDS elements are only occasionally filled with water. Open water is seen to hold the greatest value in urban design, but as SUDS elements often tend to be dry, the design should be adaptable to prevailing hydrological process and create added value in all rain situations as well as during possible dry seasons.

Secondly, if vegetated SUDS play a major role in landscape architectural design as design elements, we need more knowledge about their differences in terms of biodiversity. In principle, SUDS that sustain the function of natural processes, thus promoting structural heterogeneity of habitats, uphold biodiversity. For the needs of biodiversity, it is essential to design volumes, routes, and surfaces that enhance the water cycle as well as sustain biophysical structures, processes, and functions. This initiates a holistic approach in which the functionality of SUDS is enhanced by locating them not as individual elements or as a part of the treatment train, but in connection with the larger ecological or green network. This is closely related to enhancing local biodiversity that requires extra attention during the design phase together with a multidisciplinary approach [39].

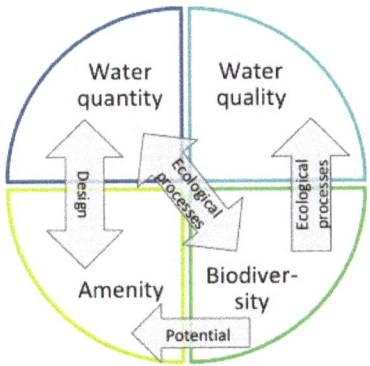

Figure 6. Mutual interconnections of the four criteria.

Thirdly, all four criteria can be assimilated into the ESS concept [14] as water quantity and quality management are strongly related to the regulation of the water cycle and purification service. Furthermore, amenity is related to the provision of cultural ecosystem services. The fourth criterion, biodiversity, is a more complicated issue. When examined in the framework of the cascade model [20], biodiversity is not an ecosystem service, but rather a requirement for it, marking this pillar as being fundamentally different from the others. However, in green stormwater infrastructure related research, biodiversity is commonly regarded as a supporting service and used together with habitat provision [14].

The assimilation of the four criteria into the ESS concept will help to understand the relation of SUDS to other systems. In order to strengthen ecosystem service provision, an understanding is required of the ecological processes and system dynamics in urban greenspaces [38]. Furthermore, sustainability advantages provided by short distances of the compact city ideal should be valued against the space requirements of ecological processes. This underlines Ahern's [2] notion that the concept of sustainability changes as cities are understood and accepted as dynamic systems.

An urban area, such as Kirstinpuisto, consists of both physical infrastructures and social structures composed by its residents. Concurrently, the area is still a catchment and also a part of the wider ecosystem, as are all urban sites [9]. As an outcome, it is an example of a social ecological system (SES) [60], in which the hydrological cycle can be combined into urban functions with the help of multifunctional SUDS. However, multifunctionality is not self-evident, but requires a focused approach [6]. The results of this paper indicate that a balanced approach is needed to consider different preconditions, interrelations, and possible outcomes in the landscape architectural design process.

SUDS elements are widely used practical implementation of GI in urban development. GI has the ability to work as a platform for different systems, such as hydrology, transportation, and tourism [6], as well as to support sustainable urban development [4]. In that framework, SUDS elements have a special role to collectively mediate local hydrology, biodiversity, and amenity values, if conditions for those parameters are created during the design process.

Kirstinpuisto is a good example of a new urban space; a former brownfield site with almost non-existent green areas will be transformed into a residential area with requirements for public urban green areas. SUDS elements are needed for its stormwater management, but it can also play a more significant role creating biodiversity and amenity values. As the benefits of new multifunctional SUDS are considered, one must be aware of the challenges with multifunctionality. Some of the expected outcomes can already be precisely measured during the design phase (such as water quantity management), or later after its realization (such as plant species richness). However, some of the outcomes will accrue through a dynamic process together with new residents, new hydrological or soil conditions, maintenance procedures, or with a changing climate.

The results of the study reinforce Jack Ahern's notions about the safe-to-fail design approach [61], in which urban landscape is understood as a system that can be guided to perform different functions. We need more understanding of the process of that guidance as well as of the intrinsic characteristics of the desired multifunctionality [6]. Especially knowledge concerning the contribution of SUDS to local biodiversity (which elements support which kind of species and habitats, and the ways it can be matched with an existing green network) is essential as SUDS is used in increasing amounts as a retrofit solution or as a part of new greenspaces with desire for multiple benefits.

6. Conclusions

This paper studied the multifunctionality of three stormwater management treatment trains that were composed of differing SUDS elements. The four criteria of SUDS design (water quantity, water quality, biodiversity, and amenity) were used to measure multifunctionality. The aim was to understand how SUDS scenarios could enhance multifunctionality of urban greenspaces as well as how this should be considered in the design process with an application to a case study area.

There has been a lack of holistic knowledge concerning the generation of multifunctionality as a part of the landscape architectural design process of stormwater management. This paper discussed and tested indicators for different criteria with the results indicating that the links and feedback between the SUDS criteria should be considered more profoundly. A deeper understanding of the interconnections between urban hydrological processes and the provision of natural functions of a site is needed to increase biodiversity and related benefits in urban greenspaces.

Furthermore, the study introduced that existing modeling tools can be utilized for the assessment of water quantity and quality criteria while such tools to assess amenity and biodiversity values delivered by SUDS elements are not available at the same level. In addition, both amenity and biodiversity depend much more on the framework where SUDS elements are adapted. These results reflect that we are more familiar with those uncomplicated features of SUDS elements that resemble a traditional pipe network. By contrast, study methods for both the assessment of complex criteria and complete understanding of the desired multifunctionality need further development.

The results confirm that multifunctionality criteria are interconnected. If biodiversity criteria have failed, it has a degenerative impact on both the amenity and water quantity management potential of the site. This suggests that if the delivery of multifunctional benefits is not considered during the design process, it is quite likely to ruin any chances of achieving goals related to multifunctionality. On the other hand, through a skillful analysis of local preconditions and with site specific design decisions, we can enhance multifunctionality.

The study can be seen as a remark to open a conversation concerning how we can assess different criteria of multifunctionality that are not commensurate by nature and not even necessarily equal. There is an obvious need to deliver more easily adaptable measuring methods for the values different

SUDS elements involve, especially concerning biodiversity. Furthermore, a fitting multicriteria analysis for SUDS elements is needed alike.

Finally, the desired provision of multifunctionality requires not only an acknowledgement of the interdependencies of its different aspects, but also a consideration of other urban functions. A careful coordination of these functions in the design process is essential, if multifunctional SUDS elements are to be successfully applied to a dense urban structure. This ultimately leads towards a system thinking approach.

Supplementary Materials: The following are available online at http://www.mdpi.com/2071-1050/11/7/1854/s1, Figure S1: Measured vs. predicted turbidity for training and testing period in ANFIS model, Figure S2: Correlation between turbidity and total suspended solids (a), concentration of chromium (b) and concentration of copper (c).

Author Contributions: Conceptualization, E.L. and A.K.; methodology, E.L., A.K., O.T., and T.K.; software, A.K.; validation, E.L., A.K., O.T., and T.K.; investigation, E.L. and A.K.; writing—original draft preparation, E.L.; writing—review and editing, A.K., O.T., and T.K.; visualization, E.L.; supervision, T.K.; project administration, E.L.; and funding acquisition, E.L.

Funding: This research was funded by the Ministry of Agriculture and Forestry as a part of the UrbanStormwaterRisk project 2016–2019.

Conflicts of Interest: The authors declare no conflict of interest. The funders had no role in the design of the study; in the collection, analyses, or interpretation of data; in the writing of the manuscript, or in the decision to publish the results.

References

1. Benedict, M.A.; McMahon, E.T. *Green Infrastructure: Smart Conservation for the 21st Century*; Sprawl Watch Clearinghouse: Washington, DC, USA, 2006.
2. Ahern, J. Urban landscape sustainability and resilience: The promise and challenges of integrating ecology with urban planning and design. *Landsc. Ecol.* **2012**, *28*, 1203–1212. [CrossRef]
3. European Environment Agency. *Green Infrastructure and Territorial Cohesion. The Concept of Green Infrastructure and Its Integration into Policies Using Monitoring Systems*; EEA Technical Report No. 18/2011; EEA: Brussels, Belgium, 2012.
4. Hansen, R.; Pauleit, S. From multifunctionality to multiple ecosystem services? A conceptual framework for multifunctionality in green infrastructure planning for Urban Areas. *Ambio* **2014**, *43*, 516–529. [CrossRef]
5. Hansen, R.; Frantzeskaki, N.; McPhearson, T.; Rall, E.; Kabish, N.; Kaczorowska, A.; Kain, J.-H.; Artmann, M.; Pauleit, S. The uptake of the ecosystem services concept in planning discourses of European and American cities. *Ecosyst. Serv.* **2015**, *12*, 228–246. [CrossRef]
6. Wang, J.; Banzhaf, E. Towards a better understanding of Green Infrastructure: A critical review. *Ecol. Indic.* **2018**, *85*, 758–772. [CrossRef]
7. UN. *New Urban Agenda*; United Nations A/RES/71/256; UN: New York, NY, USA, 2017.
8. Lennon, M.; Scott, M. Delivering ecosystems services via spatial planning: Reviewing the possibilities and implications of a green infrastructure approach. *Town Plan. Rev.* **2014**, *85*, 563–587. [CrossRef]
9. Brown, R.; Keath, N.; Wong, H.F. Urban water management in cities: Historical, current and future regimes. *Water Sci. Technol.* **2009**, *59*, 847–855. [CrossRef]
10. Fletcher, T.D.; Shuster, W.; Hunt, W.F.; Ashley, R.; Butler, D.; Arthur, S.; Trowsdale, S.; Barraud, S.; Semadeni-Davies, A.; Bertrand-Krajewski, J.-L.; et al. SUDS, LID, BMPs, WSUD and more—The evolution and application of terminology surrounding urban drainage. *Urban Water J.* **2015**, *12*, 525–542. [CrossRef]
11. C753 SuDS Manual. Available online: https://www.ciria.org/Resources/Free_publications/SuDS_manual_C753.aspx (accessed on 15 February 2018).
12. Jose, R.; Wade, R.; Jefferies, C. Smart SUDS: Recognising the multiple-benefit potential of sustainable surface water management systems. *Water Sci. Technol.* **2015**, *71*, 245–251. [CrossRef] [PubMed]
13. Mak, C.; Scholz, M.; James, P. Sustainable drainage system site assessment method using urban ecosystem services. *Urban Ecosyst.* **2017**, *20*, 293–307. [CrossRef]

14. Prudencio, L.; Null, S. Stormwater management and ecosystem services: A review. *Environ. Res. Lett.* **2018**. Available online: https://iopscience.iop.org/article/10.1088/1748-9326/aaa81a/meta (accessed on 1 December 2018). [CrossRef]
15. Zhang, K.; Chui, T. Linking hydrological and bioecological benefits of green infrastructures across spatial scales—A literature review. *Sci. Total Environ.* **2019**, *646*, 1219–1231. [CrossRef] [PubMed]
16. Hoang, L.; Fenner, R.A. System interactions of stormwater management using sustainable urban drainage systems and green infrastructure. *Urban Water J.* **2016**, *13*, 739–758. [CrossRef]
17. Roe, M.; Mell, I. Negotiating value and priorities: Evaluating the demands of green infrastructure development. *J. Environ. Plan. Manag.* **2013**, *56*, 650–673. [CrossRef]
18. Liquete, C.; Kleeschulte, S.; Dige, G.; Maes, J.; Grizzetti, B.; Olah, B.; Zulian, G. Mapping green infrastructure based on ecosystem services and ecological networks. A Pan-European case study. *Environm. Sci. Policy* **2015**, *54*, 268–280. [CrossRef]
19. Madureira, H.; Andresen, T. Planning for multifunctional urban green infrastructures: Promises and challenges. *Urban Des. Int.* **2014**, *19*, 38–49. [CrossRef]
20. Potschin, M.B.; Haines-Young, R.H. Ecosystem services: Exploring a geographical perspective. *Prog. Phys. Geogr.* **2011**, *35*, 575–594. [CrossRef]
21. Meerow, S.; Newell, J. Spatial planning for multifunctional green infrastructure: Growing resilience in Detroit. *Landsc. Urban Plan.* **2017**, *159*, 62–75. [CrossRef]
22. Hansen, R.; Olafsson, A.S.; van der Jagt, A.P.N.; Rall, E.; Pauleit, S. Planning multifunctional green infrastructure for compact cities: What is the state of practice? *Ecol. Indic.* **2019**, *96*, 99–110. [CrossRef]
23. Schifman, L.; Herrmann, D.; Shuster, W.; Ossola, A.; Garmestani, A.; Hopton, M. Situating green infrastructure in context: A Framework for adaptive socio-hydrology in cities. *Water Resour. Res.* **2017**, *53*, 139–154. [CrossRef]
24. O'Donnell, E.; Woodhouse, R.; Thorne, C. Evaluating the multiple benefits of a sustainable drainage scheme in Newcastle, UK. In *Proceedings of the Institution of Civil Engineers: Water Management*; ICE Publishing: London, UK, 2018; Volume 171, pp. 191–202.
25. Dagenais, D.; Thomas, I.; Paquette, S. Siting green stormwater infrastructure in a neighborhood to maximize secondary benefits: Lessons learned from a pilot project. *Landsc. Res.* **2017**, *42*, 195–210. [CrossRef]
26. Pappalardo, V.; La Rosa, D.; Campisano, A.; La Greca, P. The potential of green infrastructure application in urban runoff control for land use planning: A preliminary evaluation from a southern Italy case study. *Ecosyst. Serv.* **2017**, *26*, 345–354. [CrossRef]
27. Muerdter, C.P.; Wong, C.K.; Lefevre, G.H. Emerging investigator series: The role of vegetation in bioretention for stormwater treatment in the built environment: Pollutant removal, hydrologic function, and ancillary benefits. *Environ. Sci. Water Res. Technol.* **2018**, *5*, 592–612. [CrossRef]
28. Tahvonen, O. Adapting bioretention construction details to local practices in Finland. *Sustainability* **2018**, *10*, 276. [CrossRef]
29. Carpenter, C.; Todorov, D.; Driscoll, C.; Montesdeoca, M. Water quantity and quality response of a green roof to storm events: Experimental and monitoring observations. *Environ. Pollut.* **2016**, *218*, 664–672. [CrossRef] [PubMed]
30. Ferrans, P.; Rey, C.; Pérez, G.; Rodríguez, J.; Díaz-Granados, M. Effect of Green Roof Configuration and Hydrological Variables on Runoff Water Quantity and Quality. *Water* **2018**, *10*, 960. [CrossRef]
31. Echols, S.; Pennypacker, E. From Stormwater Management to Artful Rainwater Design. *Landsc. J.* **2008**, *27*, 268–290. [CrossRef]
32. Backhaus, A.; Fryd, O. The aesthetic performance of urban landscape-based stormwater management systems: A review of twenty projects in Northern Europe. *J. Landsc. Archit.* **2013**, *8*, 52–63. [CrossRef]
33. Bastien, N.; Arthur, S.; McLoughlin, M. Valuing amenity: Public perceptions of sustainable drainage systems ponds. *Water Environ. J.* **2012**, *26*, 19–29. [CrossRef]
34. Wood, E.; Harsant, A.; Dallimer, M.; de Chavez, A.; McEachan, R.; Hassall, C. Not all green space is created equal: Biodiversity predicts psychological restorative benefits from urban green space. *Front. Psychol.* **2018**, *9*. [CrossRef]
35. de Groot, R.; Alkemade, R.; Braat, L.; Hein, L.; Willemen, L. Challenges in integrating the concept of ecosystem services and values in landscape planning, management and decision making. *Ecol. Complex.* **2010**, *7*, 260–272. [CrossRef]

36. Müller, A.; Bocher, P.; Fischer, C.; Svenning, J.-C. 'Wild' in the city context: Do relative wild areas offer opportunities for urban biodiversity? *Landsc. Urban Plan.* **2018**, *170*, 256–265.
37. Aronson, M.; Lepczyck, C.; Evans, K.; Goddard, M.; Lerman, S.; MacIvor, J.; Nilon, C.; Vargo, T. Biodiversity in the city: Key challenges for urban green space management. *Front. Ecol. Environ.* **2017**, *15*, 189–196. [CrossRef]
38. Keesstra, S.; Nunes, J.; Novara, A.; Finger, D.; Avelar, D.; Kalantari, Z.; Cerda, A. The superior effect of nature based solutions in land management for enhancing ecosystem services. *Sci. Total Environ.* **2018**, *610–611*, 977–1009. [CrossRef] [PubMed]
39. Monberg, R.; Howe, A.; Ravn, H.; Jensen, M. Exploring structural habitat heterogeneity in sustainable urban drainage systems (SUDS) for urban biodiversity support. *Urban Ecosyst.* **2018**, *21*, 1159–1170. [CrossRef]
40. Halliday, B.; Matthews, T.; Iervasi, D.; Dodemaide, D.; Pickett, P.; Linn, M.; Burns, A.; Bail, I.; Lester, R. Potential for water-resource infrastructure to act as refuge habitat. *Ecol. Eng.* **2015**, *84*, 136–148. [CrossRef]
41. Kazemi, F.; Beecham, S.; Gibbs, J.; Clay, R. Factors affecting terrestrial invertebrate diversity in bioretention basins in an Australian urban environment. *Landsc. Urban Plan.* **2009**, *92*, 304–313. [CrossRef]
42. Szulczewska, B.; Giedych, R.; Borowski, J.; Kuchcik, M.; Sikorski, P.; Mazurkiewicz, A.; Stanczyk, T. How much green is needed for a vital neighbourhood? In search for empirical evidence. *Land Use Policy* **2014**, *38*, 330–345. [CrossRef]
43. Forman, T. Urban Habitat, Vegetation, Plants. In *Urban Ecology*; Cambridge University Press: Cambridge, UK, 2014; pp. 205–240.
44. Rossman, L.A. *A Storm Water Management Model User's Manual*; EPA/600/R-05/040; U.S. Environmental Protection Agency: Cincinnati, OH, USA, 2016.
45. Krebs, G.; Kokkonen, T.; Valtanen, M.; Koivusalo, H.; Setälä, H. A high resolution application of a stormwater management model (SWMM) using genetic parameter optimization. *Urban Water J.* **2013**, *10*, 394–410. [CrossRef]
46. Krebs, G.; Kokkonen, T.; Valtanen, M.; Setälä, H.; Koivusalo, H. Spatial resolution considerations for urban hydrological modelling. *J. Hydrol.* **2014**, *512*, 482–497. [CrossRef]
47. Niazi, M.; Nietch, C.; Maghrebi, M.; Jackson, N.; Bennett, B.; Tryby, M.; Massoudieh, A. Storm Water Management Model: Performance review and gap analysis. *J. Sustain. Water Built Environ.* **2017**, *3*, 04017002. [CrossRef]
48. Niemi, T.J.; Warsta, L.; Taka, M.; Hickman, B.; Pulkkinen, S.; Krebs, G.; Moisseev, D.N.; Koivusalo, H.; Kokkonen, T. Applicability of open rainfall data to event-scale urban rainfall-runoff modelling. *J. Hydrol.* **2017**, *547*, 143–155. [CrossRef]
49. Warsta, L.; Niemi, T.J.; Taka, M.; Krebs, G.; Haahti, K.; Koivusalo, H.; Kokkonen, T. Development and application of an automated subcatchment generator for SWMM using open data. *Urban Water J.* **2017**, *14*, 954–963. [CrossRef]
50. Khadka, A.; Kokkonen, T.; Niemi, T.J.; Lähde, E.; Sillanpää, N.; Koivusalo, H. Towards natural water cycle in urban areas: Modelling stormwater management designs. *Urban Water J.* **2018**. submitted.
51. Nash, J.E.; Sutcliffe, J.V. River Flow Forecasting through Conceptual Models Part I—A Discussion of Principles. *J. Hydrol.* **1970**, *10*, 282–290. [CrossRef]
52. Khadr, M.; Mohamed, E. Data-Driven Modeling for Water Quality Prediction Case Study: The Drains System Associated with Manzala Lake, Egypt. *Ain Shams Eng. J.* **2017**, *8*, 549–557. [CrossRef]
53. Shing, J.; Jang, R. ANFIS: Adaptive-Network-Based Fuzzy Inference System. *IEEE Trans. Syst. MAN Cibern.* **1993**, *23*, 665–685.
54. Memon, S.; Cristina Paule, M.; Lee, B.Y.; Umer, R.; Sukhbaatar, C.; Hee Lee, C.H. Investigation of Turbidity and Suspended Solids Behavior in Storm Water Run-off from Different Land-Use Sites in South Korea. *Desalination Water Treat.* **2015**, *53*, 3088–3095. [CrossRef]
55. Nasrabadi, T.; Ruegner, H.; Sirdari, Z.Z.; Schwientek, M.; Grathwohl, P. Using Total Suspended Solids (TSS) and Turbidity as Proxies for Evaluation of Metal Transport in River Water. *Appl. Geochem.* **2016**, *68*, 1–9. [CrossRef]
56. Gascon, M.; Mas, M.; Martinez, D.; Dadvand, P.; Forns, J.; Plasencia, A.; Nieuwenhuijsen, M. Mental Health Benefits of Long-Term Exposure to Residential Green and Blue Spaces: A Systematic Review. *Int. J. Environ. Res. Public Health* **2015**, *12*, 4354–4379. [CrossRef]

57. Francis, R.; Lorimer, J. Urban reconciliation ecology: The potential of living roofs and walls. *J. Environ. Manag.* **2011**, *92*, 1429–1437. [CrossRef]
58. Cook-Patton, S.; Bauerle, T. Potential benefits of plant diversity on vegetated roofs: A literature review. *J. Environ. Manag.* **2012**, *106*, 85–92. [CrossRef] [PubMed]
59. Ulrich, R. Human responses to vegetation and landscapes. *Landsc. Urban Plan.* **1986**, *13*, 29–44. [CrossRef]
60. Flynn, C.; Davidson, C. Adapting the social-ecological system framework for urban stormwater management: The case of green infrastructure adoption. *Ecol. Soc.* **2016**, *21*, 19. [CrossRef]
61. Ahern, J. From fail-safe to safe-to-fail. *Landsc. Urban Plan.* **2011**, *100*, 341–343. [CrossRef]

© 2019 by the authors. Licensee MDPI, Basel, Switzerland. This article is an open access article distributed under the terms and conditions of the Creative Commons Attribution (CC BY) license (http://creativecommons.org/licenses/by/4.0/).

Article

A Multicriteria Planning Framework to Locate and Select Sustainable Urban Drainage Systems (SUDS) in Consolidated Urban Areas

Sara Lucía Jiménez Ariza, José Alejandro Martínez, Andrés Felipe Muñoz, Juan Pablo Quijano, Juan Pablo Rodríguez *, Luis Alejandro Camacho and Mario Díaz-Granados

Environmental Engineering Research Centre (CIIA), Department of Civil and Environmental Engineering, Universidad de los Andes, Bogotá 111711, Colombia; sl.jimenez133@uniandes.edu.co (S.L.J.A.); ja.martinez912@uniandes.edu.co (J.A.M.); af.munoz2325@uniandes.edu.co (A.F.M.); jp.quijano116@uniandes.edu.co (J.P.Q.); la.camacho@uniandes.edu.co (L.A.C.); mdiazgra@uniandes.edu.co (M.D.-G.)
* Correspondence: pabl-rod@uniandes.edu.co (J.P.R.); Tel.: +57-1-339-4949 (ext. 2804)

Received: 28 February 2019; Accepted: 27 March 2019; Published: 17 April 2019

Abstract: The implementation of sustainable urban drainage systems (SUDS) is increasing due to their advantages, which transcend runoff control. As a result, it is important to find the appropriate SUDS locations to maximize the benefits for the watershed. This study develops a multiscale methodology for consolidated urban areas that allows the analysis of environmental, social, and economic aspects of SUDS implementation according to multiple objectives (i.e., runoff management, water quality improvements, and amenity generation). This methodology includes three scales: (a) citywide, (b) local, and (c) microscale. The citywide scale involves the definition of objectives through workshops with the participation of the main stakeholders, and the development of spatial analyses to identify (1) priority urban drainage sub-catchments: areas that need intervention, and (2) strategic urban drainage sub-catchments: zones with the opportunity to integrate SUDS due the presence of natural elements or future urban redevelopment plans. At a local scale, prospective areas are analyzed to establish the potential of SUDS implementation. Microscale comprises the use of the results from the previous scales to identify the best SUDS placement. In the latter scale, the SUDS types and treatment trains are selected. The methodology was applied to the city of Bogotá (Colombia) with a population of nearly seven million inhabitants living in an area of approximately 400 km^2. Results include: (a) The identification of priority urban drainage sub-catchments, where the implementation of SUDS could bring greater benefits; (b) the determination of strategic urban drainage sub-catchments considering Bogotá's future urban redevelopment plans, and green and blue-green corridors; and (c) the evaluation of SUDS suitability for public and private areas. We found that the most suitable SUDS types for public areas in Bogotá are tree boxes, cisterns, bioretention zones, green swales, extended dry detention basins, and infiltration trenches, while for private residential areas they are rain barrels, tree boxes, green roofs, and green swales.

Keywords: multiscale framework; runoff management; spatial analysis; SUDS location and selection; urban drainage planning; stormwater treatment train

---

1. Introduction

Increasing populations in cities and the resulting urban sprawl have been particularly marked in Latin America and the Caribbean. For example, in Colombia, the urban population has increased from 40% in 1951 to 78% in 2018 [1,2]. Unlike other countries in the region, urban growth has been concentrated in four major cities: Bogotá, Medellín, Cali, and Barranquilla [2]. As the rapid urbanization is often at the expense of the loss of valuable ecosystems and lands, serious environmental, social, and economic problems have emerged and are expected to worsen if cities fail in adopting sustainable urbanization practices. Although many concepts and definitions on sustainable urbanization have emerged, all of them refer with equal concern to environmental, governance, social, and economic sustainability [3]. In this context, sustainable urban drainage systems (SUDS) constitute an opportunity to enhance stormwater management offering multiple options for runoff control and additional benefits related with social [4], environmental [5], and economic aspects [6,7].

In the first place, SUDS reduce runoff volumes and peaks resembling the natural hydrological cycle through processes such as infiltration and detention [8–11]. Also, these systems improve the runoff quality via filtration, sedimentation, dispersion, and biological processes [10–12]. Furthermore, the presence of vegetation helps to create multifunctional spaces where runoff becomes an asset rather than a waste. As a consequence, SUDS have the potential to improve the landscape, enhance water quality, promote ecosystems connectivity, and reduce vulnerability to flooding thus helping the transition of urbanized areas to water sensitive or sponge cities [13,14].

Several types of SUDS can be implemented in public and private areas such as wet ponds, dry extended detention basins, constructed wetlands, grassed swales, bioretention zones, rain barrels, green roofs, and infiltration basins among others. Connected sets of these systems constitute stormwater treatment trains, which maximize the benefits related to runoff control. The performance of systems and trains depends on: (a) the physical, environmental and social characteristics of the emplacement; (b) the processes for runoff control, which include infiltration, detention, and conveyance; and (c) in the case of trains, the synergy between the SUDS types. For this reason, urban planning strategies involving SUDS could be developed to maximize their performance according to the watershed needs and stakeholders' perspectives. As such, a multiscale and multicriteria approach is fundamental to identifying the opportunities for SUDS implementation within a city.

Researchers have considered a variety of objectives and scales to plan for the proper location of SUDS. Objectives include runoff management, water quality improvement, and amenity generation. The most usual scales are regional, citywide, local, and microscale. Certain studies use compound indices and other GIS-based techniques to define priority areas according to hydrological and hydraulic aspects [15–19], socioeconomic and environmental aspects [17,18,20,21], and water quality issues [19]. Though, these studies have some limitations because most of the analyses correspond to the local scale and the microscale. Moreover, critical areas identified at a city scale are not used to develop specific strategies for more detailed scales. Steaming from these previous contributions in GIS applications, some other works have focused on the preferred optimal locations and configurations using benefit–cost analysis, exact optimization methodologies (e.g., linear and dynamic programming), meta-heuristics and, more recently, stochastic mixed integer linear programming that accounts for the variability of rainfall [22–27].

For example, Martin-Mikle et al. [15] defined a comprehensive methodology that includes four urban scales. However, they selected priority areas according to hydrological and hydraulic aspects only. Likewise, Garcia-Cuerva et al. [21] analyzed a watershed of 121 km^2 in North Carolina (USA) to define preferred SUDS locations and conducted a hydrological analysis of the impacts of SUDS implementation within a particular watershed sub-catchment, but they recommended areas by exclusively considering the population's socioeconomic attributes. Dagenais et al. [17] proposed a methodology in which the identification of priority zones was followed by the location of SUDS in a specific area. Nevertheless, this methodology was applied, in particular, to the local scale.

Some other studies have focused on SUDS' location assessing factors like: physical restrictions [17,28]; performance in runoff reduction, flooding mitigation, and water quality improvement [29,30]; scale, including street, neighborhood, and sub-catchment [28]; and whether the area is public or private [21,28]. The analysis conducted for the private space has generally disregarded the specific characteristics of these areas, however recent work related to permeable pavements (which can be used in a private space) considered such specific characteristics [26]. For instance,

Gogate et al. [30] established alternatives for a primarily residential area, including green roofs due to the prevalence of flat roofs, but the analysis of specific spatial constraints for leaky wells and rain gardens was absent. Instead, the authors pre-selected SUDS types by evaluating the systems suitability in a developing country based on a thorough literature review and the general characteristics related to residential and commercial land use in the area. Garcia-Cuerva et al. [21] evaluated public and private space to implement bioretention cells and rainwater harvesting systems. However, in this study, the SUDS location only considered land use (i.e., commercial, residential, institutional, and vacant land) and omitted possible site-related restrictions of these systems, such as the maximum recommended slope.

Regarding SUDS selection, the definition of the best system or set of structures have comprised two main approaches: (a) performance evaluation through models to determine runoff volume reduction [30,31], and (b) multicriteria analysis considering qualitative and/or quantitative explanatory variables mainly at local scale and microscale [32,33]. The use of models can involve a high computational cost and requires detailed information that is not always available, particularly for a preliminary evaluation. Nonetheless, it is important to define recommendations and general directions for the city over the spectrum of SUDS alternatives. Therefore, multicriteria qualitative analysis is essential to conducting preliminary analyses for SUDS selection in a specific area.

Few studies have focused their attention on connected sets of SUDS or train selection. One example is the work of Charlesworth et al. [34], who defined a management train to mitigate flood events. They categorized the city area according to recommended SUDS types considering a hierarchy for stormwater control processes—giving priority to source control and infiltration. However, development of tools to select SUDS types classified under the same control process is required to define specific alternatives according to the potential benefits of each SUDS type.

In Latin America and some developing countries, the examples of SUDS prioritization are limited and usually focus on hydraulic and hydrological aspects. For example, Mora-Melià et al. [35] identified critical points for the installation of green roofs based on flooding reports in Curicó (Chile). Likewise, Gogate and Rawal [36] outlined a methodology to recognize places to conduct artificial groundwater recharge in the city of Pune (India). On the other hand, the few studies that included SUDS selection did not consider larger spatial scales (i.e., city scale). For instance, Petit-Boix et al. [37] developed a methodology that included life cycle analysis (LCA) for the selection of SUDS for an area of 0.42 km^2 in São Carlos (Brazil). In addition, Gogate et al. [30] proposed a multicriteria analysis to select strategies of SUDS implementation in a watershed (11.71 km^2) in Pune.

Analysis at city scale is fundamental for decision-making at smaller urban scales. Additionally, due to the multiple benefits from SUDS, these systems could be compared through multicriteria analysis, which include environmental, social and economic aspects, rather than only hydraulic and hydrologic criteria enhancing the common practice in several countries. Equally important is the analysis of private areas, where it is fundamental to evaluate site-specific restrictions (e.g., slope, infiltration rate, or distance to the water table). Nonetheless, there are few examples considering these aspects in the literature, which constitute gaps for the decision-making of SUDS implementation. For this reason, the present study defines a multiscale-planning framework to identify strategic and priority urban drainage sub-catchments in consolidated urban areas, and it recommends specific SUDS types and treatment trains on public and private areas. The methodology involves analyses at three scales: citywide, local scale, and microscale. At city scale, priority and strategic areas are identified according to stakeholders' interests and characteristics of the territory by means of the analysis of georeferenced information. At local scale, public and private spaces are evaluated considering slope, infiltration rate, water table, and distance to buildings. The microscale includes a process to select SUDS types and SUDS treatment trains. The city of Bogotá (Colombia) was selected as a case for the study of the application of the proposed methodology.

2. Materials and Methods

A methodology to guide SUDS implementation is proposed at three spatial scales: (1) citywide scale, (2) local scale, and (3) microscale. This approach intends to select a location and systems according to the watershed needs and stakeholders' preferences. Figure 1 describes the proposed methodology by summarizing the main activities at each step, the required information, and the expected results.

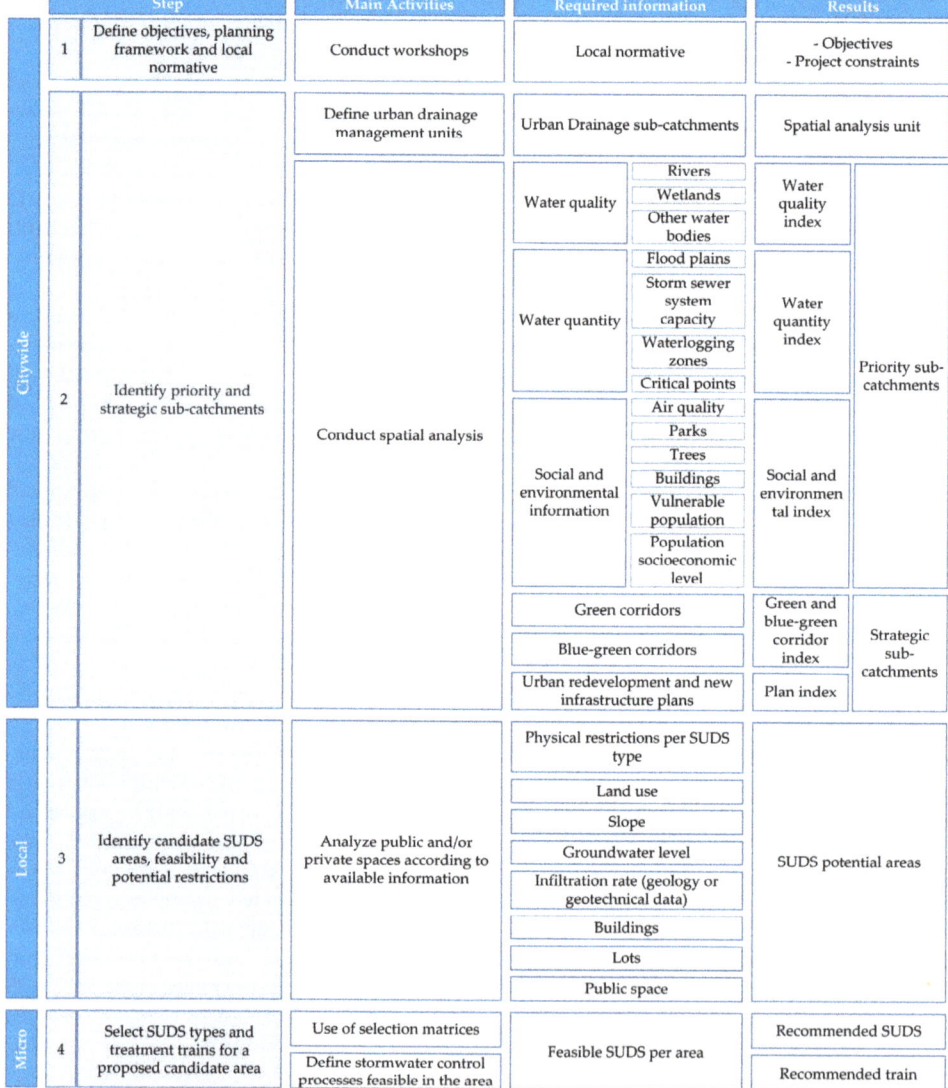

Figure 1. Multiscale methodology for sustainable urban drainage systems planning.

2.1. Citywide Scale

The main purpose of the analysis conducted at this scale was to spot urban drainage sub-catchments to address the defined citywide objectives. In this sense, two main steps are developed: (i) to define the citywide objectives using stakeholders' multicriteria perspectives, gathered by means of workshops, and (ii) to identify priority and strategic sub-catchments appraising available georeferenced information.

One important element in multicriteria decision problems is the weighting method used, which can be subjective or objective. Subjective methods base the definition of weights on preferences of the decision makers. Nevertheless, subjective weighting can have some disadvantages given that the knowledge and experience of the stakeholders may condition the results [38]. Objective weighting disregards subjective judgment and is based on mathematical procedures. Generally, the weights depend on the variability or correlation of the performance of each alternative for the evaluated criteria [38]. Both approaches are integrated into the methodology because they could generate significant differences between the territory's needs and the stakeholders' preferences in the initial phases of SUDS implementation.

2.1.1. Definition of Citywide Objectives and Their Subjective Weighting

The recognition of stakeholders' perspectives is essential when it comes to including SUDS in urban planning. Therefore, the methodology proposes workshops to identify priority aspects for different stakeholders and to improve the understanding of their vision with regards to the stormwater drainage system. The structure of the workshops is based on the soft systems methodology (SSM) developed by Checkland [39] and applied by Sánchez & Mejía [40]. This methodology deals with complex problems linked to multiple stakeholders' perceptions. For this research, the SSM involves three parts: (a) open questions about SUDS (i.e., advantages, disadvantages, components, objectives, limitations, and stakeholders' responsibilities), (b) conceptualization of the urban drainage system using a CATOWE (Customers, Actors, Transformation process, World view, Owners and Environmental constraints) analysis, and (c) closed questions about SUDS (i.e., citywide objectives, selection criteria, performance evaluation, limitations) in which the participants assign scores from zero (0) to three (3), where zero (0) means that the aspect is not applicable, one (1) that it is of low importance, two (2) moderate importance, and three (3) high importance. The delegates score objectives related to water quantity, water quality, and social aspects. The results from the workshops should be analyzed according to the existing regulations to identify shortcomings and to evaluate the stakeholders' interpretations.

2.1.2. Priority Urban Drainage Sub-Catchments

Priority urban drainage sub-catchments are areas that need an intervention due to problems related to runoff management or the characteristics of the environment. Several criteria can be analyzed to generate a qualitative index for the urban area according to three main objectives: (1) water quality improvements, (2) runoff management, and (3) amenity improvement. The relevance of these objectives depends on the stakeholders' judgment.

The analysis of the main urban water bodies guides the identification of priority urban drainage sub-catchments for water quality improvements. In this case, it is proposed to consider four commonly used water quality determinants as criteria [41]: Biological Oxygen Demand (BOD), Total Suspended Solids (TSS), Total Nitrogen (TN), and Total Phosphorus (TP). The values of these determinants are classified as high, medium, and low, where high indicates high concentrations of nutrients, solids or organic matter. A global index is defined for the rivers and wetlands according to the highest classification considering all determinants. Finally, this index is assigned to the closest urban drainage sub-catchment.

The analysis for runoff management incorporates information about storm sewer system capacity, waterlogging zones, critical points (i.e., points with insufficient hydraulic capacity), and urban river flood plains. These variables can include qualitative and quantitative information. Hence, a standardized classification from zero (0) to one (1) is defined, where 0 refers to the absence of data, 0.25 low priority, 0.5 medium-low priority, 0.75 medium priority, and 1 high priority. To draw up a water quantity index, the highest value in any of the four criteria is selected and assigned to the urban drainage sub-catchment.

Social and environmental criteria are analyzed to identify places where amenity generation is necessary due to the presence of a vulnerable population and poor environmental conditions. The selected criteria were defined according to previous studies that developed similar indices [17,18,20]. These studies identified the vulnerable population according to age [17,18,20], education [20], income [20], housing [18], and ethnic background [17,18,20]. As for the environmental variables, the analysis considers air quality, access to parks, vegetation, and impermeable surfaces among others [18,20]. The selection of the variables depends on the available information and the city characteristics. Hence, the proposed criteria are: (a) air pollutants with highest values in accordance with the local air quality index, (b) distance to parks, (c) trees per hectare or population, (d) occupied area, (e) infant and elderly population per hectare, and (f) low socioeconomic level residential areas. Regarding the distance to parks, we calculated the ratio of the number of residential lots within a radius of 300 m from parks with an area bigger than one (1) hectare to the total residential lots per unit of analysis as proposed by Ekkel and de Vries [42].

In each case, a normalized index from zero (0) to one (1) was defined. This index is obtained by subtracting the lowest value and dividing it by the range of values of each criteria [20]. For the environmental variables, in cases where a benchmark relative to city norms or international standards exists, this value is adopted as the maximum. For the distance to parks and trees per hectare, the index is defined as one minus the index, because in such cases the maximum value represents the best condition.

The average of three objective weighting methods was used to define the social index and the prioritized urban drainage sub-catchments according to water quality improvements, runoff management, and amenity improvement: (a) Entropy Method [43], (b) Criteria Importance Through Inter-criteria Correlation (CRITIC), and (c) Principal Components and Factor Analysis [44]. For CRITIC, the method was applied according to the modifications proposed by Jahan et al. [45]. A classification was made according to the percentiles 0.25, 0.5, 0.75, and 1.

2.1.3. Strategic Urban Drainage Sub-Catchments

Strategic urban drainage sub-catchments correspond to zones with the opportunity to integrate SUDS in the city, due to the presence of natural elements or future urban infrastructure works. Two main characteristics are analyzed: (a) the presence of green and blue-green corridors; and (b) urban redevelopment and new infrastructure plans. Green corridors are defined as longitudinal green spaces that can be composed of green road dividers or parks, whereas blue-green corridors are comprised of water bodies with green areas around them. Green and blue-green corridors are considered strategic because they favor superficial drainage. Also, their identification is essential to integrate multiple public spaces using treatment trains. With respect to urban development and new infrastructure plans, these can provide an opportunity to integrate SUDS in public areas.

According to the characteristics of the green and blue-green corridors, an index is calculated for every urban drainage sub-catchment. The characteristics considered for blue-green corridors analysis include the approximated total length within the urban perimeter and the area inside the unit of analysis. In this case, the total length was used as a proxy of connectivity. Green corridors are assessed according to their width, length, area, and distance to a blue-green corridor. The connectivity is evaluated through the distance to a blue-green corridor. Also, the geometric variables allow us to determine the potential intervention area. These characteristics are normalized and added to define an index for blue-green corridors and green corridors. This is done by subtracting the minimum value and dividing the result by the range of values. In the case of the distance to a blue-green corridor, the value used corresponds to one minus the index. Lastly, the two indices are added and normalized to define a combined index for the corridors. The percentiles 0.25, 0.5, 0.75, and 1 are used to classify the values into low, medium-low, medium, and high opportunity.

The main urban redevelopment and new infrastructure plans were selected and categorized according to their stage and activity. Table 1 shows the stages and activities considered. The stages correspond to the degree of progress of the plan. For example, the reserved stage means that the plan would be included in a future project. The excluded stage refers to plans that were left out of projects, and the commissioned stage refers to plans being assigned to a project. If information about the stage is absent, the plan is assumed to be in an early stage to assign the score. The activities refer to the type of intervention to be made. For instance, adaptation corresponds to infrastructure modifications, and reconstruction to the full replacement of an existent structure. Also, the plans are graded in relation to the public elements (roads, sideways, bays, among others) that are part of them, as presented in Table 2. It is worth mentioning that Table 2 does not pretend to be exhaustive, but illustrative of the main urban elements in which SUDS can be placed. Higher scores are given to the most suitable elements for SUDS implementation.

Table 1. Scores for plan stages and activities.

Stage Activity	In Progress	Commissioned	Excluded	Reserved	Suspended	Completed	No Data
Prefeasibility or feasibility studies	4	4	0	4	1	4	3
Studies and designs	2	3	0	4	1	1	2
Adaptation	0	0	0	3	0	0	0
Conservation	0	0	0	3	0	0	0
Construction	0	0	0	3	0	0	0
Diagnostic	0	0	0	0	0	0	0
Road improvement or maintenance (regular or occasional)	0	0	0	0	0	0	0
Reconstruction	0	0	0	3	0	0	0
Road rehabilitation	0	0	0	2	0	0	0
No data	0	0	0	0	0	0	0

Table 2. Scores for public elements.

Element	Score
Tree-lined roads	4
Sidewalk	4
Parking bay	4
Road	1
Bike trail	2
Bus station	1
Square	4
Main bus station	2
Pedestrian bridge	2
Vehicular bridge	1
Road divider	4
Ramp	2
Green areas	4
Cable car facilities	1

For each infrastructure plan, scores from zero (0) to four (4) are assigned according to Tables 1 and 2. A score of zero (0) refers to a non-relevant plan or element. A score of four (4) means that the element is part of the public space system, or the activity is pertinent, and the plan is in its early stages (e.g., prefeasibility studies in a reserved area). These scores are compared, and each plan is qualified with the lowest score between them. Later, the score of the set of plans that are inside every unit of analysis is added. A normalized index from zero (0) to one (1) is established according to the maximum score within an analysis unit. The values are classified according to the percentiles 0.25, 0.5, 0.75, and 1.

2.2. Local Scale

Public and private spaces are analyzed to determine the feasibility of twelve SUDS types: (1) grassed swales, (2) infiltration trenches, (3) permeable pavements, (4) wet ponds, (5) bioretention zones, (6) tree boxes, (7) sand filters, (8) constructed wetlands, (9) soakaways, (10) infiltration basins, (11) extended dry detention basins, and (12) rain barrels and cisterns. In addition, for private constructions the implementation of green roofs is considered. The selection of these SUDS types was based on an extensive literature review that included several design manuals and guidelines worldwide. Reviewed manuals and guidelines included six to fourteen SUDS types—excluding pre-treatment and other complementary structures—and the selected SUDS types correspond to the most commonly presented [46–68].

SUDS screening for the public space considers the type of space, site-specific restrictions, and spatial requirements. Type of space includes parks (P), squares (S), roads dividers (R), sidewalks (W), and parking lots (Pa). Site-specific restrictions comprise slope, distance to the groundwater level, infiltration rate (obtained from citywide geology or geotechnical datasets), and distance to foundations. Water storage capacity was not included as a restriction at this scale, thus it has to be considered when assessing the performance of the selected SUDS types. Spatial requirements cover minimum area, length to width ratio, and length. These requirements depend on the SUDS type and they are part of the feasibility evaluation of the ones with larger area requirements. In any case, the potential spaces must have a minimum area of 1 m^2. The considered restrictions are applied in all available public and private areas, and the specific values are presented in Table 3. Furthermore, the proximity to channels and pipes constitutes an additional criterion because it determines whether the suitable areas could be connected to the conventional drainage system. Also, some areas are discarded for the implementation of wet ponds and constructed wetlands because of their distance to channels and streams. The latter only evaluate the potential for connection, thus more detailed analyses are needed to assess the actual capacity of pipes, channels, and streams.

Table 3. Implementation constraints of sustainable urban drainage systems (SUDS).

Parameter	Restriction Type	Grassed Swales	Infiltration Trenches	Permeable Pavements	Wet Ponds	Bioretention Zones	Tree Boxes	Sand Filters	Constructed Wetlands	Soakaways	Infiltration Basin	Extended Dry Detention Basin
Slope (%)	Maximum	10[1]	5[1]	5[1]	15[1]	10[1]	10[1]	5[1]	15[1]	15[9]	3[4]	15[1]
	Minimum	1[11]	1[2]	0.5[3]	-	-	-	1[2]	1[5]	-	0[3]	1[2]
Distance to groundwater level (m)	Minimum	1.5[1]	3[2]	3[8]	1.3[7]	1.8[3]	1[3]	1.5[1]	1.3[7]	1[4]	1.2[7]	3[1]
Infiltration rate (mm/h)	Minimum	13[3]	7[7]	13[3]	-	7[10]	7[10]	13[7]	-	13[7]	13[7]	7[2]
Distance to foundations (m)	Minimum	4[9]	6[12]	6[12]	6[12]	6[12]	2[13]	1.5[6]	6[12]	6[12]	6[12]	6[2]
Area (m²)	Minimum	-	-	-	150[14]	-	-	-	1000[7,15]	-	45	45
Length to width ratio	Minimum	-	-	-	2:1[14]	-	-	-	3:1[4]	-	2:1[6]	2:1[6]
Width (m)	Minimum	-	-	-	8[14]	-	-	-	18	-	5	5
Length (m)	Minimum	-	-	-	20[14]	-	-	-	56	-	9	9
Public space	Type	P R	P R	S W Pa	P R	P S R W	P S R W	P R	P R	P S R W	P R	P R
Private space	Use[16]	Re C D	Re C D	Re C D	C D	Re C D	Re C D			Re C D		

(-) No data, (P) parks, (S) squares, (R) road dividers, (W) sidewalks, (Pa) parking lots, (Re) residential use, (C) commercial use, (D) public facilities. [1] [55], [2] [60], [3] [51], [4] [49], [5] [61], [6] [50], [7] [57], [8] [69], [9] [63], [10] [70], [11] [54], [12] [65]. [13] Recommendation from the local environmental agency (Secretaria Distrital de Ambiente, SDA) (2015), [14] [53], [15] [71], [16] [72].

The first step for private spaces analysis is to identify the land uses (e.g., residential, commercial, or industrial) to be considered, followed by an evaluation of the non-occupied portion of the selected land–use category, taking into account the restrictions presented in Table 3. The analysis of green roof feasibility focuses on identifying suitable constructions. Thus, two characteristics are considered: (a) presence of flat roofs and (b) a minimum area of 200 m^2—according to recommendations from Moore et al. [73]. The suitability of rainwater barrels and cisterns as rainwater harvesting (RWH) practices for capturing and storing stormwater for later use are evaluated conforming to other criteria. For the public space, the feasibility of underground cisterns depends mainly on the approximated storage volume and the distance to a pluvial drainage pipe (i.e., pipe with diameter below 0.6 m in a radius of 20 m). An area is considered suitable for cistern installation if the storage volume is above 10 m^3. For private spaces, water demand for non-potable uses and rainwater availability determines rain barrel feasibility. In this way, if there is a potential for rainwater harvesting the private area is considered suitable for a rainwater barrel. RWH has grown over the last decades as it has potential use for drought mitigation, increased demand satisfaction, reduction of stormwater runoff volumes, and pollutant loads [74–84].

2.3. Microscale

2.3.1. Site Selection

Site selection was driven by the results at citywide and local scales. The best-case scenario is when an urban drainage sub-catchment has been defined as priority and strategic, and there is available space. In this sense, the urban drainage sub-catchment rated with the highest scores for priority and strategic criteria was evaluated. After that, according to the available space, specific areas were chosen for SUDS implementation.

2.3.2. Selection of SUDS

SUDS selection depended on their performance related to multiple aspects. Thus, a qualitative matrix was defined to compare the feasible SUDS types in an area. This matrix contains criteria related to stormwater quality improvements, stormwater volume reduction, amenity, maintenance, and costs. For each criterion, three levels are defined: high, medium, and low. In the case of quality improvement, high means over 80% pollutant load reduction, moderate indicates 30% to 80% of pollutant load reduction, and low corresponds to less than 30% of pollutant load reduction [85]. Table 4 presents the defined levels corresponding to the characteristics of the different SUDS types and information reported in the literature.

Table 4. Qualification according to efficiency in pollutant removal and relevant processes.

SUDS Type	Quality Improvement						Runoff Control		Amenity			Maintenance	Cost	
	Nutrients	Metals	Bacteria	Sediment	Oil and Grease	Trash and Debris	Filtration and Sorption	Volume Control	Maximum Discharge Control	Perception Improvement	Interference with Activities on Site	Safety risks (users)	Activities and Risk of Clogging	Capital Cost	Maintenance Cost
Grassed swale	$M^{1,2,3}$	M^1	L^1	$M^{1,2,3}$	M^1	M^1	L^1	L^1	L^1	M	H	H	L	L^8	L^8
Rain barrel and cistern (RWH)	N^1	N^1	N^1	N^1	N^1	N^1	N^1	M^1	M^1	N	L	L	L	M^8	M^8
Bioretention zone [b]	$M^{1,4,5}$	$H^{1,3,4,5}$	$M^{4,5}$	$M^{4,5}$	H^1	H^1	H^1	M^1	L^1	H	H	M	M	M^8	M^8
Tree box [b]	M^1	M^a	M^a	M^1	H^1	H^1	M^1	M^1	L^1	M	H	L	M	M^8	M^8
Extended dry detention basin [b]	$L^{2,3,6}$	M^2	$M^{2,3,6}$	$M^{2,3}$	M^2	H	L^7	L	M	H	M	M	M	M^8	M^8
Infiltration trench	$M^{2,3}$	H^1	H^1	H^1	M^1	H^1	H^1	H^1	H^1	N	L	M	H	M^8	H^8
Permeable pavement [b]	L^3	M^1	M^1	H^1	H^1	M	M^1	L^1	M^1	H	H	L	M	H^8	H^8
Wet pond	$M^{2,3}$	$M^{2,3}$	$M^{2,3}$	$M^{2,3}$	M^2	H^6	L^7	L	H	L	H	H	H	H^8	H^8
Sand filter [b]	$M^{1,3}$	$M^{3,6}$	M^1	$H^{1,3}$	M^1	H^6	M^1	L^1	L^1	H	H	M	H	M^8	H^8
Constructed wetland	$M^{1,2,3,6}$	$M^{2,3,6}$	L^1	H^1	L^1	L^1	H^1	H^1	H^1	H	L	H	M	H^8	M^8
Soakaway	L^1	H^1	H^1	H^1	M^1	H^1	M^1	H^1	M^1	N	M	L	H	M	H
Infiltration basin	$M^{1,2}$	H^1	H^1	H^1	H^1	H^1	H^1	H^1	H^1	H	L	M	M	M	H
Green roof	L^1	L^1	L^1	L^1	L^1	L^1	L^1	H^1	M^1	M	L	L	M	H^8	L^8

(H) High, (M) medium, (L) low, (N) null. [a] Conditions equivalent to bioretention zones are assumed. [b] Performance related to quality improvement and runoff control can improve depending on the infiltration rate of the area. [1] [85], [2] [53], [3] [71], [4] [86], [5] [87], [6] [88], [7] [89], [8] [55].

2.3.3. Treatment Trains Selection

Five processes are identified to configure and select treatment trains: (a) infiltration, (b) detention, (c) rainwater harvesting, (d) conveyance, and (e) irrigation. Feasible relations and the sequential order among these processes are presented in Table 5. These relations result by dismissing unsuitable associations between processes and identifying processes that should be at the final stage. In this sense, it was considered that the runoff captured for later uses (e.g., rain water harvesting and irrigation) must be treated, and therefore they cannot be an initial process. Also, these relationships allow the formation of treatment trains with more than two components. For instance, for a three-stage treatment train, if the initial process is conveyance and this is followed by infiltration, according to Table 5, the final process can be rainwater harvesting or irrigation.

Table 5. Processes combinations.

Final Process Initial Process	Infiltration	Detention	Conveyance	Rainwater Harvesting (RWH)	Irrigation
Infiltration			X	X	X
Detention	X		X	X	X
Conveyance	X	X		X	X
Rainwater harvesting (RWH)					
Irrigation					

Sequential order schemes between two SUDS types are summarized in Table 6. Rows correspond to the initial component of the train and columns to the second component. The processes for stormwater control are presented in pairs. The first letter of each pair indicates the process that the initial component would perform. The second letter shows the process performed by the final component. Several combinations are presented given the different processes suitable for each SUDS type. These sequences are defined by the characteristics of the evaluated SUDS types. For instance, SUDS types used in the treatment of runoff from extended areas or several sites should be at the end of the treatment train. In this sense, systems such as extended dry detention basins, wet ponds, constructed wetlands and infiltration basins are at the end of the sequential schemes [49]. The schemes allow us to conceive trains of two, three, or more stages. For example, if the first component of a three-stage train is a grassed swale that conveys the runoff to a bioretention zone, as stated in Table 6 a feasible third element is a cistern.

To calculate a score for each feasible treatment train identified, each SUDS type is rated according to its characteristics and the stormwater control processes (see Table 7). In this manner, the score of a train is the result of the information presented in Tables 4 and 7, and the recommended trains correspond to the ones with higher scores.

Table 6. Sequential schemes between SUDS types.

FINAL / INITIAL	Grassed Swale	Rain Barrel and Cistern (RWH)	Bioretention Zone	Tree Box	Extended Dry Detention Basin	Infiltration Trench	Permeable Pavement	Wet Pond	Sand Filter	Constructed Wetland	Soakaway	Infiltration Basin
Grassed swale		C,D C,R C,Ir	C,D C,I C,Ir	C,D C,I C,Ir	C,D C,I	C,I	C,I	C,D	C,D C,I	C,D C,Ir	C,D C,I	C,D C,I
Rain barrel and cistern (RWH)	D,C					D,C D,I						
Bioretention zone	I,C D,C	C,D D,R I,R D,Ir I,Ir		C,D	C,D	I,C D,C D,I	C,D D,I	C,D	C,D	C,D		C,D
Tree box	I,C D,C	D,R I,R D,Ir I,Ir	D,Ir			I,C D,C D,I	D,I					
Infiltration trench	I,C	D,R I,R C,R D,Ir I,Ir C,Ir	C,I C,Ir	C,I C,Ir						C,Ir	C,I	C,I
Permeable pavement		D,R I,R D,Ir I,Ir										
Sand filter	I,C	D,R D,Ir I,R I,Ir				I,C						
Soakaway		D,R I,R D,Ir I,Ir	D,Ir	D,Ir							D,I	
Green roof	D,C	D,R D,Ir	D,Ir	D,Ir		D,C D,I	D,I		D,I		D,I	

The first component indicates the process related to the row and the second component the process related to the column: (I) infiltration, (D) detention, (C) conveyance, (R) rainwater harvesting, (Ir) irrigation.

Table 7. Assigned score to the evaluated processes (from 0 to 5).

Process SUDS Type	Infiltration	Detention	Conveyance	Rainwater Harvesting	Irrigation
Grassed swale	2	1	5	0	0
Rain barrel and cistern (RWH)	0	4	0	5	5
Bioretention zone	3	4	0	0	4
Tree box	3	4	0	0	4
Extended dry detention basin	3	5	0	1	1
Infiltration trench	5	3	3	0	0
Permeable pavement	5	3	0	0	0
Wet pond	0	5	0	0	0
Sand filter	3	4	0	0	0
Constructed wetland	0	5	0	0	3
Soakaway	5	3	0	0	0
Infiltration basin	5	5	0	0	0
Green roof	0	3	0	0	4

3. Case Study

The selected case study was the city of Bogotá (Colombia), which covers approximately 400 km^2 of urban area. The urban drainage system consists of a combined sewer system in the oldest urban areas and a separate system in the newest developments. Stormwater is discharged into four urban tributaries of the Bogotá River: Torca, Salitre, Fucha, and Tunjuelo rivers. Other natural elements in the urban drainage system include wetlands within the city limits (see Figure 2). The water utility of the city defined 485 urban drainage sub-catchments.

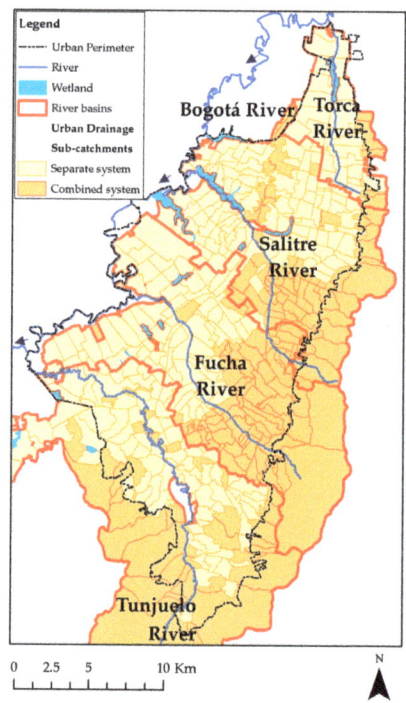

Figure 2. Main elements of the Bogotá's urban drainage system.

3.1. Citywide Information Sources

3.1.1. Water Quality Factors

Information to characterize the water quality status of urban rivers was obtained from a study conducted by Universidad de los Andes and the local environmental agency (SDA) [90]. The chosen data corresponded to the 75th percentile of measured concentrations to account for seasonal variations. A study from the water utility (EAB) and the SDA [91] was used to characterize the wetlands. The average of the reported concentration values was used.

3.1.2. Water Quantity Factors

The local planning department (SDP) classified each urban drainage sub-catchment into five levels according to their stormwater collection and transport capacity: (a) without service, (b) critical, (c) restricted, (d) moderate, and (e) high [92]. This classification was used to characterize the current urban drainage system capacity. Waterlogging zones were established according to a raster layer with different ponding areas elaborated by the local risk management institute (IDIGER). The analysis of the sewer system critical points was based on a study carried out by the local water utility (EAB) [93]. Flood plains areas were obtained from SDP data [94], which defines three risk levels: high, medium, and low.

3.1.3. Environmental and Social Factors

For the analysis of social and environmental criteria, the selected air pollutant was PM 2.5 (particulate matter with diameters that are 2.5 micrometers and smaller). The reported values in the national air quality index (ICA) and the city index (IBOCA) were assessed. These indices evidence that for most of the year, the concentrations of other air pollutants were moderate or good. Nevertheless, PM 2.5 concentrations reached an unhealthy level for sensitive groups on several occasions during 2017 [95], and more recently in early 2019. Information about PM 2.5 from the local air quality network is used [96]: 2017 time series of hourly data of 11 stations were analyzed to define the annual average for each station. If daily measures were less than 75%, data were excluded as it is set in the protocols for the city's air quality network [95]. The highest value considered for the index is 25 $\mu g/m^3$, which corresponds to the maximum allowed annual level [97].

To assess urban parks, an inventory carried out by the SDP was available. Trees per hectare were analyzed according to the tree census from the city's Botanical Garden (JBB) [98]. The occupied area within each urban plot was calculated according to the information from the city's spatial database (IDECA) [99]. The analysis of the infant (under five years) and the elderly (over sixty-five years) low-income population uses data from SISBEN (System for Identifying and Classifying Potential Beneficiaries for Social Programs in Colombia) [100]. For the identification of low-income residential areas, a classification of the city area by the SDP was considered. This classification values the characteristics of each house and its surroundings. The total area of lots rated as low or minor was calculated in every sub-catchment to conduct the analysis.

3.1.4. Strategic Urban Drainage Sub-Catchments

To identify strategic urban drainage sub-catchments, blue-green corridors and green corridors are defined from the analysis of a satellite image taken by Sentinel-2 with a resolution of 10 m per pixel. The Normalized Difference Vegetation Index (NDVI) allowed the identification of green areas considering a threshold value of 0.4. For blue-green corridors, information about channels, wetlands, ponds, rivers, riparian corridors, and preservations zones from two city databases [99,101] was used, as well as information about trees located at river rounds [98]. Regarding green corridors, the tree inventory [98] and information about the public space support the identification of linear spaces.

Renovation projects and repair works in the city were identified in the databases of the Urban Development Institute (IDU). The development plan for 2016–2020 [102] and projects supervised by the IDU [102,103] were considered. A total of 242 projects and works were evaluated including public space infrastructure, road infrastructure, and public transport infrastructure. Some of them cover various sub-catchments and comprehend different stages and elements.

3.2. Local Scale Information Sources

Public space was defined according to information from the SDP, which corresponds to georeferenced polygons of parks, squares, road dividers, sidewalks, and parking lots. Supplementary green areas were identified in an orthophoto provided by the EAB. Information about the natural and constructed drainage system was also provided by the EAB. The distance to buildings' foundations was approximated through reports by IDECA [99].

For the analysis of private space, residential use was selected because it was the predominant land-use category (i.e., approximately 40% of the city area). The analysis was conducted according to the information available for lots, uses, and buildings from IDECA [99]. Flat roofs were identified using the information available for residential use, socioeconomic level, and the number of floors in the buildings. In this sense, it was assumed that housing with more than three (3) floors had flat roofs. In addition, in low socioeconomic level areas, progressive self-constructed housing is more common, which is why these houses were presumed to have flat roofs regardless of the number of floors.

The distance to the water table and infiltration rate values were estimated from geotechnical surveys available from the geographic information system of the EAB [104]. For the distance to the groundwater level, 3384 depth measurements within the city were analyzed, whereas for the infiltration rate, the strata descriptions from 2973 geotechnical surveys were used. These descriptions were grouped into 33 classes and the permeability was defined according to: (a) the soil textural triangle from the United States Department of Agriculture (USDA); (b) the classification of Twarakavi, Šimůnek, and Schaap [105]; and (c) the saturated hydraulic conductivity estimated from the content of clay, silt, and sand.

4. Results and Discussion

4.1. Citywide Scale

4.1.1. Citywide Objectives

During 2015, workshops were held involving several stakeholders from: (a) the water utility (EAB), (b) the city environmental agency (SDA), (c) the urban development institute (IDU), (d) the risk management institute (IDIGER), and (e) researchers from public and private universities. The results were analyzed considering the local normative (Decree 528 of 2014). According to their preferences, the most important objective for implementing SUDS was stormwater quantity management. The latter was followed by storm, and thus, urban rivers water quality improvements. The objectives considered less important were the promotion of social participation, the reduction of public health risks, and reduced wrong connections in the sewer system.

These workshops allowed the identification of the main limitations for SUDS implementation in the city as a result of social, institutional, regulatory, and economic issues. Social concerns included potential negative perceptions of the communities close to SUDS projects. Institutional limitations comprised problems that resulted from the lack of interinstitutional and interdisciplinary work. Also mentioned was the lack of awareness of the role of every local institution. As regulatory limitations, the participants indicated the absence of clear policies and incentives. Technical issues were mostly associated with lack of knowledge about design, construction, operation, and maintenance of SUDS from public and private stakeholders. Additionally, the participants pointed out two economic constraints: lack of financial resources and high implementation costs.

The institutional issues were evident in the definition of the conceptual models. The stakeholders were unaware of the group of institutions involved with the design and maintenance of the city's drainage system set by local regulations. Only two of the twelve stakeholders and institutions were included in all the conceptual models that resulted from the workshops. Just one of the institutions (i.e., the SDA) mentioned entities and elements related to urban planning and recreation. On the other hand, some stakeholders evinced deficiencies in the normative, because key topics like regional interaction and cross connection issues were excluded from it.

These results are consistent with difficulties in urban stormwater management and SUDS implementation already identified in other countries. For example, Roy et al. [106] reviewed examples of stormwater management programs in Australia and the US and found technical, economical, and institutional issues. Technical issues included a lack of knowledge about the performance and requirements of the systems. Economic issues referred to a lack of information about costs. Institutional issues comprised a lack of proper regulations and interinstitutional work. Problems resulting from the absence of cooperation between institutions and regulations were also pointed out by Brown [107]. The isolated vision of stormwater management was mentioned as a problem by Dhakal and Chevalier [108]. These studies indicate the absence of improvement in this area and the negative consequences of the achievement of a sustainable system.

4.1.2. Priority Urban Drainage Sub-Catchments

Figure 3 shows the indices for water quality, water quantity, and social aspects. The priority urban drainage sub-catchments based on water quantity criteria are located mainly in the north and southwestern parts of the city. Regarding the water quality aspect, priority urban drainage sub-catchments are located mainly around the Tunjuelo, Fucha, and Salitre rivers. These results show that stormwater treatment strategies have to be implemented starting at the upper sub-catchments. According to the social index, 106 urban drainage sub-catchments were classified as a priority, which included 27% of the area. The weights for the social index that resulted from averaging the three proposed objective methods were: (a) 12% fine particulate matter levels (PM 2.5), (b) 10% distance to parks, (c) 13% trees per hectare, (d) 13% occupied area, (e) 25% low-income population under five years and over sixty years, and (f) 27% low-income residential areas. In this case, most of the urban drainage sub-catchments designed as a priority are located at the southern and southwestern parts of the city. Additional results are presented in Appendix A.

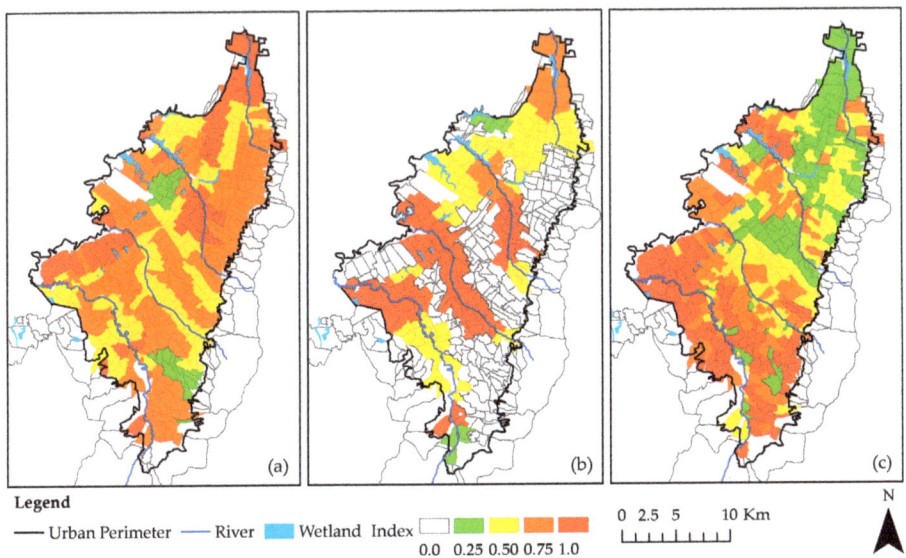

Figure 3. (**a**) Water quantity index, (**b**) water quality index, (**c**) social index.

Figure 4 presents the index that results from the analysis of water quantity, water quality, and social aspects. The results from the workshops indicate that runoff management corresponds to the stakeholders' primal concern. Subjective weights (i.e., those obtained from the workshops) were: 38% for the water quantity index, 33% for the water quality index, and 29% for the social index. Objective weights were: 27% for the water quantity index, 45% for the water quality index, and 29% for the social index. In both cases, the priority area corresponds to 29% of the analyzed area. The main difference between these two scenarios is the priority urban drainage sub-catchments along the Fucha river basin and in the north of the city. There would be more priority urban drainage sub-catchments along this river if more relevance was given to water quality. If the weight given to water quantity is higher, the north area becomes a priority. Additionally, various priority urban drainage sub-catchments are grouped in the city's southwestern part. Therefore, intervention in this part of the city is strongly recommended.

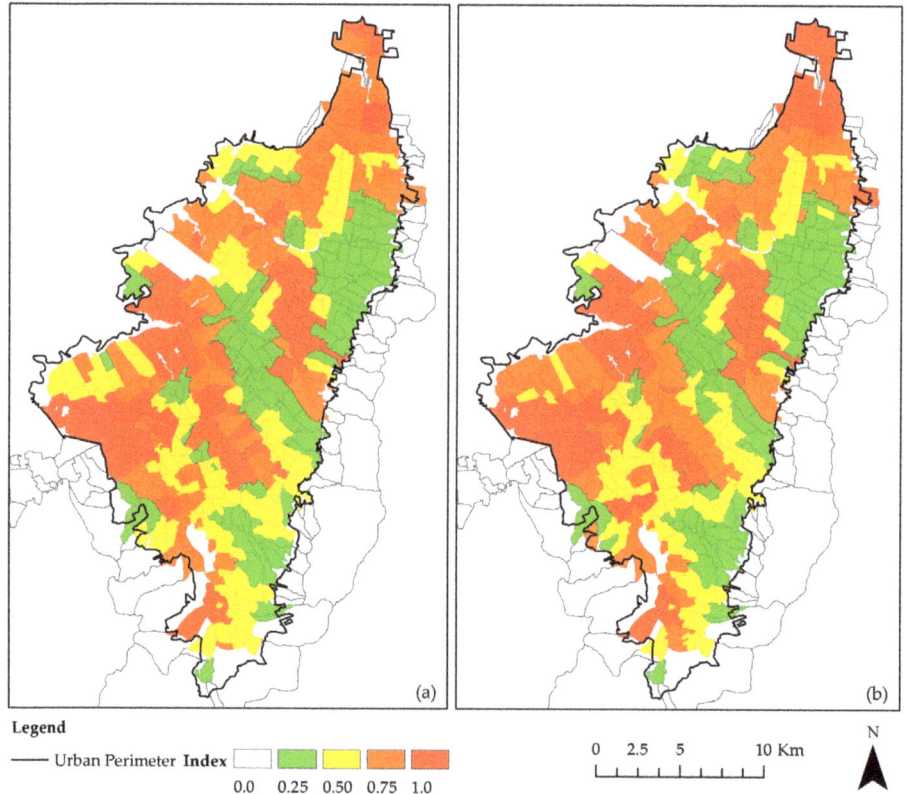

Figure 4. Priority urban drainage sub-catchments: (**a**) objective weighting, (**b**) subjective weighting.

4.1.3. Strategic Urban Drainage Sub-Catchments

The results for the analysis of corridors are summarized in Figure 5. Green corridors with a better score are located in the north of the city (Figure 5a). In particular, one corridor located along an important avenue could be an opportunity to implement SUDS. On the other hand, the main rivers of the city determine blue-green corridors. Because there are green areas adjacent to most of the Tunjuelo River, this constitutes the longest blue-green corridor. The combined index (Figure 5c) shows that there are opportunities for the joint use of the green and blue-green corridors in most of the urban drainage sub-catchments.

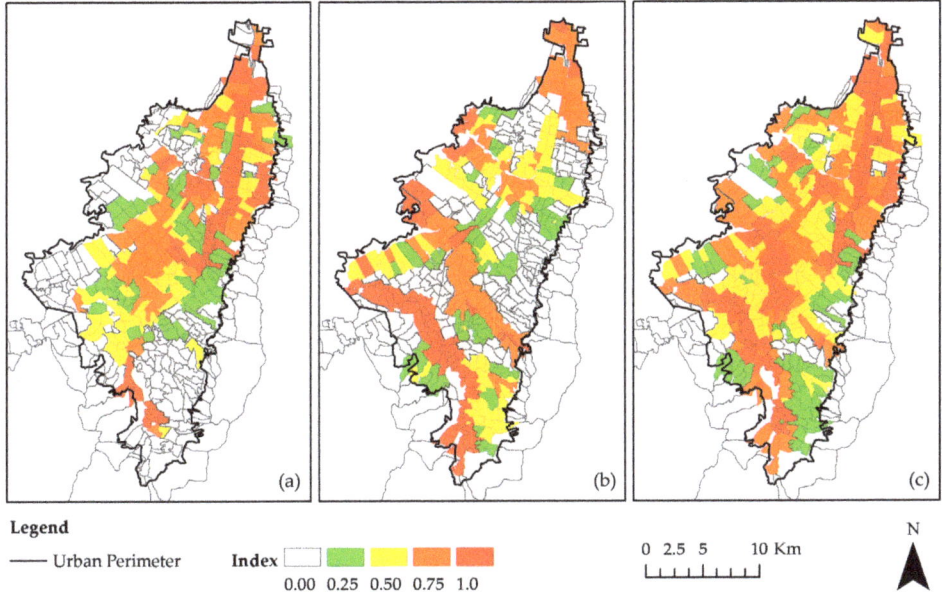

Figure 5. (a) Green corridor index, (b) blue-green corridor index, (c) green and blue-green corridor index.

Urban redevelopment and new infrastructure plans are distributed over the entire city area with a high potential for SUDS implementation (see Figure 6). However, there is a lower amount of these in the south of the city. Opportunities in the north of the city are road and public transport infrastructure that include the development of public space. A similar situation was identified in the western part of the city, which is also subject to projects for the construction of pedestrian networks, squares, and tree-lined roads. These designs are already in progress and may hinder the integration of SUDS. The southern part of the city has dispersed potential plans with a good score, which mainly constitute future public transport projects.

4.2. Local Scale

For public space, the most suitable SUDS type constitutes tree boxes, which could potentially be implemented in 58% of the public space (see Figures 7 and 8). This is because it can be implemented in several areas such as parks, squares, road dividers, and sidewalks. In contrast, infiltration basins have a low potential for implementation in the city area because of the area and minimum infiltration rate requirements. Hence, they are suitable for approximately 5.3% of the public area and 2.0% of the residential areas. Similarly, the area suitable for permeable pavements is limited to 3.2% of the public area and 8.1% of the residential area in this case study. Nevertheless, this system could be implemented in areas that were absent in the analysis. For example, narrow roads or low traffic roads may be suitable for this SUDS type and should be considered in future spatial evaluations as previous studies have identified the benefits out of implementing permeable pavements in different impervious areas due to their multifunctionality [23,109]. Permeable pavements have the potential to provide more hydrological and environmental benefits in comparison with traditional pavements. For example, in addition to managing stormwater through detention and infiltration, these systems help to reduce the heat island effect [110,111].

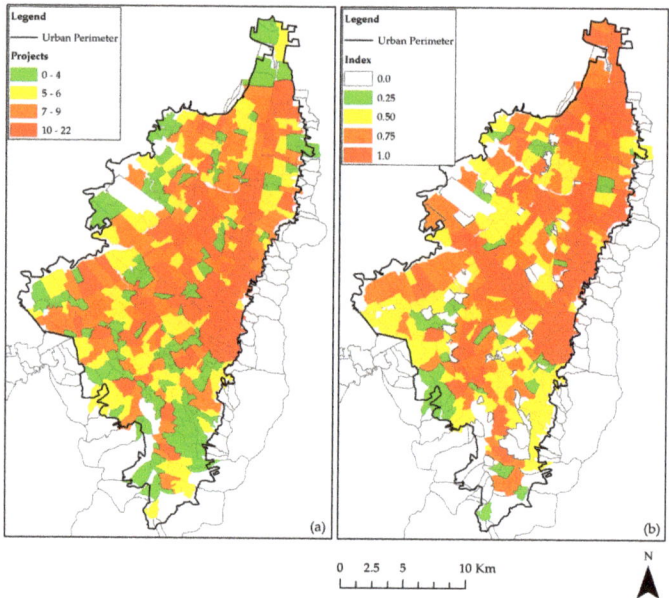

Figure 6. (**a**) Number of urban redevelopment and new infrastructure plans per urban drainage sub-catchment, (**b**) plan index.

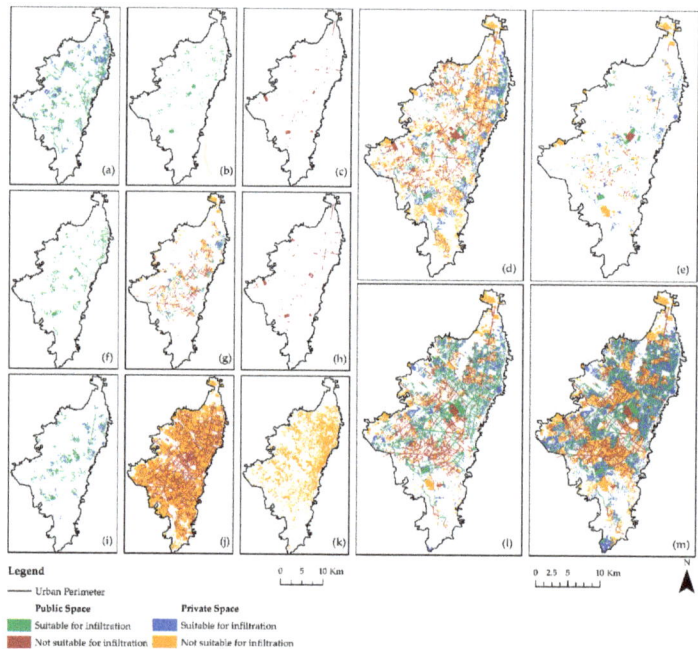

Figure 7. SUDS potential areas: (**a**) soakaways, (**b**) infiltration basins, (**c**) constructed wetlands, (**d**) grassed swales, (**e**) extended dry detention basins, (**f**) sand filters, (**g**) permeable pavements, (**h**) wet ponds, (**i**) infiltration trenches, (**j**) rain barrels and cisterns, (**k**) green roofs, (**l**) bioretention zones, and (**m**) tree boxes.

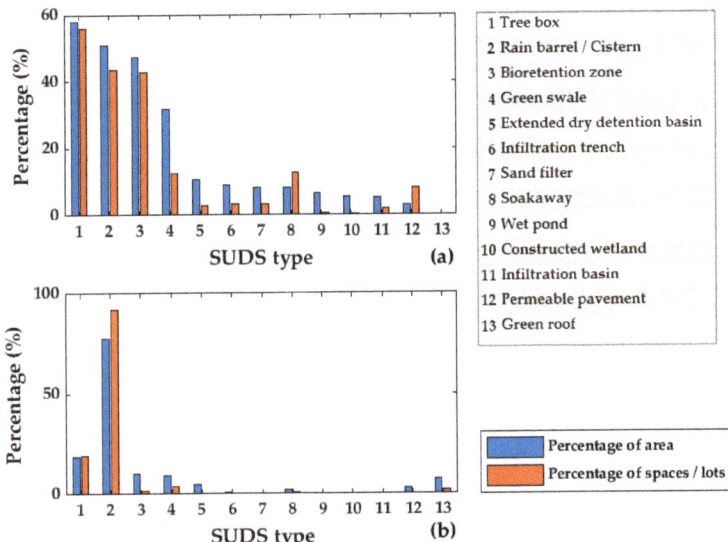

Figure 8. (**a**) Percentage of public area suitable for the evaluated SUDS types, (**b**) percentage of private residential area suitable for the evaluated SUDS types.

Figure 8 shows the percentage of the total analyzed area and the spaces suitable for the SUDS types. In this case, spaces refer to polygons of public space or lots for the private area. Wet ponds and constructed wetlands present the biggest differences between the percentage of suitable areas and spaces (see Figure 8a). Some 6.5% of the analyzed public area was found to be feasible for wet ponds, but this area corresponded to 0.6% of the number of analyzed spaces. Likewise, wetlands are suitable for 5.4% of the analyzed public area, which corresponds to 0.3% of the number of public spaces. This results from the minimum area required, limiting the implementation of this SUDS type in the southern part of the city (see Figure 7c,h). In contrast, permeable pavements are suitable in 3.2% of the public area, which is equivalent to 8.1% of the number of public spaces. There are small public spaces, mainly in the center of the city, that are feasible for this SUDS type (see Figure 7g).

Concerning private space, the most suitable SUDS type for residential use is rain barrels (see Figure 8). This is because the analysis considered flexibility in implementing this SUDS type. Nevertheless, additional restrictions related to the characteristics of the buildings could reduce the amount of suitable space. The potential area for other SUDS types is more reduced. For example, tree boxes are the second most suitable SUDS type in the residential area, but they could only be implemented in 19% of the analyzed lots. However, their implementation could bring more advantages than the rainwater barrels, particularly in terms of amenity and water quality improvement. Bioretention zones, green swales, and green roofs present notable differences between the percentage of suitable area and lots. This indicates that the opportunities for implementation concentrate in lots with large unoccupied areas.

Figure 9 presents a comparison between the suitable public and private residential areas for SUDS implementation. In each case, the value for private and public space suitability is determined according to the difference between the areas divided by the biggest area (private or public). Private residential areas have a greater potential for SUDS implementation in most of the city due to rain barrels. Figure 9a shows that 51% of the urban drainage sub-catchments have a value of over 0.80 in relation to private residential space suitability, which comprises 43% of the evaluated area. Nevertheless, other SUDS types present more benefits in terms of runoff control and amenity generation. In this sense, Figure 9b indicates that when rain barrels and cisterns are omitted, there are areas in the southwest of the city where the implementation of other kinds of SUDS is more feasible in the public space. The number of urban drainage sub-catchments with a value of suitability for private space over 0.80 changes to 14%. Figure 9c also excludes green roofs, reducing the number of urban drainage sub-catchments with that value to 11%, which corresponds to 0.26% of the evaluated area. This shows that the implementation of SUDS in private residential space needs to involve the constructed area, and the use of other types of SUDS is more feasible in public spaces. In general, public areas have a greater potential in the city center, and private residential areas in the northern and southern parts of the city.

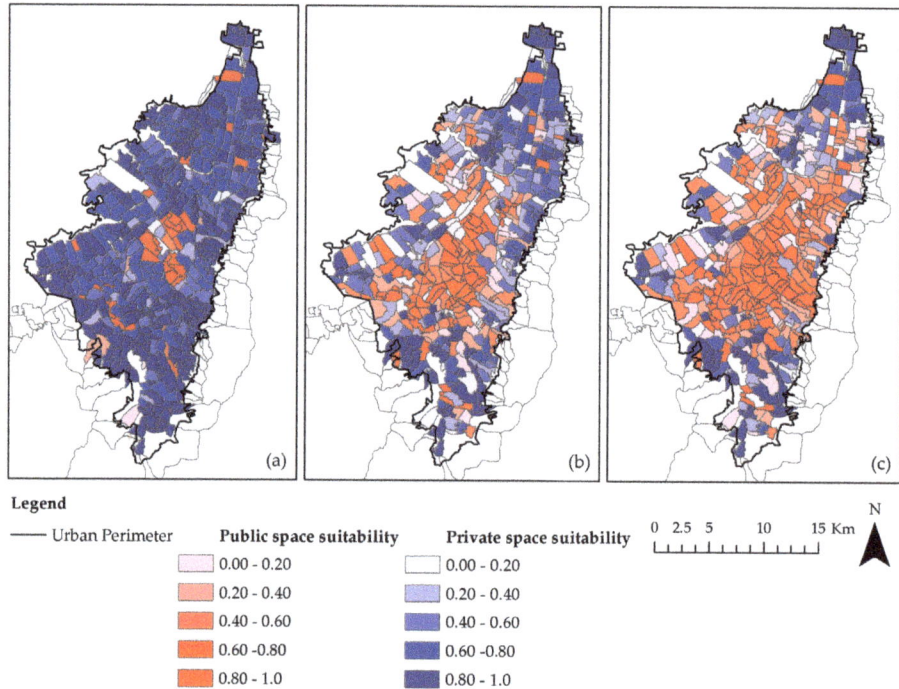

Figure 9. Public and private (residential) space comparison: (**a**) difference between total suitable area for SUDS implementation in public and private space, (**b**) difference between total suitable area for SUDS implementation in public and private space disregarding rain barrels and cisterns, (**c**) difference between total suitable area for SUDS implementation in public and private space disregarding rain barrels, cisterns, and green roofs.

4.3. Microscale

Two sites were selected to carry out the microscale analysis. The selection process included field visits to places identified in priority and strategic urban drainage sub-catchments, and from recommendations of local institutions (i.e., EAB and SDA). The first site corresponded to San Cristobal Park, in the southeast of the city (Figure 10). It is part of three urban drainage sub-catchments in the upper basin of the Fucha river. It could be a strategic area for SUDS implementation according to the analysis of green and blue-green corridors and due to its proximity to the Fucha River. The urban drainage sub-catchments in which the park is located are not prioritized, but improvements in water quantity and quality could have positive impacts downstream in prioritized urban drainage sub-catchments. The second site was a road divider located in the south of the city, referred to as the Tunal road divider. In this area, there is a future project to build a massive transport system, which constitutes an opportunity to implement SUDS. It is in an urban drainage sub-catchment where the green and blue-green corridors index is equal to one (1). Thus, even though this road divider is not in a priority urban drainage sub-catchment, it was selected for this analysis.

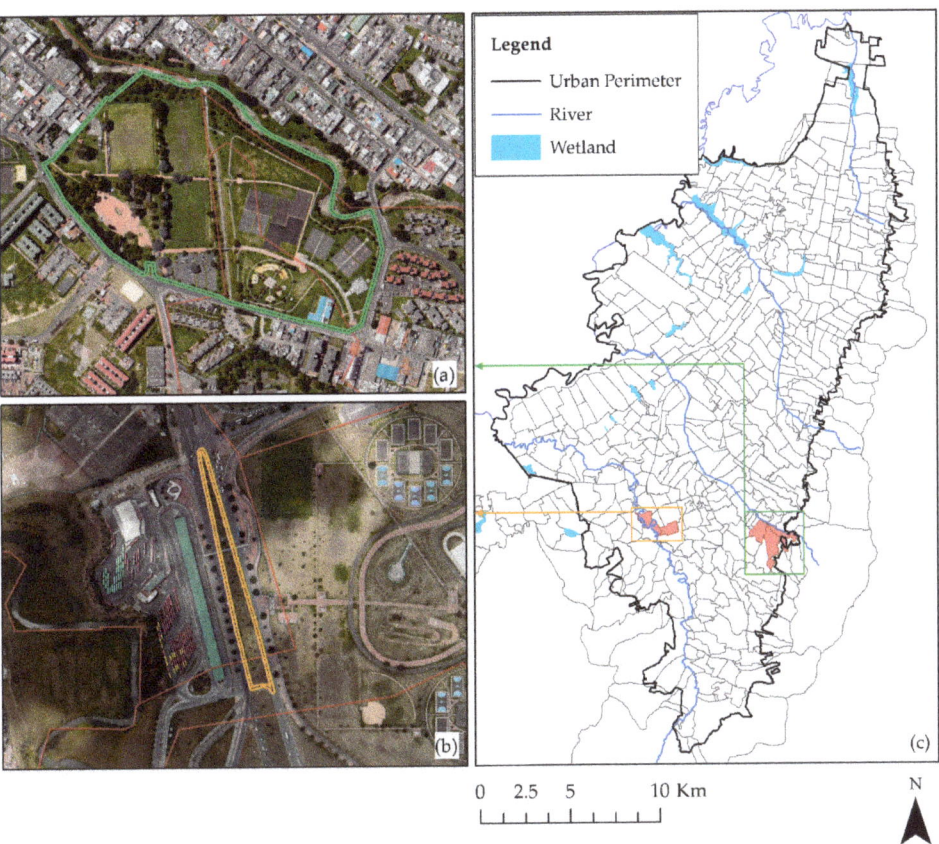

Figure 10. Selected sites: (**a**) San Cristobal Park, (**b**) the Tunal road divider, (**c**) sites location.

The proposed methodology for the treatment train selection was applied considering 1000 weights combination for five aspects to define a score from zero (0) to five (5) for each train. The suitable SUDS types for each selected site were determined by the spatial analysis conducted at local scale. In addition, in situ-evaluations led to the inclusion of other SUDS types. For San Cristobal Park, the processes analyzed were conveyance, detention, infiltration, and irrigation. For the Tunal road divider, the process of rainwater harvesting was included instead of irrigation, because it is possible to implement a cistern. Tables 8 and 9 present the most highly recommended two-stage treatment trains according to the suitable SUDS types in the selected areas. Each column presents the pairs of processes analyzed and the recommended treatment train. The number one (1) indicates the first element of the train and number two (2) indicates the second element of the train. If the SUDS types can be arranged into two different orders, both trains are shown. The weights considered for the most frequent trains are summarized and compared.

Table 8. Recommended treatment train for San Cristobal Park.

Processes	C-Ir	D-Ir	C-D	C-I
Possible trains	2	1	7	5
SUDS types			Most frequent train	
Tree box		1		
Bioretention zone	2	2		2
Grassed swale	1		1	1
Extended dry detention basin			2	2
Frequency	93%	100%	100%	79%
Higher score over 5 (most frequent train)	4.27	3.96	4.50	4.02
Evaluated weights (%)		Most frequent train weights (%)		
Quality improvement				
Runoff Control				
Amenity				
Maintenance				
Costs				

Processes: (I) infiltration, (D) detention, (C) conveyance, (Ir) irrigation. (1) First stage, (2) second stage.

Table 9. Recommended treatment train for the Tunal road divider.

Processes	C-R	D-R	C-D	I-R	C-I	I-D
Possible trains	2	3	11	3	10	3
SUDS types				Most frequent train		
Tree box						1
Bioretention zone		1	1			
Underground cistern	2	2	2	2		
Grassed swale	1		1	1	1	
Infiltration trench			2		2	2
Frequency	100%	92%	79%	100%	100%	79%
Higher score over 5 (most frequent train)	4.73	4.13	4.21	4.39	4.42	4.24
Evaluated weights (%)			Most frequent train weights (%)			
Quality improvement						
Runoff Control						
Amenity						
Maintenance						
Costs						

Processes: (I) infiltration, (D) detention, (C) conveyance, (R) rainwater harvesting. (1) First stage, (2) second stage.

The recommended train for San Cristobal Park varied depending on the process analyzed. If the processes are conveyance and detention, the recommended train is always composed by a grassed swale followed by an extended dry detention basin. For the processes of conveyance and infiltration, the recommended train is composed of a grassed swale and a bioretention zone in 79% of cases. In this case, according to Table 8, the set of water quantity weights are above the median, which indicates the importance of this aspect in recommending this train. Based on these results, a SUDS train composed by a grassed swale and an extended dry detention basin was designed and constructed as a pilot unit, which has been monitored for water quantity and quality performance since 2017. A similar case is presented in the Tunal road divider in the detention and conveyance analyses. For detention and infiltration, the recommended train is composed of a bioretention zone followed by an infiltration trench. According to the variation in the weights, this train selection follows the assignation of higher weights to water quality and lower weights to maintenance (see Table 9). These results indicate that stakeholders' preferences and project constraints are decisive in the best alternative. As in San Cristobal Park, a SUDS train conformed by tree boxes, infiltration trenches, and bioretention zones was designed in detail as another pilot case.

5. Conclusions

The development of a multiscale and multicriteria analysis is necessary to integrate the systems, scales, stakeholders, and benefits of SUDS. In this sense, the proposed methodology aims to promote a holistic approach for urban stormwater management. In addition, it seeks the inclusion of SUDS in citywide policies providing the tools to identify priority and strategic areas.

The city of Bogotá was selected to apply the proposed methodology, resulting in the identification of its advantages and limitations. In the first place, stakeholder participation in the early stages proved its importance in defining projects that responded to their concerns, and improved the city area. SUDS constitute a new approach to stormwater management in the city; thus, one of the advantages of the workshops is that they shed light on stakeholder misconceptions and gaps regarding SUDS implementation. For example, the most relevant aspects for stakeholders in Bogotá were still linked to the traditional view of the drainage system. Thus, activities like the workshops could provide an important pedagogic component. Additionally, they evince the need for institutional changes to involve social diversity and technical aspects in local regulations.

The identification of priority and strategic sub-catchments is fundamental to creating policies for SUDS implementation. In Bogotá, the evaluation of the areas according to water quantity, water quality, and social aspects indicates that the southwestern part of the city is an area that requires intervention. These interventions could be supported by the strategic sub-catchments, particularly by the sub-catchments identified through the analysis of corridors. The use of corridors provides an opportunity in every river basin to create connected spaces and give value to the runoff by improving environmental conditions.

The analysis of public and private areas according to physical constraints was a preliminary approximation that indicated the most suitable SUDS types conforming to the city characteristics. This analysis showed that the type of suitable area (i.e., public or private) varies in every urban drainage sub-catchment. On this account, regulations and incentives need to be oriented according to the potential areas. However, some of the constraints, such as the minimum infiltration rate, were estimated only roughly, meaning that site-specific analyses are still necessary to validate the results.

In Bogotá, the most suitable SUDS types for the public space were tree boxes, cisterns, bioretention zones, green swales, extended dry detention basins, and infiltration trenches. Regarding the private space, the SUDS types with more available space included rain barrels, tree boxes, green roofs, and green swales. According to the results, the constructed area is very important for runoff management in the private space. Moreover, SUDS implementation in the northern and southern parts of the city needs to include private areas due to the reduced amount of suitable public space.

Residential use was analyzed because it is the predominant type of use in the city. Nevertheless, the area available is fractionated into small spaces, limiting the suitability of many SUDS types, especially in city zones with smaller lot sizes. Furthermore, some SUDS types were not suitable for residential use. Therefore, it is recommended to conduct future studies in the city analyzing other city uses (i.e., institutional or commercial) to identify the ones with greater potential for SUDS implementation in every urban drainage sub catchment.

The purpose of the proposed methodology for train management is to simplify the identification of the most suitable train according to the processes and SUDS types whose implementation is feasible in a particular area. Nonetheless, the final recommendation can vary in accordance to the stakeholders' preferences. In this sense, it is fundamental to identify the most relevant aspects for them and the requirements of their emplacement.

Author Contributions: All the authors participated in the conceptualization and methodology. S.L.J.A., J.A.M., A.F.M. and J.P.Q. developed the formal analysis for the public space. J.P.Q. conducted the workshops with stakeholders. J.A.M. performed the formal analysis for the private spaces. Writing of the original draft and visualization was made by Jiménez. Supervision, writing, review, and editing was done by J.P.R., L.A.C. and M.D.-G.

Funding: This research was part of the project "Investigación de las tipologías y/o tecnologías de Sistemas Urbanos de Drenaje Sostenible (SUDS) que más se adapten a las condiciones de la ciudad de Bogotá D.C." funded by EAB and SDA.

Acknowledgments: The authors would like to acknowledge the support of the water utility (EAB) and the local environmental agency (SDA). In addition, we wish to thank Robert Pitt and Alexander Maestre for their advice and suggestions, as well as María Nariné Torres and other members of the CIIA research group, who shared processed information fundamental to the development of the analysis.

Conflicts of Interest: The authors declare no conflicts of interest. The funders had no role in the design of the study; in the collection, analyses, or interpretation of data; in the writing of the manuscript, and in the decision to publish the results. However, as city agencies, they participated in the developed workshops.

Appendix A

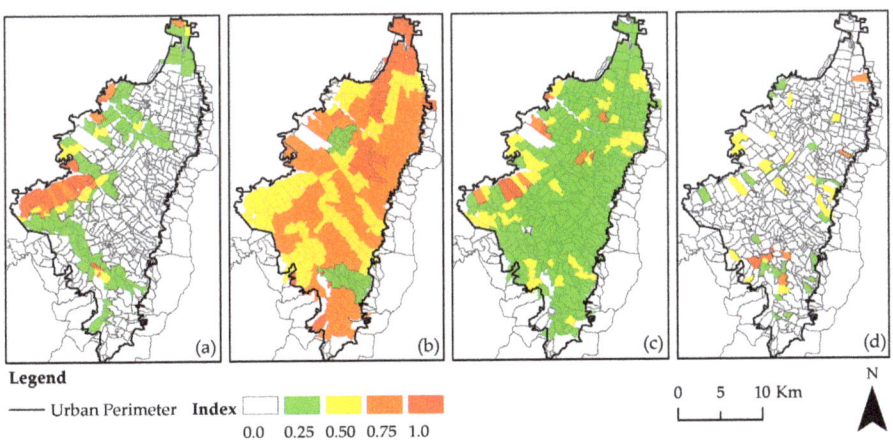

Figure A1. (**a**) Flood plains index; (**b**) storm sewer system capacity index; (**c**) ponding zones index; (**d**) critical points index.

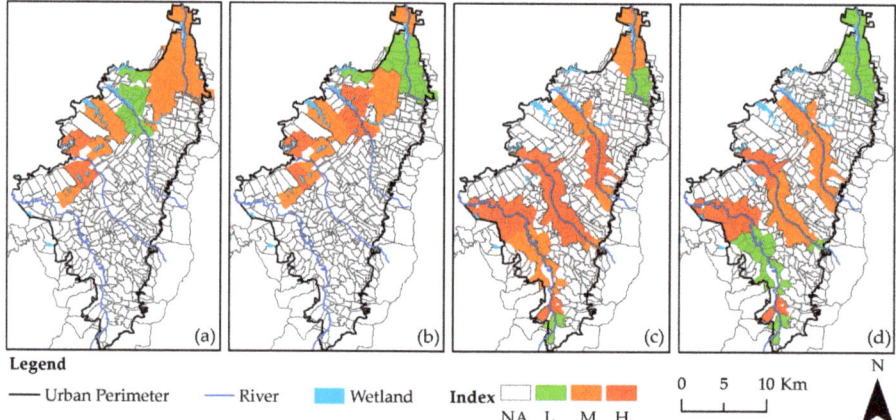

Figure A2. (**a**) Nutrient concentrations in wetlands; (**b**) TSS and BOD concentrations in wetlands; (**c**) nutrient concentrations in rivers; (**d**) TSS and BOD concentrations in rivers.

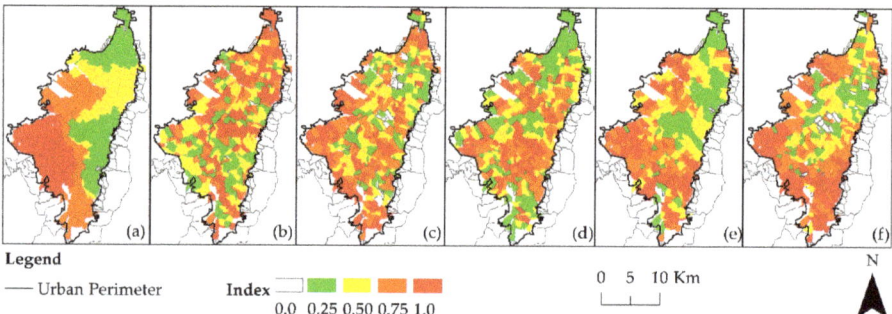

Figure A3. (**a**) Fine particulate matter levels (PM 2.5); (**b**) distance to parks; (**c**) trees per hectare; (**d**) occupied area; (**e**) population under five years and over sixty years with low economic resources; (**f**) residential areas with a low socioeconomic level.

References

1. DANE. Resultados preliminares: Censo Nacional de Población y Vivienda—CNPV 2018. Available online: https://sitios.dane.gov.co/cnpv-presentacion/src/#cuanto00 (accessed on 19 March 2019).
2. Aldana-Domínguez, J.; Montes, C.; González, J.A. Understanding the past to envision a sustainable future: A social-ecological history of the Barranquilla Metropolitan Area (Colombia). *Sustainability* **2018**, *10*, 2247. [CrossRef]
3. Shen, L.Y.; Jorge Ochoa, J.; Shah, M.N.; Zhang, X. The application of urban sustainability indicators—A comparison between various practices. *Habitat Int.* **2011**, *35*, 17–29. [CrossRef]
4. Keeley, M.; Koburger, A.; Dolowitz, D.P.; Medearis, D.; Nickel, D.; Shuster, W. Perspectives on the use of green infrastructure for stormwater management in Cleveland and Milwaukee. *Environ. Manag.* **2013**, *51*, 1093–1108. [CrossRef]
5. Poustie, M.S.; Deletic, A.; Brown, R.R.; Wong, T.; de Haan, F.J.; Skinner, R. Sustainable urban water futures in developing countries: The centralised, decentralised or hybrid dilemma. *Urban Water J.* **2015**, *12*, 543–558. [CrossRef]
6. Duffy, A.; Jefferies, C.; Waddell, G.; Shanks, G.; Blackwood, D.; Watkins, A. A cost comparison of traditional drainage and SUDS in Scotland. *Water Sci. Technol.* **2008**, *57*, 1451–1459. [CrossRef] [PubMed]
7. Ossa-Moreno, J.; Smith, K.M.; Mijic, A. Economic analysis of wider benefits to facilitate SuDS uptake in London, UK. *Sustain. Cities Soc.* **2017**, *28*, 411–419. [CrossRef]

8. De Macedo, M.B.; do Lago, C.A.F.; Mendiondo, E.M. Stormwater volume reduction and water quality improvement by bioretention: Potentials and challenges for water security in a subtropical catchment. *Sci. Total Environ.* **2019**, *647*, 923–931. [CrossRef]
9. Winston, R.J.; Dorsey, J.D.; Hunt, W.F. Quantifying volume reduction and peak flow mitigation for three bioretention cells in clay soils in northeast Ohio. *Sci. Total Environ.* **2016**, *553*, 83–95. [CrossRef]
10. Lucke, T.; Nichols, P.W.B. The pollution removal and stormwater reduction performance of street-side bioretention basins after ten years in operation. *Sci. Total Environ.* **2015**, *536*, 784–792. [CrossRef] [PubMed]
11. Braswell, A.S.; Winston, R.J.; Hunt, W.F. Hydrologic and water quality performance of permeable pavement with internal water storage over a clay soil in Durham, North Carolina. *J. Environ. Manag.* **2018**, *224*, 277–287. [CrossRef]
12. Flanagan, K.; Branchu, P.; Boudahmane, L.; Caupos, E.; Demare, D.; Deshayes, S.; Dubois, P.; Meffray, L.; Partibane, C.; Saad, M.; et al. Field performance of two biofiltration systems treating micropollutants from road runoff. *Water Res.* **2018**, *145*, 562–578. [CrossRef]
13. Fletcher, T.D.; Shuster, W.; Hunt, W.F.; Ashley, R.; Butler, D.; Arthur, S.; Trowsdale, S.; Barraud, S.; Semadeni-Davies, A.; Bertrand-Krajewski, J.-L.; et al. SUDS, LID, BMPs, WSUD and more—The evolution and application of terminology surrounding urban drainage. *Urban Water J.* **2015**, *12*, 525–542. [CrossRef]
14. Zevenbergen, C.; Fu, D.; Pathirana, A. (Eds.) *Sponge Cities: Emerging Approaches, Challenges and Opportunities. Special Issue*; MDPI: Basel, Switzerland, 2018; ISBN 9783038972723.
15. Martin-Mikle, C.J.; de Beurs, K.M.; Julian, J.P.; Mayer, P.M. Identifying priority sites for low impact development (LID) in a mixed-use watershed. *Landsc. Urban Plan.* **2015**, *140*, 29–41. [CrossRef]
16. Xu, H.; Chen, L.; Zhao, B.; Zhang, Q.; Cai, Y. Green stormwater infrastructure eco-planning and development on the regional scale: A case study of Shanghai Lingang New City, East China. *Front. Earth Sci.* **2016**, *10*, 366–377. [CrossRef]
17. Dagenais, D.; Thomas, I.; Paquette, S. Siting green stormwater infrastructure in a neighbourhood to maximise secondary benefits: Lessons learned from a pilot project. *Landsc. Res.* **2017**, *42*, 195–210. [CrossRef]
18. Meerow, S.; Newell, J.P. Spatial planning for multifunctional green infrastructure: Growing resilience in Detroit. *Landsc. Urban Plan.* **2017**, *159*, 62–75. [CrossRef]
19. Wang, Y.; Montas, H.J.; Brubaker, K.L.; Leisnham, P.T.; Shirmohammadi, A.; Chanse, V.; Rockler, A.K. A Diagnostic Decision Support System for BMP Selection in Small Urban Watershed. *Water Resour. Manag.* **2017**, *31*, 1649–1664. [CrossRef]
20. Heckert, M.; Rosan, C.D. Developing a green infrastructure equity index to promote equity planning. *Urban For. Urban Green.* **2016**, *19*, 263–270. [CrossRef]
21. Garcia-Cuerva, L.; Berglund, E.Z.; Rivers, L. An integrated approach to place Green Infrastructure strategies in marginalized communities and evaluate stormwater mitigation. *J. Hydrol.* **2018**, *559*, 648–660. [CrossRef]
22. Dearden, R.A.; Price, S.J. A proposed decision-making framework for a national infiltration SuDS map. *Manag. Environ. Qual. Int. J.* **2012**, *23*, 478–485. [CrossRef]
23. Jato-Espino, D.; Sillanpää, N.; Charlesworth, S.M.; Andrés-Doménech, I. Coupling GIS with stormwater modelling for the location prioritization and hydrological simulation of permeable pavements in urban catchments. *Water* **2016**, *8*. [CrossRef]
24. Shoemaker, L.; Riverson, J.; Alvi, K.; Zhen, J.X.; Paul, S.; Rafi, T. *SUSTAIN—A Framework for Placement of Best Management Practices in Urban Watersheds to Protect Water Quality*; National Risk Management Research Laboratory, Office of Research and Development, US Environmental Protection Agency: Cincinnati, OH, USA, 2009.
25. Tiwari, K.; Goyal, R.; Sarkar, A. GIS-based Methodology for Identification of Suitable Locations for Rainwater Harvesting Structures. *Water Resour. Manag.* **2018**, *32*, 1811–1825. [CrossRef]
26. Tuomela, C.; Jato-Espino, D.; Sillanpää, N.; Koivusalo, H. Modelling Stormwater Pollutant Reduction with LID Scenarios in SWMM. In *New Trends in Urban Drainage Modelling*; Mannina, G., Ed.; Springer International Publishing: Cham, Switzerland, 2019; pp. 96–101. ISBN 978-3-319-99866-4.
27. Cooper, D.; Calvert, J. *Ipswich Borough Council Draft Strategic Flood Risk Assessment November 2007*; Ipswich Borough Council: Ipswich, UK, 2007.
28. Kuller, M.; Bach, P.M.; Ramirez-Lovering, D.; Deletic, A. Framing water sensitive urban design as part of the urban form: A critical review of tools for best planning practice. *Environ. Model. Softw.* **2017**, *96*, 265–282. [CrossRef]

29. Zellner, M.; Massey, D.; Minor, E.; Gonzalez-Meler, M. Exploring the effects of green infrastructure placement on neighborhood-level flooding via spatially explicit simulations. *Comput. Environ. Urban Syst.* **2016**, *59*, 116–128. [CrossRef]
30. Gogate, N.G.; Kalbar, P.P.; Raval, P.M. Assessment of stormwater management options in urban contexts using Multiple Attribute Decision-Making. *J. Clean. Prod.* **2017**, *142*, 2046–2059. [CrossRef]
31. Eaton, T.T. Approach and case-study of green infrastructure screening analysis for urban stormwater control. *J. Environ. Manag.* **2018**, *209*, 495–504. [CrossRef] [PubMed]
32. Morales-Torres, A.; Escuder-Bueno, I.; Andrés-Doménech, I.; Perales-Momparler, S. Decision Support Tool for energy-efficient, sustainable and integrated urban stormwater management. *Environ. Model. Softw.* **2016**, *84*, 518–528. [CrossRef]
33. Wang, M.; Sweetapple, C.; Fu, G.; Farmani, R.; Butler, D. A framework to support decision making in the selection of sustainable drainage system design alternatives. *J. Environ. Manag.* **2017**, *201*, 145–152. [CrossRef]
34. Charlesworth, S.; Warwick, F.; Lashford, C. Decision-making and sustainable drainage: Design and scale. *Sustainability* **2016**, *8*, 782. [CrossRef]
35. Mora-Melià, D.; López-Aburto, C.S.; Ballesteros-Pérez, P.; Muñoz-Velasco, P. Viability of green roofs as a flood mitigation element in the central region of Chile. *Sustainability* **2018**, *10*, 1130. [CrossRef]
36. Gogate, N.G.; Rawal, P.M. Identification of potential stormwater recharge zones in dense urban context: A case study from Pune city. *Int. J. Environ. Res.* **2015**, *9*, 1259–1268.
37. Petit-Boix, A.; Sevigné-Itoiz, E.; Rojas-Gutierrez, L.A.; Barbassa, A.P.; Josa, A.; Rieradevall, J.; Gabarrell, X. Floods and consequential life cycle assessment: Integrating flood damage into the environmental assessment of stormwater Best Management Practices. *J. Clean. Prod.* **2017**, *162*, 601–608. [CrossRef]
38. Zardari, N.H.; Ahmed, K.; Shirazi, S.M.; Yusop, Z.B. *Weighting Methods and Their Effects on Multi-Criteria Decision Making Model Outcomes in Water Resources Management*; Springer: Cham, Switzerland, 2015; ISBN 9783319125855.
39. Checkland, P. Soft Systems Methodology: A Thirty Year Retrospective. *Syst. Res. Behav. Sci.* **2000**, *17*, 11–58. [CrossRef]
40. Sánchez, A.; Mejía, A. Learning to support learning together: An experience with the soft systems methodology. *Educ. Action Res.* **2008**, *16*, 109–124. [CrossRef]
41. Chapra, S.C. *Surface Water-Quality Modeling*; Waveland Press, Inc.: Long Grove, IL, USA, 2008; ISBN 978-1-57766-605-9.
42. Ekkel, E.D.; de Vries, S. Nearby green space and human health: Evaluating accessibility metrics. *Landsc. Urban Plan.* **2017**, *157*, 214–220. [CrossRef]
43. Deng, H.; Yeh, C.H.; Willis, R.J. Inter-company comparison using modified TOPSIS with objective weights. *Comput. Oper. Res.* **2000**, *27*, 963–973. [CrossRef]
44. Nicoletti, G.; Scarpetta, S.; Boylaud, O. *Summary Indicators of Product Market Regulation with an Extension to Employment Protection Legislation*; ECO Working Paper No. 226; OECD: Paris, France, 1999.
45. Jahan, A.; Mustapha, F.; Sapuan, S.M.; Ismail, M.Y.; Bahraminasab, M. A framework for weighting of criteria in ranking stage of material selection process. *Int. J. Adv. Manuf. Technol.* **2012**, *58*, 411–420. [CrossRef]
46. Dylewski, K.L.; Brown, J.T.R.; LeBleu, C.M.; Eve, F. *Brantley Low Impact Development Handbook for the State of Alabama*; Alabama Department of Environmental Management: Auburn, AL, USA, 2014.
47. Luoni, S.; Amos, C.A.; Breshears, K.; Huber, J.; Jacobs, C.; Reyenga, S.M.; Komlos, L.; Guzman, D.; Roark, B.; Lewis, S.; et al. *Low Impact Development: A Design Manual for Urban Areas*; University of Arkansas Community Design Center: Fayetteville, AR, USA, 2010; ISBN 9780979970610.
48. Wilson, S.; Bray, B.; Neesam, S.; Bunn, S.; Flanagan, E. *Sustainable Drainage. Cambridge Design and Adoption Guide*; Cambridge City Council: Cambridge, UK, 2009.
49. Woods Ballard, W.; Wilson, S.; Udale-Clarke, H.; Illman, S.; Scott, T.; Ashley, R.; Kellager, R. *The SuDS Manual*; CIRIA: London, UK, 2007; ISBN 978-0-86017-697-8.
50. Urban Drainage and Flood Control Distric. *Urban Storm Drainage. Criteria Manual. Volume 3—Best Management Practices*; Water Resources Publications, LLC: Denver, CO, USA, 2010; ISBN 1-887201-66-1.
51. City of Edmoton. *Low Impact Development Best Management Practices Design Guide*; City of Edmoton: Edmoton, AB, Canada, 2011.

52. Lawson, K.; Callow, P.; Shepherd, L.; Goodyear, K.; Presland, V.; Wright, P.; Morris, P.; Hughes, P.; Downs, C.; Dawson, P. *Sustainable Drainage Systems (SUDS). Design and Adoption Guide*; Essex County Council: Essex, UK, 2012.
53. Revitt, M.; Ellis, B.; Scholes, L. *Report 5.1. Review of the Use of Stormwater BMPs in Europe*; Middlesex University: Middlesex, UK, 2003.
54. City of Los Angeles. *Development Best Management Practices Handbook*; City of Los Angeles: Los Angeles, CA, USA, 2011.
55. Strecker, E.; Sheffield, A.; Cristina, C.; Leisenring, M. *Stormwater BMP Guidance Tool. A Stormwater Best Management Practices Guide for Orleans and Jefferson Parishes*; Bayou Land RC&D & Louisiana Public Health Institute: New Orleans, LA, USA, 2010.
56. Department of Environmental Resources Prince George's County. *Low-Impact Development Design Strategies. An Integrated Design Approach*; Department of Environmental Resources Prince George's County: Largo, MD, USA, 1999.
57. Center for Watershed Protection. *Maryland Stormwater Design Manual. Volumes I & II*; Maryland Department of the Environment: Baltimore, MD, USA, 2000.
58. The Low Impact Development Center. *Mount Rainier Urban Green Infrastructure Master Plan*; The Low Impact Development Center: Beltsville, MD, USA, 2013.
59. Philadelphia Water Department. *Stormwater Management Guidance Manual, Version 3*; Philadelphia Water Department: Philadelphia, PA, USA, 2015.
60. Riverside County Flood Control and Water Conservation District. *Design Handbook for Low Impact Development Best Management Practices*; Riverside County Flood Control and Water Conservation District: Riverside, CA, USA, 2011.
61. City of Santa Rosa. *Storm Water. Low Impact Development Technical Design Manual*; City of Santa Rosa & The County of Sonoma: Santa Rosa, CA, USA, 2011.
62. Fernández, B.; Muñoz, J.F.; Varas, E.; Fernández, T.; Destéfano, C.; Pizarro, G.; Rengifo, P.; Benítez, D.; Díaz, M.E.; Courar, P.; et al. *Técnicas Alternativas para Soluciones de Aguas Lluvias en Sectores Urbanos. Guía de Diseño*; Ministerio de Vivienda y Urbanismo: Santiago, Chile, 1996.
63. Toronto and Region Conservation Authority; Credit Valley Conservation Authority. *Low Impact Development Stormwater Management Planning and Design Guide*; Toronto and Region Conservation Authority: Toronto, ON, Canada, 2010.
64. Melbourne Water. *WSUD Engineering Procedures: Stormwater*; CSIRO Publishing: Collingwood, VIC, Australia, 2005; ISBN 0-643-09092-4.
65. Virginia Department of Transportation. *BMP Design Manual of Practice*; Virginia Department of Transportation: Richmond, VA, USA, 2013.
66. Department of Water & Swan River Trust. Structural Controls. In *Stormwater Management Manual for Western Australia*; Department of Water Government of Western Australia: Perth, Australia, 2007; ISBN 978-1-921094-61-3.
67. Massachusetts Department of Environmental Protection. Structural BMP Specifications for the Massachusetts Stormwater Handbook. In *Stormwater Handbook Volume 2*; Massachusetts Department of Environmental Protection: Boston, MA, USA, 2008.
68. Blick, S.A.; Kelly, F.; Skupien, J.J. *Stormwater Best Management Practices Manual*; New Jersey Department of Environmental Protection Division of Watershed Management: Trenton, NJ, USA, 2004.
69. Faha, L.; Faha, M.; Milligan, B. *Low Impact Development Approaches Handbook*; Clean Water Services: Tualatin, OR, USA, 2009.
70. Department of Defense USA. *Unified Facilities Criteria (UFC): Low Impact Development Manual*; Department of Defense USA: Washington, DC, USA, 2010.
71. Debo, T.N.; Reese, A.J. *Stormwater Management*, 2nd ed.; Lewis Publishers: Boca Ratón, FL, USA, 2003; ISBN 1566705843.
72. Jia, H.; Yao, H.; Tang, Y.; Yu, S.L.; Zhen, J.X.; Lu, Y. Development of a multi-criteria index ranking system for urban runoff best management practices (BMPs) selection. *Environ. Monit. Assess.* **2013**, *185*, 7915–7933. [CrossRef] [PubMed]
73. Moore, S.L.; Stovin, V.R.; Wall, M.; Ashley, R.M. A GIS-based methodology for selecting stormwater disconnection opportunities. *Water Sci. Technol.* **2012**, *66*, 275–283. [CrossRef]

74. Aladenola, O.O.; Adeboye, O.B. Assessing the potential for rainwater harvesting. *Water Resour. Manag.* **2010**, *24*, 2129–2137. [CrossRef]
75. Campisano, A.; Butler, D.; Ward, S.; Burns, M.J.; Friedler, E.; DeBusk, K.; Fisher-Jeffes, L.N.; Ghisi, E.; Rahman, A.; Furumai, H.; et al. Urban rainwater harvesting systems: Research, implementation and future perspectives. *Water Res.* **2017**, *115*, 195–209. [CrossRef] [PubMed]
76. Zhang, X.; Hu, M. Effectiveness of rainwater harvesting in runoff volume reduction in a planned industrial park, China. *Water Resour. Manag.* **2014**, *28*, 671–682. [CrossRef]
77. Coombes, P.J.; Argue, J.R.; Kuczera, G. Figtree Place: A case study in water sensitive urban development (WSUD). *Urban Water* **2000**, *1*, 335–343. [CrossRef]
78. Ghisi, E.; Montibeller, A.; Schmidt, R.W. Potential for potable water savings by using rainwater: An analysis over 62 cities in southern Brazil. *Build. Environ.* **2006**, *41*, 204–210. [CrossRef]
79. Jones, M.P.; Hunt, W.F. Performance of rainwater harvesting systems in the southeastern United States. *Resour. Conserv. Recycl.* **2010**, *54*, 623–629. [CrossRef]
80. Herrmann, T.; Schmida, U. Rainwater utilisation in Germany: Efficiency, dimensioning, hydraulic and environmental aspects. *Urban Water* **2000**, *1*, 307–316. [CrossRef]
81. Rahman, A.; Keane, J.; Imteaz, M.A. Rainwater harvesting in Greater Sydney: Water savings, reliability and economic benefits. *Resour. Conserv. Recycl.* **2012**, *61*, 16–21. [CrossRef]
82. Steffen, J.; Jensen, M.; Pomeroy, C.A.; Burian, S.J. Water supply and stormwater management benefits of residential rainwater harvesting in U.S. cities. *J. Am. Water Resour. Assoc.* **2013**, *49*, 810–824. [CrossRef]
83. United States Environmental Protection Agency. *Rainwater Harvesting: Conservation, Credit, Codes, and Cost*; United States Environmental Protection Agency: Washington, DC, USA, 2013.
84. Ward, S.; Memon, F.A.; Butler, D. Performance of a large building rainwater harvesting system. *Water Res.* **2012**, *46*, 5127–5134. [CrossRef] [PubMed]
85. Boston Water and SewerCommission; Geosyntec Consultants. *Stormwater Best Management Practices: Guidance Document*; Boston Water and Sewer Commission: Boston, MA, USA, 2013.
86. Liu, Y.; Engel, B.A.; Flanagan, D.C.; Gitau, M.W.; Mcmillan, S.K.; Chaubey, I. Science of the Total Environment A review on effectiveness of best management practices in improving hydrology and water quality: Needs and opportunities. *Sci. Total Environ.* **2017**, *601–602*, 580–593. [CrossRef]
87. Liu, J.; Sample, D.J.; Bell, C.; Guan, Y. Review and Research Needs of Bioretention Used for the Treatment of Urban Stormwater. *Water* **2014**, *6*, 1069–1099. [CrossRef]
88. Fletcher, T.; Duncan, H.; Poelsma, P.; Lloyd, S. *Stormwater Flow and Quality, and The Effectiveness of Non-Proprietary Stormwater Treatment Measures—A Review and Gap Analysis. Technical Report*; Cooperative Research Centre for Catchment Hydrology: Melbourne, VIC, Australia, 2004.
89. Venner, M.; Strecker, E.; Leisenring, M.; Pankani, D.; Taylor, S. *NCHRP 25-25/83: Current Practice of Post-Construction Structural Stormwater Control Implementation for Highways*; National Cooperative Highway Research Program: Lakewood, CO, USA, 2013.
90. Rodríguez Susa, M.S.; Porras, L.S.; Martínez León, A.J.; Ramírez Zamudio, N. *Calidad del Recurso Hídrico de Bogotá (2012–2013)*; Universidad de los Andes, Facultad de Ingeniería, Departamento de Ingeniería Civil y Ambiental, Ediciones Uniandes. Alcaldía Mayor, Secretaría Distrital de Ambiente: Bogotá, Colombia, 2014; ISBN 9789587740479.
91. Empresa de Acueducto Alcantarillado y Aseo de Bogotá (EAB); Secretaria Distrital de Ambiente (SDA). *IX Fase del Programa de Seguimiento y Monitoreo de Efluentes Industriales y Afluentes al Recurso Hídrico de Bogotá*; Empresa de Acueducto Alcantarillado y Aseo de Bogotá (EAB): Bogotá, Colombia, 2010.
92. Secretaría Distrital de Planeación (SDP). *Sistema de alcantarillado—Mapa No 20*; Alcaldía Mayor de Bogotá D.C.: Bogotá, Colombia, 2013.
93. IEH GRUCON S.A. *Recopilación y Análisis de Información Requerida para la Consolidación de la Base de Datos de Conocimiento de los Puntos Críticos del Alcantarillado de Bogotá*; EAB: Bogotá, Colombia, 2011.
94. Secretaría Distrital de Planeación (SDP). *Amenaza de inundación por desbordamiento—Mapa Borrador No 04*; Alcaldía Mayor de Bogotá D.C.: Bogotá, Colombia, 2013.
95. Red de Monitoreo de Calidad del Aire de Bogotá (RMCAB). *Informe Anual de Calidad del Aire en Bogotá*; Secretarí-a Distrital de Ambiente: Bogotá, Colombia, 2017.
96. Red de Monitoreo de Calidad del Aire de Bogotá (RMCAB). Multi Station Report. Available online: http://201.245.192.252:81/ (accessed on 1 July 2018).

97. Ministerio de Ambiente y Desarollo Sostenible. *Resolución 2254 de 2017 Ministerio de Ambiente y Desarrollo Sostenible. Por la cual se adopta la norma de calidad del aire ambiente y se dictan otras disposiciones*; Ministerio de Ambiente y Desarollo Sostenible: Bogotá, Colombia, 2017.
98. Jardín Botánico de Bogotá José Celestino Mutis Visor de Información Geográfica—SIGAU. Available online: http://sigau.jbb.gov.co/SigauJBB/VisorPublico/VisorPublico (accessed on 1 July 2018).
99. Infraestructura de Datos Espaciales (IDECA) Mapa de Referencia IDECA. Available online: https://www.ideca.gov.co/es/encuestamapa-de-referencia-ideca (accessed on 15 June 2018).
100. Infraestructura de Datos Espaciales (IDECA) Mapas Bogotá. Available online: http://mapas.bogota.gov.co/# (accessed on 1 July 2018).
101. Secretaría Distrital de Ambiente (SDA) Visor Ambiental. Available online: http://www.secretariadeambiente.gov.co/visorgeo/#submenu-capas (accessed on 30 July 2018).
102. Instituto de Desarrollo Urbano (IDU); Infraestructura de Datos Espaciales (IDECA) Seguimiento de Proyectos—SIGIDU. Available online: http://idu.maps.arcgis.com/apps/webappviewer/index.html?id=6950db8fa2d440ffbb3946c468eaae4a (accessed on 25 June2018).
103. Instituto de Desarrollo Urbano (IDU); Infraestructura de Datos Espaciales (IDECA) Visor de Proyectos. Available online: http://opendata.idu.gov.co/visor_proyectos/ (accessed on 25 June 2018).
104. EAB SISGEO. Available online: http://gme.acueducto.com.co/sisgeo/ (accessed on 15 Jun 2015).
105. Twarakavi, N.K.C.; Šimůnek, J.; Schaap, M.G. Can texture-based classification optimally classify soils with respect to soil hydraulics? *Water Resour. Res.* **2010**, *46*. [CrossRef]
106. Roy, A.H.; Wenger, S.J.; Fletcher, T.D.; Walsh, C.J.; Ladson, A.R.; Shuster, W.D.; Thurston, H.W.; Brown, R.R. Impediments and Solutions to Sustainable, Watershed-Scale Urban Stormwater Management: Lessons from Australia and the United States. *Environ. Manag.* **2008**, *42*, 344–359. [CrossRef]
107. Brown, R.R. Impediments to integrated urban stormwater management: The need for institutional reform. *Environ. Manag.* **2005**, *36*, 455–468. [CrossRef]
108. Dhakal, K.P.; Chevalier, L.R. Managing urban stormwater for urban sustainability: Barriers and policy solutions for green infrastructure application. *J. Environ. Manag.* **2017**, *203*, 171–181. [CrossRef]
109. Jato-Espino, D.; Charlesworth, S.M.; Bayon, J.R.; Warwick, F. Rainfall-runoff simulations to assess the potential of SUDS for mitigating flooding in highly urbanized catchments. *Int. J. Environ. Res. Public Health* **2016**, *13*, 149. [CrossRef]
110. Kayhanian, M.; Li, H.; Harvey, J.T.; Liang, X. Application of permeable pavements in highways for stormwater runoff management and pollution prevention: California research experiences. *Int. J. Transp. Sci. Technol.* **2019**. [CrossRef]
111. Liu, Y.; Li, T.; Peng, H. A new structure of permeable pavement for mitigating urban heat island. *Sci. Total Environ.* **2018**, *634*, 1119–1125. [CrossRef] [PubMed]

© 2019 by the authors. Licensee MDPI, Basel, Switzerland. This article is an open access article distributed under the terms and conditions of the Creative Commons Attribution (CC BY) license (http://creativecommons.org/licenses/by/4.0/).

Article

Accelerated Exploration for Long-Term Urban Water Infrastructure Planning through Machine Learning

Junyu Zhang [1,2], Dafang Fu [1,2,*], Christian Urich [1,3] and Rajendra Prasad Singh [1,2]

1. Joint Research Centre for Water Sensitive Cities, Southeast University-Monash University Joint Graduate School (Suzhou), Southeast University, Suzhou 215123, China; junyu.zhang@seu.edu.mn (J.Z.); christian.urich@monash.edu (C.U.); rajupsc@seu.edu.cn (R.P.S.)
2. Department of Civil Engineering, Southeast University, #2Sipailou, Nanjing 210096, China
3. Department of Civil Engineering, Monash University, Clayton, VIC 3800, Australia
* Correspondence: fdf@seu.edu.cn; Tel.: +86-185-5182-4285

Received: 15 October 2018; Accepted: 1 December 2018; Published: 5 December 2018

Abstract: In this study, the neural network method (Multi-Layer Perceptron, MLP) was integrated with an explorative model, to study the feasibility of using machine learning to reduce the exploration time but providing the same support in long-term water system adaptation planning. The specific network structure and training pattern were determined through a comprehensive statistical trial-and-error (considering the distribution of errors). The network was applied to the case study in Scotchman's Creek, Melbourne. The network was trained with the first 10% of the exploration data, validated with the following 5% and tested on the rest. The overall root-mean-square-error between the entire observed data and the predicted data is 10.5722, slightly higher than the validation result (9.7961), suggesting that the proposed trial-and-error method is reliable. The designed MLP showed good performance dealing with spatial randomness from decentralized strategies. The adoption of MLP-supported planning may overestimate the performance of candidate urban water systems. By adopting the safety coefficient, a multiplicator or exponent calculated by observed data and predicted data in the validation process, the overestimation problem can be controlled in an acceptable range and have few impacts on final decision making.

Keywords: urban planning; water infrastructure; adaptation planning; artificial neural network; multi-layer perception

1. Introduction

Long-term strategic planning on urban infrastructures is often obsessed with future uncertainties such as the state of the world (e.g., economic situation, climate) or state of the city (e.g., population growth). These uncertainties are not statistical in nature which makes them hard to predict. One of the most convincing examples is the "Shrinking City" event in Dresden since 1990, where 7 predictions have been made during 15 years to predict the population growth and guide the city planning but none of them turned out to be right [1,2].

To deal with this issue, computational tools have been developed to look into more future scenarios and offer more reliable plans, such as Adaptation tipping points [3], Robust decision making [4], Info-gap [5]. The adaptation tipping points offered shifting between different strategies and plans but no guarantee of success adaptation due to lack of system performance evaluation. The robust decision-making and info-gap both aim to explore as much future as possible and evaluate the robustness of candidate plans by trade-off on the target.

As an improvement exploring planning tools have been developed to model the performance of different infrastructure plans under different scenarios, such as Adaptive policy making [6], Adaptation pathways [7] and Dynamic adaptive policy making [8]. The adaptation pathways are able to simulate the dynamic of different infrastructure and the adaptation among them under relatively small range of future scenarios. Meanwhile, the adaptive policy making looks into wide range of future scenarios without lack of infrastructure adaptation. As the improvement of them the dynamic adaptive policy making tries to consider both but could only work out plans for independent strategies.

The limitation of the current tools is they are not able to evaluate the adaptation of a real-world combined system (centralized + decentralized) as such simulation is excessively time-consuming. More precisely, one of the major challenges on reducing the time consumption in such exploration planning tools is the robustness problem. The more detailed designs to be modelled (especially spatial distributed decentralized systems) and the more scenarios to be considered, the more time it will take, the more robust the plan can be.

Unfortunately, there are only few methods or tools that could reduce the exploration time while maintain the exploration range. This problem is being addressed in this paper by integrating the neural network method (multi-layer perceptron) with an explorative model that simulates possible urban infrastructure adaptation, to study the feasibility of using machine learning to reduce the computational time in such exploration.

In recent years, Artificial Neural Networks (ANNs), as a data-drive, self-adaptive and non-linear forecasting tool was applied in various fields such as natural resource management [9–11], pattern recognition [12,13], medical diagnosis [14] and decision making [15,16]. As a matter of factor, the methods and its derivative tool are often used in short-term decision makings or predictions (event scale) rather than long-term planning (strategy scale). To cope with the exploration model, the machine learning algorithm was designed and trained to predict urban water infrastructure performance for individual events while the decision on planning was made based on microscopic strategy performance distribution.

In this paper, the above accelerated explorative long-term planning method was proposed and tested. The following works have been conducted: (1) a comprehensive statistical trial-and-error analysis method is proposed and tested to avoid local optimization of network structure. (2) a neural network was integrated in the explorative adaptation planning to significantly reduce the simulation time, performance was tested and analyzed; (3) a correction method was proposed and tested to minimize the overestimation problem of the designed exploration framework.

2. Methods

2.1. Site Description and the Exploration

The case was carried out in Scotchman's Creek catchment, locates at the southeast of Melbourne CBD. The catchment is mostly located within Monash City council but a part of the catchment (6%) is situated within Whitehorse City council. It has an area of approximately 10.36 km^2 and a population of approximately 25,000 residents.

The council started to introduce rainwater tanks to households since 2005 to deal with the unpredictable rainfall events (e.g., reduce peak flow during highly intensive rainfall event, store rain water during drought season). Although the council tried to set up a progressive goal of rainwater tank uptake rate in the area, there were several obstacles in making such a plan: (1) The spatial distribution of rainwater tanks will largely influence the flood resistance in the catchment resulting from them. Thus, the promoting of higher rainwater tank uptake rate cannot be easily determined compared to upsizing pipe systems; (2) The population growth in the area could infect the construction of houses and buildings which increases the impervious surfaces in the catchment as well as the opportunity for uptake rainwater tanks; (3) The flood-resistance robustness of the combined drainage system (under different rainwater tank uptake ratio and pipe system capacity) was unclear.

Thus, a long-term (2015–2035) evolution of the urban development, climate change and water infrastructure adaptation were simulated by DAnCE4Water (Dynamic Adaptation for enabling City Evolution for Water) [17,18] to set up a robust plan of progressive goals for both rainwater take uptake ratio and drainage pipe system upsizing. With the initial city scenario established based on the real-world catchment in 2015, DAnCE4Water ran in a 5-year interval to simulate the transformation of the city and assess the urban water system performance with different drainage infrastructure updates under all possible development scenarios.

The development scenario consists of two parameters: the population growth rate (PGR) and the climate change factor (CCF). The 5-year population growth rate is ranged in [0.03,0.06] which calculated based on the maximum annual growth rate (0.012 per year) in the area according to the 1990–2015 census data from the Australian Bureau of Statistics. DAnCE4Water would replace old buildings and construct new ones according to the increased population through its urban development module (UDM) [17,18]. The 5-year climate change factor is a coefficient used to magnify the 5-year designed storm. Initialized to 1.00, CCF is assumed to change every 5 years within three rates: 0.95X, 1.00X or 1.05X.

Three drainage update options were tested in this paper: (1) business as usual, (2) uptake rainwater harvesting tanks and (3) upsize drainage pipes. "Business as usual (BAU)" maintained the existing infrastructures from the previous step. The more BAU was taken, the less contribution would be done in reducing flooded junctions. "Uptake rainwater harvesting tank (RWHT)" increased the current probability of households installing rainwater harvesting tanks by 5%. The more RWHT was taken, the more decentralized systems would be built to reduce the runoff and peak flow. "Upsize drainage system (PIPE)" upgrades the drainage network, which was divided into 4 groups according to their diameters. Each upgrade enlarged one group of pipes, from the large one to the small one. The more PIPE was taken, the higher capacity of the drainage network would be.

The exploration randomly selected a PGR, a CCF and a drainage infrastructure update within the available range and applied to the base city scenario. The UDM would then generate a future scenario of the city while the performance of the combined system (the number of flooded junctions in the catchment area along the drainage network) would be evaluated by SWMM. The result city scenario was saved as the base city scenario for the next 5-year decision (see Figure 1).

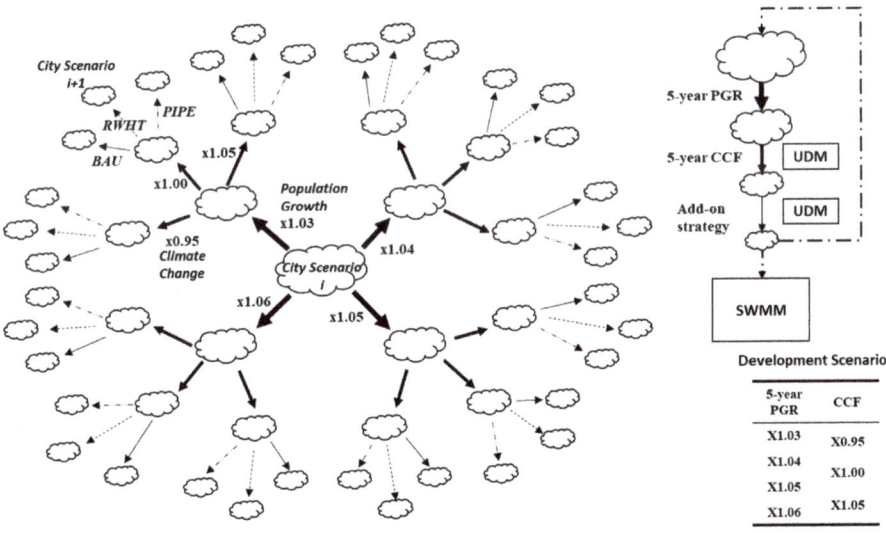

Figure 1. Designed exploration of the Scotchman's Creek catchment area.

The result scenarios were classified by the drainage infrastructure status (e.g., how many steps of BAU, RWHT and PIPE were adopted respectively). The corresponding distribution of system performance (flooded junctions) for each status was calculated. As only one strategy was taken in each decision step, the status contains the year information as well. If the number of flooded junctions of a status was below the target (110 in 2020, 100 in 2025, 90 in 2030 and 80 in 2035, which is 100%, 91%, 82%, 73% of the flooded junctions in 2015) in over 95% of the cases, the status would be consider "robust." The "robust" statuses were connected in a time line to form a drainage infrastructure implementation pathway as the long-term plan in this case study.

To compare the proposed acceleration exploration method, the plan was first explored through the above traditional exploration. The 20-year planning took 2.93 million simulations including 1.73 million explorations with uniformed input values and 1.2 million with random input values for the last two decision steps. The uniformed input values were listed in Figure 1, with 36 scenarios in 2020 (4 PGRs * 3 CCFs * 3 add-on strategies), 36^2 in 2025, 36^3 in 2030 and 36^4 in 2035. The random explorations selected result scenarios in 2025 and 2030, PGR and CCF within range of [1.03,1.06] and [0.95,1.05]. The whole exploration took 1 year and 4 months with 32 instances in the DAnCE4Water cloud server while the result was saved in a SQLite database containing the input values and output values for every simulation.

2.2. The Accelerated Exploration and ANN Design

The proposed accelerated exploration started with a normal exploration and paused when a certain amount of simulation had been finished. These simulations would be used as the training set to train an ANN while the exploration continued. The exploration then stopped when another certain amount simulation had been finished. These extra simulations would be used for validation. The ANN would be trained with different structures and settings and tested on the validation simulations. The errors of the validation would be used to choose the best structure and setting, and the ANN would do the rest of exploration by predicting with the scheduled PGR, CCF and add-on strategies (as the normal exploration) but skipping the UDM and SWMM process.

The results in the reference exploration (the scenarios as well as the evaluated system performance) were classified into three sets: the training set (size: 0.1%, 1% or 10%), the validation set (size: 10%) and the test set (size: the remaining data).

The training set was used to train the network (e.g., weights) while the validation set was for adjusting the structure of the network (e.g., number of nodes) [4]. The test set was used to assess the performance of a trained and validated network. In most literature [14,19–25], as the network structure are usually pre-defined or tested by trial-and-error, the validation sets are usually disused or replaced by the test sets. Under such substitution, the performance of the network is only meaningful for certain sets (the 'test sets'), which have been optimized during the training, rather than for the untrained data which we expect more precise predictions.

2.2.1. Type of ANN

There are several groups of networks such as Feedforward Networks (e.g., Multi-layer Perceptron [26], the Probabilistic Neural Network [27], the Dynamic Neural Network [28]), Recurrent Networks (e.g., Elman Network [29], Autoregressive Networks [30]), Polynomial Networks (e.g., Ridge Polynomial Networks [31], Function Link Network [32]), Modular Networks, Support Vector Machine and so forth. [33].

Among these extensive types of ANNs and their derivations, The multi-layer perceptron (MLP), a feedforward multilayer network with non-linear node functions, is the most commonly encountered one [33,34]. Practically, MLP shows successful generalization capability, effectiveness and efficiency in forecasting time series [10,11,19,23], as well as great compatibility coping with different optimization methods or existing models [19,35]. Although MLP is usually the better choice or at least the same performance with respect to other proposal networks [33], there remain certain delimitations that have a remarkable impact on the training accuracy and efficiency. Such aspects include the structure of the network, the activation function of nodes, the existence of bias units, the quality and quantity of training and validation datasets, the choice of training algorithm and parameters and so forth. In this paper, the MLP network will be adopted while the design process of these aspects will be investigated and adapted to the case study. The network will be established using PyBrain [36], a modular Machine Learning Library for Python.

2.2.2. The Structure of MLP Network

The MLP usually consists of nodes(units) arranged in three types of layer: the input layer, the hidden layer(s) and the output layer. As Figure 2 shows, each node (unit) has its own output value y and is connected by real-valued weights w to all (and only) the nodes of the subsequent layer. For the ith node in the lth layer n_{il}, let S_{il} be the set of nodes that connect to n_{il}, $f(x)$ be the activation function of n_{il}, the output value is calculated using Formula (1):

$$y_{n_i^l} = f(\sum_{n_j^m \in S_i^l} w_{ji}^{ml} y_{n_j^m}) \qquad (1)$$

where $y_{n_i^l}$ is the output value the ith node in the lth layer; w_{ji}^{ml} is the weight of the connection between this node and the jth node in the mth layer; $y_{n_j^m}$ is the output value of the jth node in the mth layer; $f(x)$ be the activation function of this node.

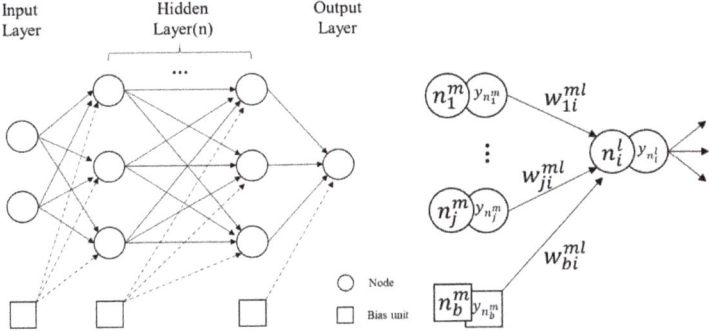

Figure 2. Structure and value propagation of MLP.

The input layer receives the input data while the output of output layer refers to the predicted results. Thus, both only requires only 1 layer to fulfill the task. The number of nodes in these layers are determined according to the number of input variables and target variables [37]. In some cases, the input and output variables are linearly normalized to (0,1) or (−1,1), to avoid computational problems or to meet algorithm requirement [24,38,39]. In this study, such methods were not applied because: (1) with the exploration continues, the input variables will always exceed the range of the existing records while the output variable also has the chance. (2) the weights may undo the scaling.

The number of hidden layers and its nodes has a significant impact on MLP training [37,40]. Simple networks maybe less accurate in learning the problem while complex networks may take excessively long training time. one hidden layer is usually sufficient in most cases [14,19–25,33,41–43] while sometimes multiple hidden layers shows better learning on certain problems [35].

The number of nodes in hidden layer is usually determined through trial-and-error method [19,23,43]. The range of attempts is usually within 1 to 20 [14,19–25], or 3 times the number of input variables [43]. The best number of nodes was the one having the smallest mean-square error (MSE) and root-mean-square error (RMSE) and the highest correlation coefficient (r) for the validation data set. [11]

In this paper, the designed MLP consists 1 input layer, 1 hidden layer and 1 output layer. There will be 5 nodes in the input layer representing climate change factor, population, the number of decision take for BAU, RWHT and PIPE within the 20 years and 1 node in the output layer referring to the flooded junctions. No variables will be normalized. The number of nodes in the hidden layer will be determined within 1 to 20 through trail-and-error method.

2.2.3. The Activation Functions

The role of activation function (AF) in MLP is to non-linearize the linear combination of weights and node values passing through from the previous layer. Practically, there are three types of AFs: (1) the analytic AFs, which are classic functions such as Gaussian, Sigmoid and Tanh; (2) the fuzzy AFs, which has faster convergence in training; and (3) the adaptive AFs, which improves the nonlinear response of the network [40]. Although the fuzzy AFs perform better on specific problems [44], there is little evidence on the advantage of such AFs in practice. On the other hand, the adaptive AFs also suffer from a more complex and error-prone training algorithm [40]. Thus, only classic analytic AFs are considered in this study.

For nodes in the hidden layer, most commonly used AFs are the logistic sigmoid function [34,38,41], the tanh function [35,43,45]. These two functions are similar in shape while different in output ranges (sigmoid: [0,1], tanh: [−1,1]). For the output layer, most researchers adopt linear function [11,35,41,45].

In this paper, the log-sigmoid function has been used for the hidden layer nodes while linear function has been applied in the output layer to test their performance on handling random noise.

2.2.4. Bias Unit

The bias unit is an extra set of nodes added to all layers but the output layer, which helps to get a better and quicker learning of the network. The output value of a bias unit is fixed value while the weights of connection from the bias unit to the subsequent nodes are still adjustable. The addition of bias unit introduces a threshold value that may influence the activation of the subsequent nodes [24,37], or, from another perspective, helps to move the AF in the subsequent nodes along the x-axis for better learning results. Thus, in most cases, bias units always contribute positively to the network.

2.2.5. Learning Algorithm and Parameter Setting

The traditional and most commonly used training method for MLP is the two-step error-backpropagation method [14,19,24]. Firstly, the input vector is fed into the input layer, propagating forward through hidden layer(s) to the output layer. Then, the error is calculated in the gradient descent and propagated backward from the output layer through the hidden layer(s) to the input layer, which modifies the weights for every connection between nodes. The training repeats until the network's overall error are less than a predefined learning rate, or until the number of maximum epochs is reached. Learning rate is a damping factor applied to weights correction during training [40], indicating the amount that the weights are updated. Epoch is a measure of the number of times all of the training vectors are used once to update the weights. Obviously, when dealing with huge datasets, it is super time consuming if all the weights are recomputed for each training vector. Thus, there is also a batch-learning term for the backpropagating method, which feeds multiple training samples in one forward/backward pass. The number of samples in one pass is called batch size while such one forward/backward process is count as one iteration.

As the original backpropagation method is likely to be slow [41], improved strategies such as Second-order On-Line training methods have been developed. Although these second-order training algorithms are likely to converge significantly faster than first-ordered backpropagation [37], they require more complex data preprocessing as well as more storage and computational costs. Luckily, there are also several improved first-order backpropagation methods. The most commonly used is the Backpropagation with Momentum [22,24], which significantly speed up the training process. The momentum is an inertial factor applied to the weights during the back propagate process, which aims to maintain the direction of weight changing [40]. The addition of momentum accelerates convergence where the learning quality is good while precisely reduces the number of oscillations where bad [37].

The settings of training parameters are more likely to be empirical and case-dependent. In most cases, the start/fixed learning rate will be in the range of [0.01,0.3] [21,22,25,34] while the end learning rate within [0.00013,0.001] [19,21]. The number of epochs usually depends on the training data size and the computational capacity, ranging from 200 to 15,000 [19,21,22,24,34,35,42]. Momentum is typically set to 0.9 [22], although the optimal value might be task-specific [21,24,34].

The designed network structure and learning parameters are shown in Table 1. All combinations of structure and learning parameters were tested with the first 0.1% of data and validated with the following 0.05% data. After the best structure was determined, the network was again tested with different size of training set size to find the best application pattern. The validation set size is half of the training set. The best performing structure and application patter were applied to the case study to study the feasibility of ANN in supporting long-term planning.

Table 1. Designed Neural Network Parameters.

Type	Structure			Activation Function	Bias Units	Learning Settings	
	Name	Layer	Node				
MLP	input	1	5	-	True	training size [1]	0.1%, 1%, 10%
						batch size	1
	hidden	1	1–20	sigmoid	True	learning rate	0.01, 0.1, 0.3
						learning rate decay	1.0
	output	1	1	linear	False	momentum	0.1–0.9
						epoch	500, 1000, 5000

[1] Training size is the percentage of total data used as the training set, tested after the ANN structure being determined.

2.3. Trial and Error

The performance of learning results was assessed by the root-mean-square error (RMSE), which is a commonly used index in machine learning [14,20,21,34]. The lower RMSE it is, the better prediction the module makes [19].

RMSE is defined as the absolute value of the estimated error between the predicted result and the observed result, calculated by:

$$RMSE = \sqrt{\frac{\sum_{i=1}^{n}(O_i - P_i)^2}{n}} \qquad (2)$$

where O_i is the observed result; P_i is the predicted result.

As the unit of RMSE is case-dependent, the correlation coefficient (r) [14,20,21,34] was adopted to compare the training performance with other studies.

$$r = \frac{\sum_{i=1}^{n}(P_i - \overline{P})(O_i - \overline{O})}{\sqrt{\sum_{i=1}^{n}(P_i - \overline{P})^2 \sum_{i=1}^{n}(O_i - \overline{O})^2}} \qquad (3)$$

where O_i is the observed result; P_i is the predicted result; $\overline{O}$ is the mean value of the observed result; $\overline{P}$ is the mean value of the predicted result.

Practically, as the decision in long-term infrastructure implementation planning is not scenario-based but strategy-based, the distribution of predict results for each strategy combination should be more convincible than RMSE. Thus, the prediction distribution of outputs was also adopted in this study as the other performance indicator

3. Results and Discussion

3.1. ANN Structure and Training Parameters

As mentioned in the previous section, all combinations of structure (number of hidden nodes) and learning parameters (learning rate, momentum and number of epochs) were tested with the first

0.1% of all data (training size = 0.1) and validated with the following 10% of data. For each parameter, the distributions of RMSE for each candidate value under all possible combinations are shown in Figure 3.

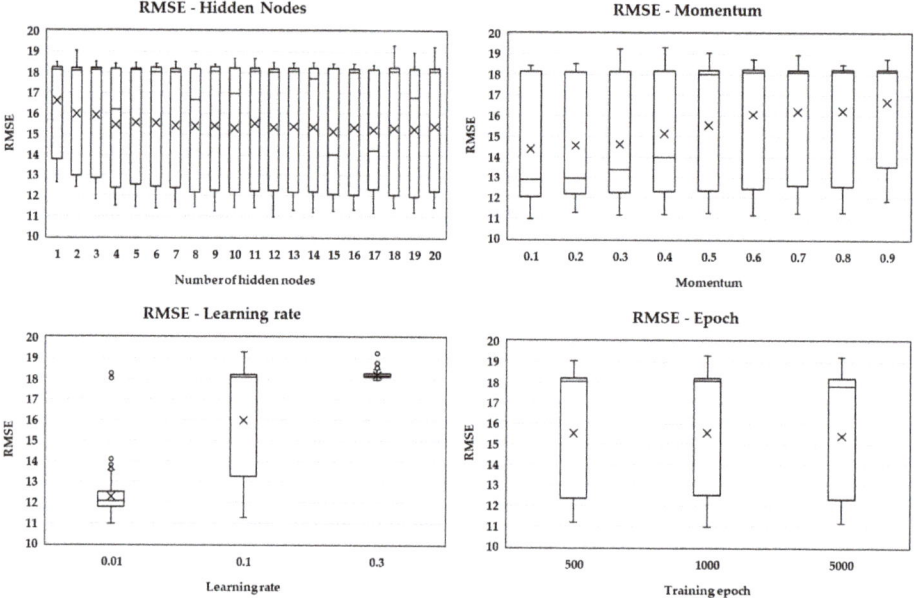

Figure 3. RMSE Distributions under different manipulated variables.

By adopting ANN(MLP) in urban water infrastructure performance prediction, the RMSE of such method ranges from 10.97–19.33 nodes with the observed flooded junctions ranging from 20 to 146. For the number of hidden nodes, setting 1 node caused the highest average RMSE (16.62) which may due to the strongest linearity of the network. With the number of hidden nodes rises to 4 nodes, the average RMSE drops gradually to 15.46 where the non-linearity starts to develop effect. From 4 nodes to 20 nodes, the average RMSE keeps stable within (15.13,15.56). Although there is no significant difference in the average RMSE with the number of hidden nodes changing, the distributions of RMSE still have dramatic and irregular variations. These distributions are characterized by the minimum, maximum, Q1, Q3 and mid-values, which indicates 100%, 75%, 50%, 25%, 0% chance of getting a higher RMSE than the given value, respectively. Thus, the lower these values are, the better performance of the network we will get.

As shown in Table 2, the MLP network with 15 nodes was always in the top 5 well-performed structure and has significant advantages in low mid-value compared to others. The 17 nodes network is slightly better than the 15 nodes one on minimum, Q3 and maximum as well as slightly poor on Q1 and mid-value. Thus, the network of 15 and 17 hidden nodes are selected as the candidate structure for the following studies.

Following the same process, the rest parameters are then determined: momentum = 0.1, learning rate = 0.01, epoch = 5000.

The candidate network was again tested with different size of training set size to find the best application pattern (see Table 3). The result indicates that network with 15 nodes performs better than the 17 nodes one under the select learning parameter, which is within 3 times the number of input variables [38]. Training with the first 10% data will have a significant improvement in reducing the RMSE while maintaining an acceptable time-saving capacity (reduce 80% of the time).

The best performing structure and application pattern (Table 3) were then applied to the case study. The overall RMSE for the whole observed data and the predicted data is 10.5722 and the detailed performance of MLP prediction is shown in Figure 4. The overall RMSE is slightly higher than the validation result (9.7961).

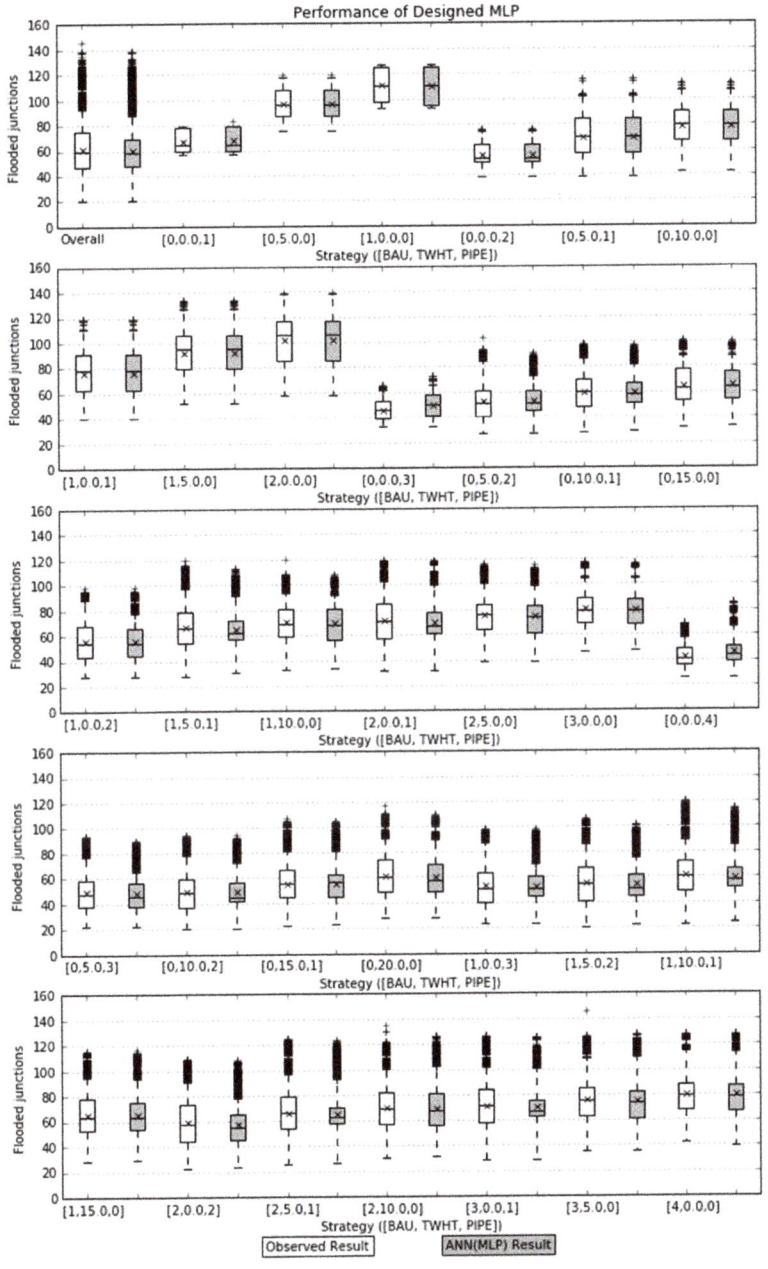

Figure 4. ANN performance for different strategy combinations (supported by Matplotlib [46]).

Table 2. Comparison of performance distribution for different number of hidden nodes.

	1st	RMSE	2nd	RMSE	3rd	RMSE	4th	RMSE	5th	RMSE
Min	12	10.97	14/17	11.17	19	11.18	9	11.25	15	11.26
Q1	19	11.95	18	12.04	16	12.05	15	12.08	13	12.16
Mid	15	14.02	17	14.20	8	16.67	19	16.79	10	16.98
Q3	17	18.15	5	18.17	10/12	18.19	13/14	18.20	15	18.21
Max	17	18.37	8	18.39	9	18.41	6	18.42	15/16	18.43

Table 3. ANN performance under different training set sizes.

	Training Size	Hidden Nodes	Learning Rate	Momentum	Epoch	RMSE
Validation set	0.001	15	0.01	0.1	5000	11.5051
	0.01					11.8653
	0.1					9.7961
	0.001	17				12.2593
	0.01					12.5760
	0.1					11.9862
Test set	0.1	15	0.01	0.1	5000	10.5722

The correlation coefficient (r) of the test set was 0.821, which was preferable compared to rs in the other close applications of ANN (flood discharge: 0.683–0.851 [47], open-channel junction velocity field: 0.035–0.884 [48], drought effects on surface water quality:0.819–0.922 [49], BOD in river: 0.505–0.821 [19]).

Taking account of the tremendous amount of data in this case study, the above result suggested the proposed statistical trial-and-error method for determining network parameters is feasible and reliable on selecting the best structures.

3.2. Performance on Supporting Long-Term Planning

To analyze the performance variations of different implementation strategy combinations for the urban water system in the case study, boxplots are again used while the upper end of the whiskers is set to 95th percentile (Figure 4). In other word, the probability of a certain system performing better than this upper end is 95%. Thus, the accuracy on the 95th percentile and Q3 is practically more important than that of mid-value, Q1 and minimum.

For strategies containing only rainwater tanks ([0,5.0,0], [0,10.0,0], [0,15.0,0] and [0,20.0,0]), the first two combinations are all included in the training set and share the same distribution with the observed results. For the latter two strategies, the 95th percentile errors are −0.24% and −1.26% respectively while the Q3 errors being −2.28% and −5.68%. This suggests the designed MLP network is effective and has relatively good performance in predicting strategies with spatial randomness. The performance of purely decentralized systems may have stronger and more linear relation with the rainfall events and urban permeability (related to buildings/population), which makes the prediction of these purely decentralized strategies better than mix strategies.

For the same reason, the purely business as usual strategies also have good predictions: for [3,0.0,0], Q3 = −0.22% and 95th = 0.18%; for [4,0.0,0], Q3 = −0.77% and 95th= −0.88%. As no additional systems were implemented in these scenarios, the designed network performs well in generalizing the relation between water system performance and rainfall events and urban permeability.

For the overall performance, the MLP result has similar minimum, Q1 and mid-value compared to the observed result (min: 20, 20; Q1: 48.1, 47.0; mid: 58.3, 60.0). Whereas the predicted values have a narrower range (20.0–88.44) than the observed ones (20–93) despite the outliers. Such phenomena indicate that the prediction in the high-value events (poorly performed water system in practice) tend to aggregate to the Q3. This suggests that, from an overview perspective, the adoption of ANN supported planning may raise the chance of overestimating the performance of urban water systems.

To make this proposed method applicable and reliable in practice, the error distributions of the result are investigated to solve the overestimating problem. As shown in Figure 5, all errors of Q3 lie between (−10.56%,8.76%) and 95th percentile between (−18.91%,14.95%). The majority of these errors are negative, indicating universal overestimations of the urban water system.

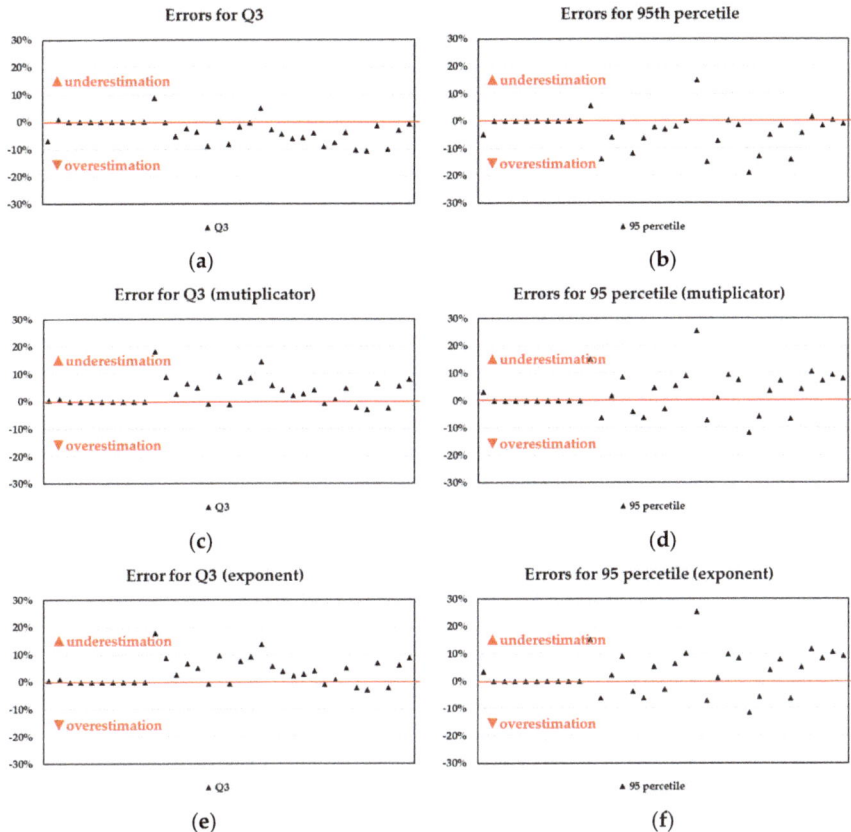

Figure 5. Error distribution of MLP predicted result and corrected result ((**a**,**b**) observed errors for 95th percentile and Q3; (**c**,**d**) corrected errors for 95th percentile and Q3 by multiplication; (**e**,**f**) corrected errors for 95th percentile and Q3 by exponent).

As Table 4 shows, the adoption of safety coefficient could effectively raise the error from negative to positive (from overestimation to under estimation) while slightly enlarge the standard deviation of the errors.

Table 4. Mean ± SD error of adopting the safety coefficient.

	Observed Error	Multiplicator	Exponent
Q3	−2.29% ± 4.28%	3.38% ± 4.73%	3.43% ± 4.72%
95th percentile	−3.13% ± 6.34%	2.63% ± 7.15%	2.96% ± 7.32%

As these errors are related to the network structure and its final status, a safety coefficient, which comes from the validation process, is adopted to adjust the final output of the network. By investigating the observed data and the predicted data in the validation set, a multiplicator or exponent can be calculated out and applied for the test set. As the 95th percentile is the dominant factor of this case study, the safety coefficient also comes from the 95th percentile of the validation (multiplicator:1.0910, exponent:1.0272).

The result of correction is shown in Figure 5. There is no obvious difference between correction with multiplicator and exponent. The corrected errors of Q3 lied in −3.05% to 18.24% (multiplicator) and −2.96% to 17.87% (exponent) while that of the 95th percentile in −11.69% to 25.41% (multiplicator) and −11.60% to 25.36% (exponent).

As shown in Table 5, the accelerated exploration identified all robust drainage infrastructure status in the reference exploration while overestimated three. The corrected accelerated exploration identified most robust drainage infrastructure status in the reference exploration while underestimated one. The underestimated one has no influence on the plan generation as there is no connectable route in the previous decision year. Thus, the correction is essential and effective to raise the robustness of the proposed accelerated exploration.

Table 5. Robust progressive goal for Scotchman's Creek.

	Reference Exploration	Accelerated Exploration	Corrected Accelerated Exploration
2020	[0,0,1][1]	[0,0,1]	[0,0,1]
2025	[0,0,2]	[0,0,2]	[0,0,2]
2030	[0,0,3] [0,5,2] [0,10,1] [0,15,0] [1,0,2] -	[0,0,3] [0,5,2] [0,10,1] [0,15,0] [1,0,2] [1,5,1]	[0,0,3] [0,5,2] [0,10,1] - [1,0,2] -
2035	[0,0,4] [0,5,3] [0,10,2] - -	[0,0,4] [0,5,3] [0,10,2] [1,5,2] [2,0,2]	[0,0,4] [0,5,3] [0,10,2] - -

[1] [BAU,RHWT(%),PIPE].

Notably, for 95th percentile, the majority of errors are controlled within ±10%. The two outliers represent the two pure strategies of upgrading pipes, [0,0.0,3] and [0,0.0,4]. Although there are great errors on these two strategies (underestimation of water system), the origin system performance of them is good enough that the errors have no influence on identifying them as good strategies (not influencing decision). This error also indicates that different from purely decentralized strategies, such purely centralized strategies which have only relations with rainfall events, do not have a preferable prediction at all.

Such a result indicates that when using the MLP to predict a black box problem, such as the urban water system in the case study, there should be at least two related input factors for each variable (the candidate infrastructure, e.g., pipe, rwht) to ensure reliable prediction.

4. Conclusions

In this study, an accelerated exploration planning method was proposed by integrating the neural network method (multi-layer perceptron) with an explorative model (DAnCE4Water), to significantly reduce the simulation time of generating a robust long-term water system adaptation plan. The proposed method was applied to a case study in Scotchman's Creek, Melbourne, Australia. Results showed the proposed method can cut down 80% of the simulation time while offering the same plan.

Instead of modifying the network parameters, the network structure and settings in this paper were determined through a comprehensive statistical trial-and-error analysis (evaluating for all possible parameter combination). With 10% of the training data, the validation error (10% data) was 9.7961, the overall prediction RMSE was 10.5722 (80% data) and the correlation coefficient (r) was 0.821. This suggests that the ANN could have stable and reliable with good designed network and low proportion of training data. It also emphasis the necessity of network design which did take time in the trial-and-error analysis but having promising return in total time saving and accuracy.

The ANN showed diverse capacity on predicting the performance of different type of flood-resisting strategies. The estimation of purely decentralized strategies (scenarios with RWHT only) and purely BAU strategies is far more accurate than that of mixed strategies. Meanwhile, the purely centralized strategies (scenarios with PIPE only) had the worst prediction. Considering the input variables related to the strategy, it is obvious that the performance estimation would be more accurate if more flood-related input variables are related to it (two for RWHT and BAU while one for PIPE). Thus, more flood-related input variables should be considered (for each strategy) in future studies.

The proposed exploration method raised the chance of overestimating the performance of urban drainage systems ($-3.13\% \pm 6.34\%$ flooded junctions than observed). By adopting the safety coefficient, a multiplicator or exponent calculated by observed data and predicted data in the validation process, the overestimation problem was controlled in an acceptable range and have very limited impacts on final decision making ($2.63\% \pm 7.15\%$ flooded junctions than observed). Such correction is effective in practice as the real-world goal for planning is either above or below a certain target. Instead of reducing the error which is a tough task, the correction shifts the error along one direction (to more underestimate side) to ensure the reliability of the given plan. As the error came from the method, the safety coefficient calculated by the validation data could be reasonable to some extent.

Although the proposed accelerated exploration method was proved to be efficiency in time saving (saved 80% of exploration) and effective (offered similar decisions after correction), there are still several aspects requires further studies. (1) The training set used in this study followed a "real-world exploration" time sequence, which means there were much more simulations in the later decision steps than in the earlier steps. Such setting may have influence on the network performance. Further studies have to be conducted on the composition of the training set to ensure efficient and effective training; (2) Further investigation in the cause of the universal overestimation have to be conducted to optimize the algorithm or training pattern. (3) More case studies should be carried out to further validate and improve the proposed accelerated exploration method.

Author Contributions: Conceptualization, J.Z., D.F., C.U. and R.P.S.; Methodology, J.Z. and C.U.; Software, J.Z.; Validation, J.Z.; Formal Analysis, J.Z.; Data Curation, J.Z. and C.U.; Writing-Original Draft Preparation, J.Z.; Writing-Review & Editing, R.P.S.; Visualization, J.Z.; Supervision, D.F.; Project Administration, D.F.; Funding Acquisition, D.F.

Funding: The research was co-funded by National Key R&D Program of China (Grant No. 2018YFC0809900) and the Priority Academic Program Development of the Jiangsu Higher Education Institution.

Acknowledgments: Thanks to Professor Dafang Fu and Christian Urich for the creative leadership and for motivation to finish the research. Thanks to Chenli Wu, who continuously supports us to finish the work.

Conflicts of Interest: The authors declare no conflict of interest.

References

1. Moss, T. 'Cold spots' of Urban Infrastructure: 'Shrinking' Processes in Eastern Germany and the Modern Infrastructural Ideal. *Int. J. Urban Reg. Res.* **2008**, *32*, 436–451. [CrossRef]
2. Wiechmann, T.; Pallagst, K.M. Urban shrinkage in Germany and the USA: A Comparison of Transformation Patterns and Local Strategies. *Int. J. Urban Reg. Res.* **2012**, *36*, 261–280. [CrossRef] [PubMed]
3. Kwadijk, J.C.J.; Haasnoot, M.; Mulder, J.P.M.; Hoogvliet, M.M.C.; Jeuken, A.B.M.; van der Krogt, R.A.A.; van Oostrom, N.G.C.; Schelfhout, H.A.; van Velzen, E.H.; van Waveren, H.; et al. Using adaptation tipping points to prepare for climate change and sea level rise: A case study in the Netherlands. *Wiley Interdiscip. Rev. Clim. Chang.* **2010**, *1*, 729–740. [CrossRef]

4. Lempert, R.J.; Groves, D.G.; Popper, S.W.; Bankes, S.C. A General, Analytic Method for Generating Robust Strategies and Narrative Scenarios. *Manag. Sci.* **2006**, *52*, 514–528. [CrossRef]
5. Ben-Haim, Y. *Info-Gap Decision Theory: Decisions under Severe Uncertainty*; Academic Press: Cambridge, MA, USA, 2006.
6. Walker, W.E.; Rahman, S.A.; Cave, J. Adaptive policies, policy analysis, and policy-making. *Eur. J. Oper. Res.* **2001**, *128*, 282–289. [CrossRef]
7. Haasnoot, M.; Middelkoop, H.; Offermans, A.; Beek, E.V.; Deursen, W.P.A.V. Exploring pathways for sustainable water management in river deltas in a changing environment. *Clim. Chang.* **2012**, *115*, 795–819. [CrossRef]
8. Haasnoot, M.; Kwakkel, J.H.; Walker, W.E.; Ter Maat, J. Dynamic adaptive policy pathways: A method for crafting robust decisions for a deeply uncertain world. *Glob. Environ. Chang.* **2013**, *23*, 485–498. [CrossRef]
9. Mustafa, M.R.; Rezaur, R.B.; Saiedi, S.; Isa, M.H. River suspended sediment prediction using various multilayer perceptron neural network training algorithms—A case study in Malaysia. *Water Resour. Manag.* **2012**, *26*, 1879–1897. [CrossRef]
10. Singh, S.; Reddy, C.S.; Pasha, S.V.; Dutta, K.; Saranya, K.R.L.; Satish, K.V. Modeling the spatial dynamics of deforestation and fragmentation using Multi-Layer Perceptron neural network and landscape fragmentation tool. *Ecol. Eng.* **2017**, *99*, 543–551. [CrossRef]
11. Ruben, G.B.; Zhang, K.; Bao, H.; Ma, X. Application and Sensitivity Analysis of Artificial Neural Network for Prediction of Chemical Oxygen Demand. *Water Resour. Manag.* **2017**, *32*, 273–283. [CrossRef]
12. Ripley, B.D. *Pattern recognition and neural networks*; Cambridge University Press: Cambridge, UK, 2009; pp. 233–234.
13. Kumar, R.; Singh, B.; Shahani, D.T. Recognition of single-stage and multiple power quality events using Hilbert–Huang transform and probabilistic neural network. *Electr. Power Compon. Syst.* **2015**, *43*, 607–619. [CrossRef]
14. Sun, W.Z.; Jiang, M.Y.; Ren, L.; Dang, J.; You, T.; Yin, F.F. Respiratory signal prediction based on adaptive boosting and multi-layer perceptron neural network. *Phys. Med. Biol.* **2017**, *62*, 6822–6835. [CrossRef] [PubMed]
15. Ivey, R.; Bullock, D.; Grossberg, S. A neuromorphic model of spatial lookahead planning. *Neural Netw.* **2011**, *24*, 257–266. [CrossRef] [PubMed]
16. Erdem, U.M.; Hasselmo, M. A goal-directed spatial navigation model using forward trajectory planning based on grid cells. *Eur. J. Neurosci.* **2012**, *35*, 916–931. [CrossRef] [PubMed]
17. Urich, C.; Rauch, W. Exploring critical pathways for urban water management to identify robust strategies under deep uncertainties. *Water Res.* **2014**, *66*, 374–389. [CrossRef]
18. Urich, C.; Sitzenfrei, R.; Kleidorfer, M.; Bach, P.M.; McCarthy, D.T.; Deletic, A.; Rauch, W. Evolution of urban drainage networks in DAnCE4Water. In Proceedings of the 9th International Conference on Urban Drainage Modelling, Belgrade, Serbia, 4–6 September 2012.
19. Raheli, B.; Aalami, M.T.; El-Shafie, A.; Ghorbani, M.A.; Deo, R.C. Uncertainty assessment of the multilayer perceptron (MLP) neural network model with implementation of the novel hybrid MLP-FFA method for prediction of biochemical oxygen demand and dissolved oxygen: A case study of Langat River. *Environ. Earth Sci.* **2017**, *76*, 503. [CrossRef]
20. Fan, X.; Wang, L.; Li, S. Predicting chaotic coal prices using a multi-layer perceptron network model. *Resour. Policy* **2016**, *50*, 86–92. [CrossRef]
21. Mirici, M.E. Land Use/Cover Change Modelling in a Mediterranean Rural Landscape Using Multi-Layer Perceptron and Markov Chain (Mlp-Mc). *Appl. Ecol. Environ. Res.* **2018**, *16*, 467–486. [CrossRef]
22. Saeidi, S.; Mohammadzadeh, M.; Salmanmahiny, A.; Mirkarimi, S.H. Performance evaluation of multiple methods for landscape aesthetic suitability mapping: A comparative study between Multi-Criteria Evaluation, Logistic Regression and Multi-Layer Perceptron neural network. *Land Use Policy* **2017**, *67*, 1–12. [CrossRef]
23. Feng, X.; Li, Q.; Zhu, Y.; Hou, J.; Jin, L.; Wang, J. Artificial neural networks forecasting of PM 2.5 pollution using air mass trajectory based geographic model and wavelet transformation. *Atmos. Environ.* **2015**, *107*, 118–128. [CrossRef]

24. Lopez, M.E.; Rene, E.R.; Boger, Z.; Veiga, M.C.; Kennes, C. Modelling the removal of volatile pollutants under transient conditions in a two-stage bioreactor using artificial neural networks. *J. Hazards Mater.* **2017**, *324*, 100–109. [CrossRef] [PubMed]
25. Abderrahim, H.; Chellali, M.R.; Hamou, A. Forecasting PM10 in Algiers: Efficacy of multilayer perceptron networks. *Environ. Sci. Pollut. Res. Int.* **2016**, *23*, 1634–1641. [CrossRef] [PubMed]
26. Rumelhart, D.E.; Hinton, G.E.; Williams, R.J. Learning internal representation by back-propagation of errors. *Nature* **1986**, *323*, 533–536. [CrossRef]
27. Enke, D.; Thawornwong, S. The use of data mining and neural networks for forecasting stock market returns. *Expert Syst. Appl.* **2005**, *29*, 927–940. [CrossRef]
28. Guresen, E.; Kayakutlu, G.; Daim, T.U. Using artificial neural network models in stock market index prediction. *Expert Syst. Appl.* **2011**, *38*, 10389–10397. [CrossRef]
29. Lee, T.S.; Chen, I.F. Forecasting exchange rates using feedforward and recurrent neural networks. *J. Appl. Econ.* **1995**, *10*, 347–364.
30. Kodogiannis, V.; Lolis, A. Forecasting Financial Time Series using Neural Network and Fuzzy System-based Techniques. *Neural Comput. Appl.* **2002**, *11*, 90–102. [CrossRef]
31. Ghazali, R.; Hussain, A.J.; Al-Jumeily, D.; Merabti, M. *Dynamic Ridge Polynomial Neural Networks in Exchange Rates Time Series Forecasting*; Springer: Berlin, Germany, 2007.
32. Hussain, A.J.; Knowles, A.; Lisboa, P.J.G.; El-Deredy, W. Financial time series prediction using polynomial pipelined neural networks. *Expert Syst. Appl.* **2008**, *35*, 1186–1199. [CrossRef]
33. Ramos, E.G.; Martínez, F.V. A Review of Artificial Neural Networks: How Well Do They Perform in Forecasting Time Series? *Analítika Revista Análisis Estadístico* **2013**, *6*, 7–15.
34. Pham, T.D.; Yoshino, K.; Bui, D.T. Biomass estimation of Sonneratia caseolaris (l.) Engler at a coastal area of Hai Phong city (Vietnam) using ALOS-2 PALSAR imagery and GIS-based multi-layer perceptron neural networks. *GISci. Remote Sens.* **2016**, *54*, 329–353. [CrossRef]
35. Zadkarami, M.; Shahbazian, M.; Salahshoor, K. Pipeline leakage detection and isolation: An integrated approach of statistical and wavelet feature extraction with multi-layer perceptron neural network (MLPNN). *J. Loss Prev. Process Ind.* **2016**, *43*, 479–487. [CrossRef]
36. Schaul, T.; Bayer, J.; Wierstra, D.; Sun, Y.; Felder, M.; Sehnke, F. PyBrain. *J. Mach. Learn. Res.* **2010**, *11*, 743–746.
37. Ba, A.J.S. *Second-Order Methods for Neural Networks*; Springer: Berlin, Germany, 1997; pp. 201–203.
38. Piotrowski, A.P.; Napiorkowski, M.J.; Napiorkowski, J.J.; Osuch, M. Comparing various artificial neural network types for water temperature prediction in rivers. *J. Hydrol.* **2015**, *529*, 302–315. [CrossRef]
39. Zhang, G.; Eddy Patuwo, B.; Hu, M.Y. Forecasting with artificial neural networks: The state of the art. *Int. J. Forecast.* **1998**, *14*, 35–62. [CrossRef]
40. Laudani, A.; Lozito, G.M.; Riganti Fulginei, F.; Salvini, A. On Training Efficiency and Computational Costs of a Feed Forward Neural Network: A Review. *Comput. Intell. Neurosci.* **2015**, *2015*, 818243. [CrossRef] [PubMed]
41. Bayram, S.; Ocal, M.E.; Laptali Oral, E.; Atis, C.D. Comparison of Multi Layer Perceptron (Mlp) and Radial Basis Function (Rbf) for Construction Cost Estimation: The Case of Turkey. *J. Civil Eng. Manag.* **2015**, *22*, 480–490. [CrossRef]
42. Pham, B.T.; Tien Bui, D.; Prakash, I.; Dholakia, M.B. Hybrid integration of Multilayer Perceptron Neural Networks and machine learning ensembles for landslide susceptibility assessment at Himalayan area (India) using GIS. *Catena* **2017**, *149*, 52–63. [CrossRef]
43. Talebi, N.; Nasrabadi, A.M.; Mohammad-Rezazadeh, I. Estimation of effective connectivity using multi-layer perceptron artificial neural network. *Cogn. Neurodyn.* **2018**, *12*, 21–42. [CrossRef]
44. Tang, J.; Deng, C.; Huang, G.B. Extreme Learning Machine for Multilayer Perceptron. *IEEE Trans. Neural Netw. Learn. Syst.* **2016**, *27*, 809–821. [CrossRef]
45. Humphrey, G.B.; Maier, H.R.; Wu, W.; Mount, N.J.; Dandy, G.C.; Abrahart, R.J.; Dawson, C.W. Improved validation framework and R-package for artificial neural network models. *Environ. Model. Softw.* **2017**, *92*, 82–106. [CrossRef]
46. Hunter, J.D. Matplotlib: A 2D Graphics Environment. *Comput. Sci. Eng.* **2007**, *9*, 90–95. [CrossRef]
47. Seckin, N. Modeling flood discharge at ungauged sites across Turkey using neuro-fuzzy and neural networks. *J. Hydroinform.* **2011**, *13*, 842–849. [CrossRef]

48. Sharifipour, M.; Bonakdari, H.; Zaji, A.H. Comparison of genetic programming and radial basis function neural network for open-channel junction velocity field prediction. *Neural Comput. Appl.* **2018**, *30*, 855–864. [CrossRef]
49. Safavi, H.R.; Malek Ahmadi, K. Prediction and assessment of drought effects on surface water quality using artificial neural networks: Case study of Zayandehrud River, Iran. *J. Environ. Health Sci. Eng.* **2015**, *13*, 68. [CrossRef] [PubMed]

© 2018 by the authors. Licensee MDPI, Basel, Switzerland. This article is an open access article distributed under the terms and conditions of the Creative Commons Attribution (CC BY) license (http://creativecommons.org/licenses/by/4.0/).

Article

Scalable Green Infrastructure—The Case of Domestic Private Gardens in Vuores, Finland

Outi Tahvonen [1,2]

1. School of Arts, Design and Architecture, Aalto University, Otakaari 1, FI-00076 Aalto, Finland; outi.tahvonen@aalto.fi; Tel.: +358-50-5289850
2. Bioeconomy Research Unit, Häme University of Applied Sciences (HAMK), Lepaantie 129, FI-14610 Lepaa, Finland

Received: 3 November 2018; Accepted: 30 November 2018; Published: 3 December 2018

Abstract: The planning, implementation, and everyday use of the built environment interweave the green and grey components of urban fabric tightly together. Runoff from grey and impermeable surfaces causes stormwater that is managed in permeable surfaces that simultaneously act as habitats for vegetation. Green infrastructure (GI) is one of the concepts that is used to perceive, manage, and guide the components of urban green spaces. Furthermore, GI pays special attention to stormwater management and urban vegetation at several scales at the same time. This study concentrated on scalable GI in domestic private gardens. A set of garden designs in Vuores, Finland were analyzed and developed by Research by Design. The aim was to study how garden scale choices and designs can enhance GI at the block and neighbourhood scales to rethink design practices to better integrate water and vegetation throughout the scales. As a result, we propose a checklist for designers and urban planners that ensures vegetation-integrated stormwater management to enhance habitat diversity in block scale and possibility to use blocks of private plots for ecological networks. The prerequisite for garden designers is to be capable to balance between water, vegetation, and soil, and their processes and flows in detail the scale.

Keywords: garden design; scalable green infrastructure; systems thinking

1. Introduction

Ecosystem services support the well-being and health of urban residents. These benefits build up in a network of different kinds of urban green spaces that, together, can be considered an urban green infrastructure (GI). In other words, the urban fabric and its GI elements provide essential and nature-based benefits for residents as ecosystem services [1]. This approach includes a default definition of GI that comprises all shades of green in the urban context, including both public and private, and planned and unplanned urban vegetation, regardless of the land ownership or planned function. Therefore, GI and its shades of green penetrate all the land use categories.

However, the definition of GI is complex as the concept is applied to different purposes and scales. At its largest scale, the EU [2,3] perceives GI on a pan-European scale as a network joining the Natura 2000 areas that provide connections for fauna and appropriate patches for them to live in. At a smaller scale, detailed GI elements might concentrate on the techniques of green walls and roofs or best management practices in stormwater management [4]. Furthermore, discipline-specific definitions and uses make GI a multifaceted concept [5,6]. In the context of urban drainage management, GI is considered as networks of decentralized stormwater management practices, while landscape architects and urban ecologists use GI for describing networks of green spaces and landscape ecology [7]. According to Fletcher and others [5]: "A central tenet of green infrastructure is, of course, the use of vegetated systems to deliver desired ecosystem services". These approaches stress the connection of water and vegetation within GI.

While the definitions of the concept of GI depend on the used scale [8] and discipline [5], certain common attributes define its nature. GI is multifunctional, scalable, connective, and resilient [8,9]. Multifunctionality reflects the ecological, technical, and sociocultural functions that exist simultaneously in one space, such as buffering of climatic extremes, biomass productions, provision of habitats and biodiversity, species movement routes or opportunities for social interaction and nature experience. This division of multifunctionality to three main components, ecological, economic and sociocultural functions, relate the whole concept to sustainable development and its triple bottom line [10,11].

Urban planning deals with these attributes in all land use categories, including commercial, industrial, residential, and traffic areas, as opposed to just parks and conservation areas. While the share of the green component of the total surface of high-density areas is limited on its own, it can be integrated into buildings and constructions as well as green roofs and walls [12]. In addition, different land uses generate different concentrations of pollution in runoff, so considering multiple land uses simultaneously might complicate the design process [13]. From the perspective of GI, low density housing (LDH) is one of the most diverse land use categories. The GI of LDH comprises small areas managed by owners, and the needs and habits of gardens vary as time passes. These separate, small areas form a coherent gardenscape [14].

LDH and the garden matrix formed in the area cover a significant share of an urban area. According to Loram and others [15], the gardens of low density housing cover 22% of the surface area of examined towns and cities in the UK, while according to Mathieu and others [16], these constitute 36% of a town in New Zealand. The share of the gardens in LDH areas of total urban green spaces has been found to amount to 35–47% [15] or even over 50% [16]. It is assumed that the share of the garden area of LDH will continue to increase because of ongoing urbanization [17].

The characteristics of domestic gardens are determined based on plot sizes and the layout of buildings and parking spaces within the plot, as impervious surfaces prevent vegetation from growing. The ratio between impervious and pervious surfaces on a plot depends on the density, period of construction, and building types in the area [18]. The layout of this grey and impermeable proportion of a plot defines both the accumulation of stormwater and areas that may infiltrate and allow ground soil-based growth of vegetation. Furthermore, water and vegetation are interwoven through soil or growing media. The characteristics of soil determine both the hydraulic conductivity of water, the water storage, and the capillary action to bring water up the roots of vegetation, but also nutrient and water provision for the needs of vegetation [8]. Few studies have described the nature and extent of impermeable and permeable surfaces at a garden scale. Lawn is the most commonly used surface, covering 55–60% of the surface area [16,19]. The prevalence of pavement and asphalt has also been investigated, and a 13% increase was noted in their proportions in Leeds, UK over the previous 30 years [20].

Therefore, areas with LDH constitute a diverse gardenscape that serves as part of the urban ecological network and provides the same ecosystem services as other urban green spaces. It can therefore improve the air quality and microclimate as well as human health and wellbeing, contribute to stormwater management, and play a part in flood control [21].

This study examines how garden design can be used to improve the role of the gardens of low density housing as part of the GI and the effects of this on the block and neighbourhood scales. The main driver in this study is to explore the opportunities for developing GI from a perspective of garden design. The research data is based on the standard practices of the design process of the Research by Design method as well as choices made in an area with LDH in Finland. The research questions are as follows: How can garden designs that combine vegetation and stormwater management enhance GI at the garden scale? How is this improved design practice on the scale of plots reflected at the scales of the entire block and neighbourhood?

2. Theoretical Background: Planning and Design of Scalable Stormwater and Vegetation Systems

In the context of GI planning and design, scalability can be perceived at both the scales used in the design and the links between these as well as at a temporal scale. In the present paper, scalability primarily refers to spatial links between different scales.

2.1. Garden Scale

Plot-specific garden design brings together the needs of garden users and the conditions provided by a plot. In this context, the conditions consist of the layout formed by the placement of buildings in relation to the streets and the arrangements for entrances and car parking on the plot. This layout determines the need for passageways and, as a result, often also includes the extent and placement of impervious surfaces on the plot. In turn, the actual vegetation on the plot will be located in the areas that are free from impervious surfaces, although some vegetation may also be planted between the hard surfaces for purposes such as screening the yard from outsiders or improving the comfort of entryways.

From a garden design perspective, vegetation plays a number of different roles. While vegetation is one of the key elements for spatial design, it differs from other design elements, such as terrain shapes or structures, as it is living and changes constantly. In addition to creating spatial features, plants can serve as space dividers, frames to a view, or ornaments; produce biodiversity and a habitat for fauna as planting systems; and improve the microclimate; or provide screening to residential spaces. In addition to these goals, the selection of plants is determined by availability, factors related to growth potential at the design site, and hardiness [22,23].

Vegetation and water are the most fundamental and central elements of GI [8]. In the context of scalable GI, the smallest unit of vegetation is an individual plant, whose viability is based on the availability of water and nutrients at the growth site. If a growth site does not provide the conditions necessary for a plant to grow, these must be improved by means such as irrigation or fertilizing, or the plant's growth will be stunted or the plant may die [24]. However, the water centric approach to this small scale GI element concentrates on plants capability to minimize urban runoff. Ossola and others [25] studied how an increase in habitat complexity minimizes the urban runoff. They found three main factors: an increase in canopy density and volume, preservation of surface litter, and maintenance of the soil macropore structure. These factors apply to the plant scale.

When examining GI, particularly as a tool combining stormwater management and vegetation, two main approaches can be observed: vegetation integrated best management practices and tools stressing the extent of different surfaces. The Green Factor (GF) or similar tools give scores at the design stage to different surfaces and their proportions of designed area in order to improve the capacity of plots to generate urban green spaces. For example, the volume of growing media under a surface material can be a GF scoring criterion. While this is not a stormwater structure as such, it describes the water infiltration and retention potential under the surface materials [26,27]. However, stormwater management is more commonly based on sustainable urban drainage systems (SuDSs) that emulate the processes of the natural water cycle [28,29]. SuDSs provide a more or less standard toolbox of constructions with relatively well-known functions in order to manage the quantity or quality of stormwater. However, there are several approaches to categorize SuDS, and for instance Charlesworth and others [30] categorized SuDS into five device groupings (adapted in Figure 1). SuDS-based design has recently highlighted an aim of combining stormwater management with amenities and puts more emphasis on biodiversity [31]. This combines SuDS with urban vegetation. However, it is notable that not all SuDSs contain vegetation or rely on the processes of plant growth in stormwater management (Figure 1). This observation was supported by Wootton-Beard and others [32] as they claimed that urban design and planning require biology as well as engineering.

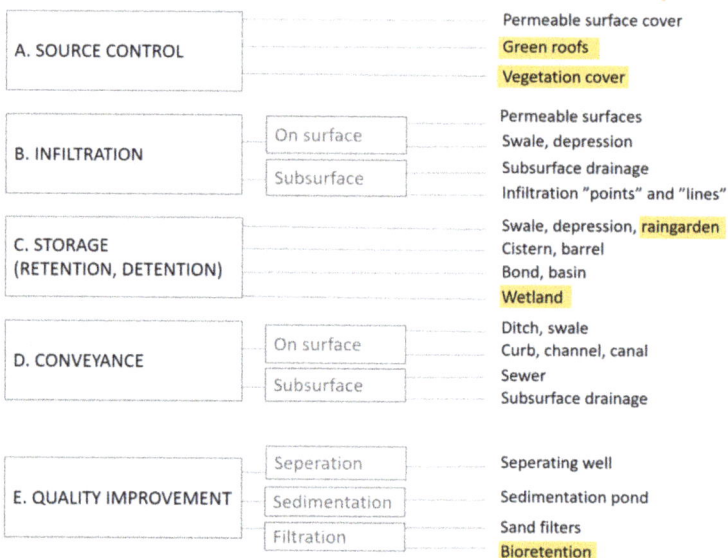

Figure 1. Sustainable urban drainage systems (SuDSs) devise grouping (in left) describes the general functions of stormwater management practices. These functions emulate the processes of the natural water cycle. Technical details of individual SuDS (in right) and their primary function define how they belong to different SuDS devise grouping. SuDS that contain pivotal and functional roles of vegetation are marked in yellow. (SuDS devise grouping adapted from [30], SuDS examples adapted from local practices described in [33]).

2.2. Scaling Up

In the water system, in contrast with separate SuDSs, stormwater management may also be designed as treatment trains. In these trains, a single SuDS is not assumed to solve the challenges concerning quantity, quality, or amenity, but instead, is perceived as an individual part of a larger solution [31]. Designing the trains also allows a better perspective of the different management practices in the whole design area to be obtained. As a result, the stormwater management of the upper parts of a watershed can be implemented with methods that reduce the volume of generated stormwater, while the approaches used at the lower parts of the system can be expected to level flood peaks and flows. However, the design of this treatment train must be viewed separately from flood passage design, as the treatment chain aims to solve the challenge of stormwater management in several consecutive sections. Therefore, an individual SuDS is not required to provide the most efficient solution possible, but rather, the tasks of stormwater management can be divided between the different parts of the treatment train.

Plot-scale treatment trains consist a set of SuDS placed in sequential order along the gradient. If it is not allowed to provide runoff or drained water from plots, then the treatment train consists only the SuDS inside the plot. However, the approach of treatment train applies also to up scaled water systems in blocks and neighbourhoods. At these scales the main focus is on different purposes and functions, or SuDS groupings according to Charleswoth [30], for the parts of the entire water management system.

When scaling up to watersheds or sub-watersheds, studies have been shown that the percentage of impervious surface area predicts the condition of the receiving water body [34]. With a higher proportion of impervious surfaces in the watershed, more problems are caused in receiving waterbodies by contaminants, erosion, and changes to temperature and flow rate [35]. Indeed, in urban planning, the Total Impervious Area (TIA) has been used as one of the indicators for the ecological impacts of planned construction and for estimation of pollutant loads from different land use categories [36]. Nonetheless, there are some weaknesses associated with the use of TIA in studies, which Brabec and others [34] have identified to include variation and a lack of clarity over which part of an impervious surface is directly connected to drainage system. As a result, the concept of the Effective Impervious Area (EIA) has been introduced alongside TIA. EIA only includes the impervious surfaces that are directly hydraulically connected to the drainage system. The concept does not include those impervious areas whose surface runoff is directed to areas covered with vegetation. However, the EIA has not been established as a standard indicator for planning and related steering, and the studies using the concept have mainly used it to describe existing neighbourhoods, focusing on plot-specific observations and aerial photographs [34,37].

In a plant system, the next scale up from an individual plant is a group of plants or a plant community. This may be a monocultural mass planting in a built environment or a habitat comprising various species in several overlapping layers. Recently multi-layer vegetation has been noted to be a key factor in supporting biodiversity [38,39].

The planning of urban ecological networks involves the identification of urban green spaces as patches, corridors, and matrices. Traditionally, the backbone for these networks has consisted of public green areas, such as parks, green spaces around streets, protective green zones, and conserved areas. In recent discussions, however, attention has been focused on the matrix between these patches and corridors, the exact part of GI that this study concerns [14]. When considering the urban green as a whole on a city scale, it is important to note that it plays a variety of roles in addition to the ecological one. These roles include curbing the urban heat island phenomenon, providing an environment for commuting and recreation, and fostering the equal availability of so-called green services to different residential areas [40].

3. Materials and Methods

This study explored the garden scale choices by first identifying a set of state-of-the-art garden designs and then developing and re-designing these garden designs to better serve GI by scaling them up to the block and neighbourhood scales. This development at the garden scale was carried out as an iterative design process during re-designing and upscaling.

The method followed the Research by Design (RbD) method, which explores practical design processes through several iterative and scientific reflective cycles [41], and systematically combines research inquiry and design thinking [42]. RbD, as one of the qualitative methods, aims not to gather numerical data, but focuses on the human element on how vegetation and stormwater management could be integrated during the design process in scales of gardens. According to Glanville [43], RbD combines both the research object and the means of carrying out the study. Here, the object was a set of garden designs that simultaneously serve as the means of carrying out the role of garden design in the context of GI in LDH.

This study applied the idea of grounded theory (GT) for analyzing the data produced in the design process of RbD. GT provides a general and non-discipline specific methodology that was used to analyze the iterative part of this study to reveal the conceptual context and linkages of vegetation-integrated stormwater management. Furthermore, GT allows a wide range of data collection methods.

On a city scale, urban green spaces, biodiversity, and green infrastructure are often studied by remote sensing or from satellite images that show the existing situation. In this study, garden designs were used to present a view of how things ought to be "instead of how things [actually] are" in accordance with Simon's [44] description of the difference between natural science and design.

3.1. The Context

The data of this study comprised 24 garden designs from the Vuores neighbourhood in Tampere, Finland, which served as the location for a national housing fair in 2013 (Figure 2). The gardens were designed and constructed simultaneously in the same area, and they followed the same design guidelines. The gardens can be considered to reflect the views of professional designers on the practical application of the main theme of the fair, sustainable stormwater management. The gardens in the fair area also play a significant role in creating an idea of a functional and ecological garden that meets today's standards among detached house constructors, as Finland's national housing fair is annually visited by nearly 100,000 people. According to surveys, visitors have reported getting ideas for their garden as one of the main reasons for visiting the fair [45].

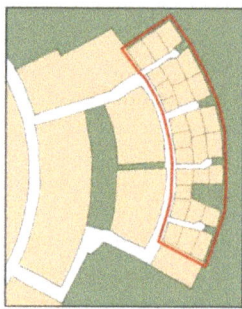

Figure 2. Vuores is a new development south of Tampere, Finland. This study concentrated on private domestic gardens and their garden designs in this area (marked in red). These plots are located between a large park/urban forest and multi-storey buildings.

3.2. The Process

This study examined designs and designing. The practical design work involved finding a balance between a number of factors (presented in Section 2.1), of which stormwater management or creating potential for biodiversity are only two examples.

First, the analysis of a set of existing garden designs concentrated on how the elements of stormwater management and vegetation existed and situated, and how they were integrated into the designs. Furthermore, the intended functions of these elements were mapped as it was the backbone of conventional design process. Then, in the second phase the garden designs were re-designed to improve water and vegetation integration, meanwhile the original layout and functions in plot scale were respected. These improved designs were further developed by considering their input first to block and then to neighbourhood scales. This scaling up and down provides an iterative design process that was repeated once for each plot. It was originally developed as garden scale designs, however, the outputs of these upper scales are also reported in this study (Figure 3).

In this study, RbD was used to provide several re-designing loops to ensure and develop designers approach to integrate vegetation with water. These loops were analyzed by coding and categorizing designs, that follows the applied methods in grounded theory (GT). Open coding was used to identify, name, and describe the development of designs. In coding we mapped all the main changes in the set of improved designs, meaning that the information in drawings was switched to written form. There were 2–8 coded changes or observations per design. These codes were then organized under categories describing more general themes, and they are presented in the section Vuores but also in the theory section. Our findings present inductively produced knowledge of designers' possibilities to integrate vegetation and water in plot scale. The theory concerning this finding is presented in the Section 2, but the core category, soil-vegetation-water system, is presented in Section 5.

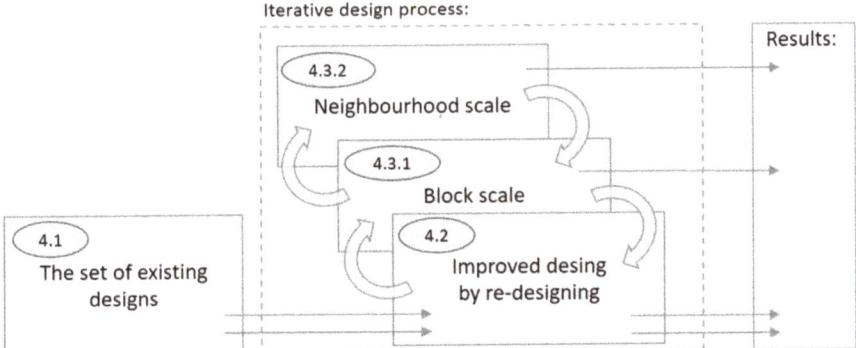

Figure 3. The used method, Research by Design (RbD), focused on the iterative process of re-designing the set of garden designs. This developed the garden scale designs by scaling up to the block and neighbourhood levels. The results of this study were based on the outcome of the garden scale development process, but findings are also presented on the block and neighbourhood scales. Numbers in the figure refer to section numbering in this paper.

The data included all available 24 garden designs in Vuores marked in red in Figure 2. This data seemed to be wide enough as the same categories started to appear in analysis and therefore the saturation of this data was achieved.

Although the research material was based on extremely practice-oriented work and its results, we consider this study to be an important addition to scientific research where the primary focus related to LDH has previously been on examining existing areas or investigating a single functional aspect. As noted by Harrison-Atlas and others [46], carefully defined studies that bridge the gap between science and practice are needed in the context of sustainability.

4. Results

Whether consisting of carefully prepared design documents or a series of separate choices made by an owner, the solutions related to the vegetation and stormwater management on a garden scale are defined in a garden design. In this section, we first analyze garden designs prepared by professional designers, and subsequently improve the integration of water and vegetation by re-designing these on plot, block, and neighbourhood scales.

4.1. Analysis of a Set of Existing Garden Designs

In garden design, decisions are made on the form and style of the overall layout, the location and sizing of different features, and the use of space dividers to separate different parts of the garden. The space may be divided into spaces using structures, planting areas, terrestrial elevation, or a variation in surface materials. While all of these elements were seen in the gardens in the Vuores housing fair site, the proportion of sealed surface was higher than in typical gardens. Paved pathways were used to support visitor movement during rainy days at the fair. In general, the design area was made for the everyday use of families. The Finnish housing fair concept did not adopt the show garden style with diverse and ornate plantings that is common in countries such as the United Kingdom.

Our analysis of the garden designs revealed, in this case, the difficulty of combining stormwater management with vegetation. In Vuores, plot sizes ranged between 454 and 935 m^2, and the floor area ratio was 0.35. These numbers depict the relatively high density of LDH in the Finnish developments. While opportunities for stormwater management have been provided in master planning, the garden scale solutions have primarily handled vegetation and stormwater management as distinct systems. For instance, gutters and water retention may even isolate vegetation from the SuDS. Moreover, narrow planting strips located in the middle of delineated paving may end up relying fully on irrigation water. At the time of the fair, stormwater management had only recently been introduced to the public discussion in Finland, and the main focus in the fair area was on presenting individual and, at times, rather isolated solutions and products. Stormwater management methods integrated in vegetation mostly consisted of rain gardens and the infiltration of small amounts of water at the edges of lawns [47].

In this set of designs, vegetation served five different main purposes. First, plants were used for property boundaries as both cut hedgerows and freely growing plant masses. Vegetation was also used as an element for separating the spaces and functions within the plot, in which case the elements usually consisted of shrubs or perennials. Some of the vegetation also appeared to serve an ornamental purpose. In some of the gardens, plants also contributed to food production in green houses and vegetable gardens, a task that relies on annual plants and their intensive growth during a single growing season. Lawns were the fifth use of vegetation; they were used to determine the shape of spaces, even if not otherwise demarcating the area. None of the garden designs retained the original vegetation of the plot. Figure 4a presents a schematic drawing of the types of vegetation and their locations and describes the overall arrangements of the gardens in the fair area.

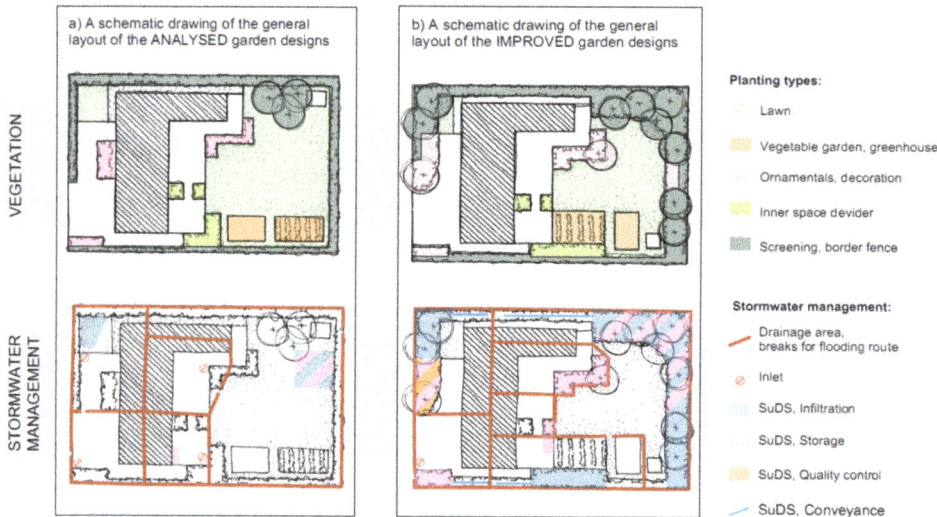

Figure 4. (**a**) A schematic drawing describing the arrangement of plotS, the volumes and locations of different planting types, and stormwater management; (**b**) The same schematic design after improving the integration of vegetation into stormwater management shows the change in vegetation's roles.

4.2. Improved Garden Designs

The following step included examining the opportunities for better integration of stormwater and vegetation when redesigning the gardens. The starting point was the general principles of the original design, and the aim was to retain the functions, styles, form, and space dividers used in the design (Figure 4b).

The first step in the design process was to refine the size of the planting areas according to their functional type. This led to enlarged planting areas which played a key role in property boundaries. Similarly, the inner space dividers located in the middle of the hardscapes were enlarged to better provide the required soil volume to improve both the infiltration capacity and the storage of water for the use of vegetation. Ornamental plantings also partly served as space dividers on the plots, especially when combined to raise beds or other constructions. For these, the utilisation of runoff must be more carefully considered as a raised planter may be totally separated from the soil by structures or capillary gaps. In practice, this first step means ensuring there is sufficient volume of growing media for water retention and plant growth.

The second step appeared to concentrate on the re-evaluation of the placement of the different planting types in relation to the runoff sources. Planting types with high water demand were located close to the downspouts and outer edges of impermeable surfaces to better benefit the available runoff. The designs revealed that ornamental plantings, in particular, if not growing in raised beds, and inner space dividers could benefit stormwater integrated growing conditions. For residents, these planting types are, in any case, part of the essential vegetation for gardening as a hobby. Of all planting types, greenhouses and vegetable gardens require the most water. Paradoxically, these types were usually placed at the most remote part of the plot, at the back of the yard, in the original designs. However, these plantings require a consistent supply of water to yield crops, and therefore, water storage in containers or barrels is needed.

The third step of the re-design process appeared to consist of defining a stormwater treatment train. The re-design process aimed to integrate the planting types and their water demands into the treatment trains. The single SuDSs in the original designs were transformed into multi-phase treatment trains. The aforementioned utilisation of the ornamental plantings or inner space dividers emerged as a central development. However, a challenge arose in this context due to the local recommendations which state that infiltration should occur at a distance of at least 3 m, and preferably 6 m from a building. Moreover, in Finland, ground frost sheets are used next to buildings at a 1.5-m distance from the wall base for ground frost insulation purposes, which sets limits for planting vegetation on the sides of buildings.

The re-design process revealed that the treatment train seems to form a linear set of separate SuDSs. This happens when designing starts solely with stormwater management. However, when designing is integrated with vegetation, it also expressly concerns extensive surfaces, such as large planting areas or entire lawns. In fact, the supply of water to these areas can be managed as extensive surface runoff that evenly crosses pavement borders. In an LDH plot, paved surface areas are primarily so small that no problematic erosion forms at the lawn borders. The situation may be different, however, if the water is initially directed to a certain point using kerbstones. A similar difference in approaches is also apparent in planning the management of water from a downspout (a spot-like release) or from paving used in the garden (as a wide front runoff). The utilisation of surfaces as part of the treatment train as water resources for vegetation was one of the key changes made to the original designs. This means that impervious surfaces should be perceived as water-generating areas and the vegetation surface should be perceived as an equal water-using area, even if it is not named as a method of SuDS. Therefore, all vegetation covered surfaces should be perceived as part of the stormwater management train, in which the slope and the material of the surface determine its effectiveness in stormwater management.

According to the examined garden designs, the placement of infiltrating SuDSs on the plots was based on, firstly, the avoidance of non-permitted infiltration areas and, secondly, the sizing of SuDSs. Moreover, in cold climates, snow, snow piling sites, and melted water on top of frozen ground require careful placement and sizing.

The practices of stormwater management including infiltration always require water flows to be perceived as both surface runoff and surface layer runoff. An examination of the water movement to the foundations of buildings and structures in relation to the drainage and frost insulation required revealed that any planting areas placed at the centres of paved areas must be carefully designed. This is due to the fact that sub-surface drainage systems intended to keep the base of a wall or pavement dry can easily be overburdened by the irrigation water used in an adjacent planting area. Another problem of subsurface drainage systems is that they are usually maintenance-intensive and prone to clogging issues [48]. Similarly, construction layers with big grain size cause the surrounding growing media to dry, in which case the volume of the growing media must be increased. In practice, this results in the planting areas in the middle of pavements and narrow stripes expanding.

As a whole, the integration of stormwater and vegetation in LDH plots appears to work well due to the relatively low water volumes. If a plot receives runoff outside its borders or if there is an uncommonly large impervious area, the potential for plot-specific stormwater management is naturally reduced. The design process that integrates vegetation with stormwater management needs to start with form and functions like any design process. Planting types are determined by the actual functions and spaces of a garden, and then plant water availability is ensured by appropriate runoff routes, infiltration, and storage. This vegetation integrated stormwater design creates treatment trains between different planting types and ensures that stormwater does not cause problems to constructions, garden use, or, if ponding occurs for a considerably long time, vegetation. It is of utmost importance to also include vegetated areas, such as mass plantings and lawns, instead of merely focusing on band-like substitutes for ditches.

4.3. Scaling Up

The plot scale designs were improved in stages. This gradual and iterative work progressed initially at the scale of blocks and subsequently, included the entire low density housing (LDH) area. This upscaling was used to examine the significance of plot-specific choices at higher scales.

4.3.1. Blocks

At the scale of blocks, even more emphasis is put on the placement of buildings and parking spaces than at the plot level. This is due to the fact that the building masses and their elevations form a block-specific micro watershed dividing front and back yards from each other. At the same time, this placement, combined with roof shapes, determines the volume of water accumulated from roofs to the part of the plot where the water must be managed. This also determines the amount of space available for stormwater management, and therefore also the set of suitable SuDSs.

At the block scale, re-designing revealed an opportunity for a so-called shared growing media volume which emerges at the borders of plots, as opposite planting areas are adjacent to each other. This is noteworthy, as growing media volume was one of the challenges observed at the plot scale. Utilising shared growing media volume naturally requires the planting areas to be located at the same section of the plot border, and there should also be no changes expected in the neighbours' plot use.

The block scale can also be used when working on large planting areas where plant communities (man-made habitats) can be developed. These habitats can emerge at the centres of blocks when water management and vegetation are located in the same area. In the blocks examined in this study, a stormwater flood route based on the locations of building masses and their elevations and a related vegetation area had already been created at the centre of the block at the planning stage. The design at the block level also included the use of this vegetation area for safe infiltration at a sufficient distance from buildings, and a possibility, to provide a harmonious forest stand and a resulting increase in crown closure on the block. This could allow the creation of larger vegetation-covered patches with multi-layer vegetation to support biodiversity on the block scale.

In addition to the slightly obvious definition for the multi-layer, eutrophic vegetation areas, this idea for habitat construction includes the examination of other built environment habitat types (Figure 5). Second, walkways and the sides of buildings, which are kept dry to ensure accessibility or healthy structures, create a dry growth environment on, and at the immediate vicinity of, these surfaces. As a result, the placement of buildings and walkways may form dry habitats across the borders of individual plots at the block level. At the same time, these areas between buildings tend to be the ones where inhabitants wish to use vegetation to create protective screening between plots and to the street. This produces third habitat type at the block level, where vegetation is planted on naturally dry spots in the middle of hard surfaces. The growth of sufficient media to retain water and nutrients must be ensured for this habitat type, and an adequate water supply must be provided for the planted vegetation. The fourth habitat type at the block level is comprised of vegetable patches that require regular moisture. While some plots may not include these, there are good grounds for placing these at the borders of plots adjacent to neighbours' patches to ensure the necessary humidity conditions and equal levels of light.

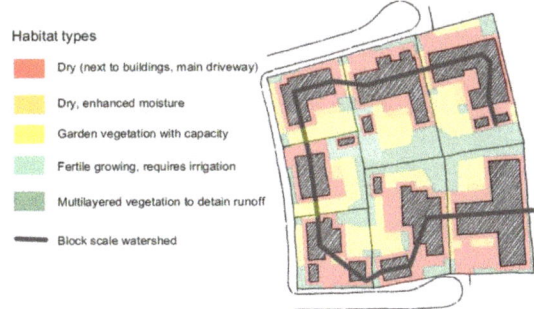

Figure 5. Adjacent plots formed five different habitat types for blocks. The moisture conditions in these habitats are based on the areas of construction layers and sub-surface drainage with irrigation dependent vegetation, with those with a high infiltration capacity with multi-layered vegetation in the centre of the block.

The fifth habitat type was open surfaces with low levels of vegetation—typically lawns and the planting areas commonly placed at lawn borders. At the block level, these lawnscapes are located in front of buildings and, particularly, next to patios. Even though lawns are rarely perceived as a part of stormwater management, the block-level examination revealed that they are located between water-producing hard surfaces and the eutrophic biotypes that need the most water, and they must therefore be perceived as part of the treatment trains.

4.3.2. Neighbourhoods

In addition to blocks consisting of plots, the GI of neighbourhoods comprises public parks and street networks. On the neighbourhood scale, vegetation is divided into trees planted alongside streets in a band-like formation or areas of plants around streets and vegetation patches in parks. Vegetation plays similar roles in parks as on the plots. However, in this area, vegetation is primarily perceived as forests and groves, meadows and other open spaces, or gardenesque sections of parks.

The layout of a neighbourhood divides the GI into the private green areas of blocks and the public green areas of parks and streets. Therefore, the layout of a neighbourhood defines what kind of GI continuum is created for people's physical activities and as a habitat for fauna. While urban planning is primarily concerned with the construction of the biophysical environment, functional connections, such as streams of water and nutrients, also affect the design of the GI, particularly at the neighbourhood scale.

Neighbourhood scale GI planning can utilise wooded patches growing in blocks as a kind of stepping stone passing through the area. This allows the lush parts of blocks to supplement broken ecological connections, support the landscape ecology patches located nearby, or create new connections. The shared growing media volumes of blocks may also be connected to park zones, thus providing possibilities for connections to the micro-organisms in the soil.

On this scale, watershed divides emerge as a result of the building masses in blocks and the elevations and inclinations of the street system. As such, street areas and kerbs serve as flood paths. However, water from the streets will primarily flow to the sewer system, as the ratio between pervious and impervious surfaces does not primarily favour SuDSs. The potential for urban green areas in stormwater management is determined by the scaling of the cross-section of the street area in urban planning. If the dimensions of streets allow it, a green street can provide a band-like connection through the street network in the form of trees planted alongside the street. On the streets along which plots are located, the stormwater management approaches are focused on water infiltration and increasing the delay in water flow (Figure 6).

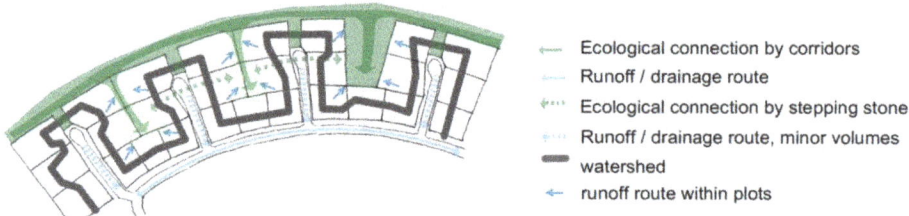

Figure 6. Neighbourhood scale defined flooding routes and vegetation patterns that may support the ecological network.

5. Discussion

The purpose of this study was to describe how the GI of an area with LDH can be developed by first, improving the garden scale designs to better integrate water and vegetation and second, scaling up from plot scale garden designs to habitats at the block scale and ecological networks at the neighbourhood scale. This bottom-up, decentralized approach follows Keeley's [49] claim about the need to develop the practices of GI planning. The results indicate that while combining stormwater management with the planting types typically used in garden design appears to work, this requires the recognition of their level of water demand. On the block scale, vegetation should make use of shared growing media between neighbours and rely the diversity of habitats that form from block scale arrangements of green and gray components. This block scale arrangement may form cohesive vegetation by shared soil volumes and smooth stormwater infiltration in the lowest corner. These habitats with multi-layer vegetation are determined in design at the scale of the entire neighbourhood, which includes the creation of a network of ecological corridors, patches, and matrices. Nonetheless, all types of habitats, from dry to water-absorbing plantings, should be appreciated in order to avoid inappropriate infiltration in areas that are drained with the means of constructions and their foundations.

Vegetation integrated stormwater management and, especially, the use of multi-layered vegetation, generates two simultaneous benefits. First, multi-layered vegetation provides a design element for defining a space and its edges. For this purpose, it is essential to have multi-layered vegetation. This space forming role of SuDSs is not too often discussed, and the guidelines seem to concentrate mainly on the nutrient removal capacity of vegetation, water tolerance, or presence of native species. Second, multi-layered vegetation has recently been mentioned in several studies as the key component of biodiversity [38,39,50]. Furthermore, this potential for biodiversity is proposed to especially rely on residential areas [39,51]. Figure 7 sums up our proposal for a designer's checklist to work with scalable GI that starts on plot scale designs.

Based on this study, there appears to be room for development in the design practices if the aim is to improve the GI of LDH. Vegetation integrated stormwater management requires constant assessment of the amount of water needed by vegetation and its capacity to tolerate ponding. However, this integration cannot be carried out without consideration of the surrounding environment and its moisture conditions in the foundations of constructions. Therefore, vegetation integrated stormwater management is based on stormwater management whereby treatment trains through vegetation-covered areas allow water to be infiltrated and stored in the growing media, thus allowing runoff be conducted slowly and as a wide front across planting areas and lawns in addition to other SuDSs. The main difference with this approach and traditional SuDS descriptions is that water is perceived as a resource that is necessary for plant growth and, additionally, the flows of water are perceived as surface layer runoff instead of only as surface runoff. This approach requires the understanding of both water and vegetation as well as the flows formed by the soil that conveys these.

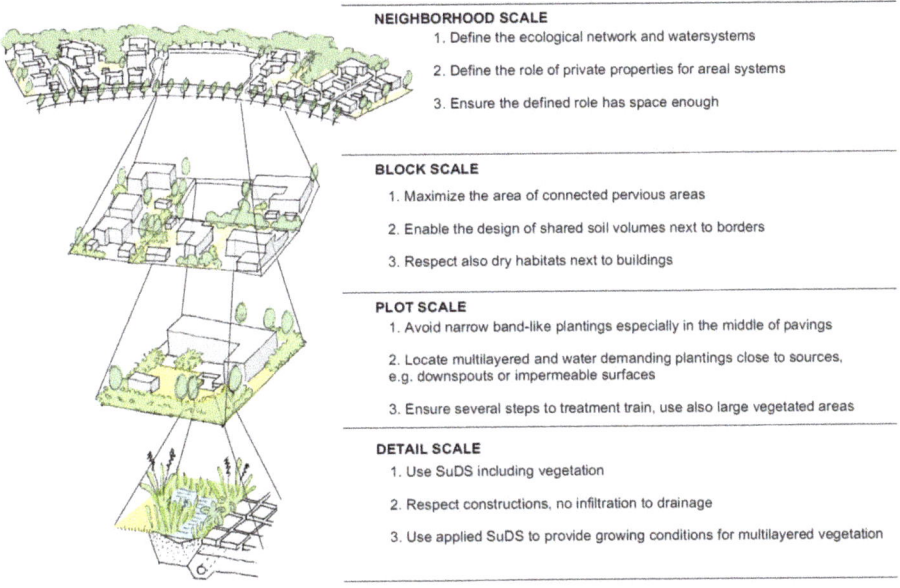

Figure 7. Proposed checklist for designers to work with scalable green infrastructure (GI) in low density housing.

In this system, soil is the interface between vegetation and water that enables water to filtrate, be retained, infiltrate, and rise due to capillary actions. In turn, vegetation absorbs the available water for its growth and releases water to the atmosphere. The decomposition of dead leaves and litter forms organic matter (OM) that contains nutrients needed for growth, and OM improves the water-holding capacity in soil that supports the availability of water to vegetation between rain events. OM supports the living conditions of micro-organisms, thereby improving biodiversity in the soil. In addition, the development of a root system supports water infiltration.

This core system of GI does not correspond to the traditional planting design process that includes the selection of plant species, but rather, is concerned with seeking a balance between soil, vegetation, and water. This system (a) can be found in some form on all surfaces of a built environment and (b) functions in constant interaction with the ways that people use areas and manage their gardens. Based on the results of this study, this system of water, vegetation and soil was identified to be a key factor in the design of vegetation integrated stormwater management. This finding is in line with the claims that the provision of ecosystem services builds on hydrologically active surfaces [52] and vegetated surfaces [53].

The proportion of sealed surfaces and their foundations limit the soil volume that is available for the system of water, vegetation, and soil. The smaller the space left for vegetation is, the more vulnerable the GI's CS is, and there might be a need to support this system by using fertilizers or irrigation. This brings up the question of what the minimum space for a self-sustaining GI core system is. If soil is considered solely as a filter through which stormwater infiltrates, the opportunity to provide soil water for vegetation is lost. The purpose is not to drown the plants with excessive water but to make sure that the soil holds available water for vegetation to withstand drought between rain events.

6. Conclusions

Garden scale GI can be enhanced by integrating stormwater management to vegetation, and this enhanced GI at plot scale affects also block and neighbourhood scales. This integration requires garden designers to have the knowledge of the interconnected system of water, vegetation, and soil and its on-going processes in the detail scale. This knowledge is essential when designing both good growing conditions for vegetation and technical safety for buildings and constructions. This integrative designing demands balancing between proportions of green and grey, impervious and pervious surfaces, to place the areas of water demand and runoff generation in relation to each other. Furthermore, designing must consider water flows not only on surface but also in surface layer next to construction foundations. This integrative approach needs to be the aim already in the early steps of design process. The careful design of separate vegetation or water systems will not suffice on its own.

Plot scale integration of stormwater and vegetation can provide improved growing conditions that serve for the continuum of different water demanding habitat types. Furthermore, it stresses the role of plots every square meter for stormwater management, not only the set of separate SuDS. This integrative approach starts from plot scale and the set of decisions in garden designs. However, urban planners need to realize its potential in block and neighbourhood scales as the outcome may improve biodiversity potential in the whole residential area and that returns back to residents as ecosystem services.

Funding: This study was financially supported through Maiju and Yrjö Rikala's Garden Foundation.

Conflicts of Interest: The authors declare no conflict of interest. The founding sponsors had no role in the design of the study; in the collection, analyses, or interpretation of data; in the writing of the manuscript, and in the decision to publish the results.

References and Notes

1. Davies, C.; Lafortezza, R. Urban green infrastructure in Europe: Is greenspace planning and policy compliant? *Land Use Policy* **2017**, *69*, 93–101. [CrossRef]
2. Supporting the Implementation of Green Infrastructure Final Report. 2016. Available online: http://ec.europa.eu/environment/nature/ecosystems/docs/green_infrastructures/GI%20Final%20Report.pdf (accessed on 1 December 2018).
3. Liquete, C.; Kleeschulte, S.; Dige, G.; Maes, J.; Grizzetti, B.; Olah, B.; Zulian, G. Mapping green infrastructure based on ecosystem services and ecological networks: A Pan-European case study. *Environ. Sci. Policy* **2015**, *54*, 268–280. [CrossRef]
4. Ahern, J. Green Infrastructure for cities. The spatial dimensions. In *Cities of the Future towards Integrated and Landscape Management*; Novotny, V., Ed.; IWA Publications: London, UK, 2007; pp. 267–283.
5. Fletcher, T.D.; Shuster, W.; Hunt, W.F.; Ashley, R.; Butler, D.; Arthur, S.; Trowsdale, S.; Barraud, S.; Semadeni-Davies, A.; Bertrand-Krajewski, J.L.; et al. SUDS, LID, BMPs, WSUD and more—The evolution and application of terminology surrounding urban drainage. *Urban Water J.* **2015**, *12*, 525–542. [CrossRef]
6. Wright, H. Understanding green infrastructure: The development of a contested concept in England. *Local Environ.* **2011**, *16*, 1003–1019. [CrossRef]
7. Benedict, M.A.; McMahon, E.T. Green Infrastructure: Smart conservation for the 21st century. *Renew. Resour. J.* **2002**, *20*, 12–17.
8. Pitman, S.D.; Daniels, C.B.; Ely, M.E. Green infrastructure as life support: Urban nature and climate change. *Trans. R. Soc. South Aust.* **2015**, *139*, 97–112. [CrossRef]
9. Kambites, C.; Owen, S. Renewed prospects for green infrastructure planning in the UK. *Plan. Pract. Res.* **2006**, *21*, 483–496. [CrossRef]
10. Mell, I.C. Green Infrstructure: Concepts and planning. *Forum eJ.* **2008**, *8*, 69–80.

11. Hansen, R.; Olafsson, A.S.; van der Jagt, A.P.N.; Rall, E.; Pauleit, S. Planning multifunctional green infrastructure for compact cities: What is the state of practice? *Ecol. Indic.* **2019**, *96*, 99–110. [CrossRef]
12. Koc, C.B.; Osmond, P.; Peters, A. A Green Infrastructure Typology Matrix to Support Urban Microclimate Studies. *Procedia Eng.* **2016**, *169*, 183–190. [CrossRef]
13. Tu, M.C.; Smith, P. Modeling Pollutant Buildup and Washoff Parameters for SWMM Based on Land Use in a Semiarid Urban Watershed. *Water Air Soil Pollut.* **2018**, *229*, 4. [CrossRef]
14. Tahvonen, O.; Airaksinen, M. Low-density housing in sustainable urban planning – Scaling down to private gardens by using the green infrastructure concept. *Land Use Policy* **2018**, *75*, 478–485. [CrossRef]
15. Loram, A.; Tratalos, J.; Warren, P.H.; Gaston, K.J. Urban domestic gardens (X): The extent & structure of the resource in five major cities. *Landsc. Ecol.* **2007**, *22*, 601–615.
16. Mathieu, R.; Freeman, C.; Aryal, J. Mapping private gardens in urban areas using object-oriented techniques and very high-resolution satellite imagery. *Landsc. Urban Plan.* **2007**, *81*, 179–192. [CrossRef]
17. Dewaelheyns, V.; Rogge, E.; Gulinck, H. Putting domestic gardens on the agenda using empirical spatial data: The case of Flanders. *Appl. Geogr.* **2014**, *50*, 132–143. [CrossRef]
18. Whitford, V.; Ennos, A.R.; Handley, J.F. City form and natural process indicators for the ecological performance of urban areas and their alpplication to merseyside, uk.pdf. *Landsc. Urban Plan.* **2001**, *57*, 91–103. [CrossRef]
19. Gaston, K.J.; Warren, P.H.; Thompson, K.; Smith, R.M. Urban domestic gardens (IV): The extent of the resource and its associated features. *Biodivers. Conserv.* **2005**, *14*, 3327–3349. [CrossRef]
20. Cameron, R.W.; Blanuša, T.; Taylor, J.E.; Salisbury, A.; Halstead, A.J.; Henricot, B.; Thompson, K. The domestic garden—Its contribution to urban green infrastructure. *Urban For. Urban Green.* **2012**, *11*, 129–137. [CrossRef]
21. Kelly, D.A. How Does Your Garden Flow? The Impact of Domestic Front Gardens on Urban Flooding. *Br. J. Environ. Clim. Chang.* **2016**, *6*, 149–158. [CrossRef]
22. Dee, C. *Form and Fabric in Landscape Architecture A Visual Introduction*; Spon Press: New York, NY, USA, 2001.
23. Booth, N.K.; Hiss, J.E. *Residential Landscape Architecture: Design Process for the Private Residence*, 6th ed.; Prentice Hall: Upper Saddle River, NJ, USA, 2012.
24. Nouri, H.; Beecham, S.; Kazemi, F.; Hassanli, A.M. A review of ET measurement techniques for estimating the water requirements of urban landscape vegetation. *Urban Water J.* **2013**, *10*, 247–259. [CrossRef]
25. Ossola, A.; Hahs, A.K.; Nash, M.A.; Livesley, S.J. Habitat Complexity Enhances Comminution and Decomposition Processes in Urban Ecosystems. *Ecosystems* **2016**, *19*, 927–941. [CrossRef]
26. Keeley, M. The green area ratio: An urban site sustainability metric. *J. Environ. Plan. Manag.* **2011**, *54*, 937–958. [CrossRef]
27. Szulczewska, B.; Giedych, R.; Borowski, J.; Kuchcik, M.; Sikorski, P.; Mazurkiewicz, A.; Stańczyk, T. How much green is needed for a vital neighbourhood? In search for empirical evidence. *Land Use Policy* **2014**, *38*, 330–345. [CrossRef]
28. Kellagher, R. *The SUDS Manual*; CIRIA: London, UK, 2007.
29. "CIRIA SuDS Manual 2015".
30. Charlesworth, S.; Warwick, F.; Lashford, C. Decision-making and sustainable drainage: Design and scale. *Sustainability* **2016**, *8*, 782. [CrossRef]
31. Bastien, N.; Arthur, S.; Wallis, S.; Scholz, M. The best management of SuDS treatment trains: A holistic approach. *Water Sci. Technol.* **2010**, *61*, 263–272. [CrossRef]
32. Wootton-Beard, P.C.; Xing, Y.; Durai Prabhakaran, R.T.; Robson, P.; Bosch, M.; Thornton, J.M.; Ormondroyd, G.A.; Jones, P.; Donnison, I. Review: Improving the Impact of Plant Science on Urban Planning and Design. *Buildings* **2016**, *6*, 48. [CrossRef]
33. Eskola, R.; Tahvonen, O. *Hulevedet Rakennetussa Viherympäristössä*; Hämeen Ammattikorkeakoulu: Hämeenlinna, Finland, 2010.
34. Brabec, E.; Schulte, S.; Richards, P.L. Impervious surfaces and water quality: A review of current literature and its implications for watershed planning. *J. Plan. Lit.* **2002**, *16*, 499–514. [CrossRef]
35. Arnold, C.L.; Gibbons, C.J. Impervious surface coverage: The emergence of a key environmental indicator. *J. Am. Plan. Assoc.* **1996**, *62*, 243–258. [CrossRef]
36. Kuusisto-Hjort, P.; Hjort, J. Land use impacts on trace metal concentrations of suburban stream sediments in the Helsinki region, Finland. *Sci. Total Environ.* **2013**, *456–457*, 222–230. [CrossRef]

37. Yang, B.; Li, S. Green Infrastructure Design for Stormwater Runoff and Water Quality: Empirical Evidence from Large Watershed-Scale Community Developments. *Water* **2013**, *5*, 2038–2057. [CrossRef]
38. Goddard, M.A.; Dougill, A.J.; Benton, T.G. Scaling up from gardens: Biodiversity conservation in urban environments. *Trends Ecol. Evol.* **2010**, *25*, 90–98. [CrossRef] [PubMed]
39. Threlfall, C.G.; Ossola, A.; Hahs, A.K.; Williams, N.S.G.; Wilson, L.; Livesley, S.J. Variation in Vegetation Structure and Composition across Urban Green Space Types. *Front. Ecol. Evol.* **2016**, *4*, 1–12. [CrossRef]
40. Fairbrass, A.; Jones, K.; McIntosh, A.; Yao, Z.; Malki-Epshtein, L.; Bell, S. *Green Infrastructure for London: A Review of the Evidence*; The Engineering Exchange: London, UK, 2018.
41. Roggema, R. Research by Design: Proposition for a Methodological Approach. *Urban Sci.* **2016**, *1*, 2. [CrossRef]
42. De Jong, T.; van der Voordt, T. Criteria for scientific study and design. In *Ways to Study and Research*; De Jong, T., van der Voordt, T., Eds.; Delft University Press: Delft, The Netherlands, 2002; pp. 19–30.
43. Glanville, R. Research Design and Designing Research. *Des. Issues* **1999**, *15*, 80–91. [CrossRef]
44. Simon, H.A. *The Sciences of the Artificial*; MIT Press: Cambridge, MA, USA, 1969.
45. Tahvonen, O. Impervious coverage in Finnish single-family house plots management and creating urban green spaces. *Archit. Res. Finl.* **2018**, *2*, 180–194.
46. Harrison-Atlas, D.; Theobald, D.M.; Goldstein, J.H. A systematic review of approaches to quantify hydrologic ecosystem services to inform decision-making. *Int. J. Biodivers. Sci. Ecosyst. Serv. Manag.* **2016**, *12*, 160–171. [CrossRef]
47. Tahvonen, O. Water for Vegetation—Knowledge Base for an Integrated Approach to Sustainable Stormwatern Management in Site Scale. In In Proceedings of the Eclas 2014 Conference, Landscape: A Place for Cultivation, Porto, Portugal, 21–23 September 2014; pp. 331–333.
48. Tu, M.C.; Traver, R. Clogging impacts on distribution pipe delivery of street runoffto an infiltration bed. *Water* **2018**, *10*, 1045. [CrossRef]
49. Keeley, M. Green Roofs Incentives: Tried and True Techniques from Europe. In In Proceedings of the Second Annual Green Roof for Healthy Cities Conference, Portland, OR, USA, 2–4 June 2004.
50. Beninde, J.; Veith, M.; Hochkirch, A. Biodiversity in cities needs space: A meta-analysis of factors determining intra-urban biodiversity variation. *Ecol. Lett.* **2015**, *18*, 581–592. [CrossRef] [PubMed]
51. Mumaw, L.; Bekessy, S. Wildlife gardening for collaborative public-private biodiversity conservation. *Australas. J. Environ. Manag.* **2017**, *24*, 242–260. [CrossRef]
52. Stone, B. *The City and the Coming Climate: Climate Change in the Places We Live*; University of Cambridge Press:: Cambridge, MA, USA, 2012.
53. Abunnasr, Y.F. Climate Change Adaptation: A Green Infrastructure Planning Framework for Resilient Urban Regions. Ph.D. Thesis, University of Massachusetts Amherst:, Amherst, MA, USA, 2013; pp. 1–243.

© 2018 by the author. Licensee MDPI, Basel, Switzerland. This article is an open access article distributed under the terms and conditions of the Creative Commons Attribution (CC BY) license (http://creativecommons.org/licenses/by/4.0/).

Article

A GIS-Based Framework Creating Green Stormwater Infrastructure Inventory Relevant to Surface Transportation Planning

Xiaofan Xu, Dylan S. P. Schreiber, Qing Lu and Qiong Zhang *

Department of Civil and Environmental Engineering, University of South Florida, 4202 E. Fowler Avenue, ENB 118, Tampa, FL 33620, USA; xiaofanxu@mail.usf.edu (X.X.); dschreiber@mail.usf.edu (D.S.P.S.); qlu@usf.edu (Q.L.)
* Correspondence: qiongzhang@usf.edu; Tel.: +1-813-974-6448

Received: 30 October 2018; Accepted: 8 December 2018; Published: 11 December 2018

Abstract: The stormwater runoff that carries pollutants from the land adjacent to road transportation systems may impair the water environment and threaten the ecosystem and human health. A proper management approach like green stormwater infrastructure (GSI) can help control flooding and the runoff pollutants. One barrier for GSI analysis relevant to system-level surface transportation planning is the lack of the inventory of GSI in many U.S. cities. This study aims to develop a GIS-based framework for creating GSI inventory in a time and labor efficient way, different from the traditional survey-based method. The new proposed framework consists of three steps, including road categorization, GSI mapping, and GSI type identification using the GIS data, high-resolution land-cover image, and Google Earth street view pictures. The new approach was tested in Philadelphia, Pennsylvania and also applied in Tampa, Florida. The results showed that the new GIS-based framework can achieve similar accuracy to the survey-based method while saving time and labor. The GSI inventory created in the study demonstrated the usefulness of the proposed framework for analyzing the status of GSI implementation and identifying gaps for future planning in terms of potential locations and underrepresented GSI types.

Keywords: stormwater management; green infrastructure; geographic information system; mapping method

1. Introduction

National Pollutant Discharge Elimination System (NPDES) regulates that transportation authorities are responsible for managing the stormwater runoff that carries pollutants from the land adjacent to road transportation systems. The proper stormwater management can help control flooding and the runoff pollutants that may impair the water environment and threaten the ecosystem and human health [1,2]. Green infrastructure is a stormwater management approach with many economic and human health benefits including flood mitigation, erosion control, improved water quality, groundwater recharge, mitigated effect of urban heat islands, reduced energy demands for cooling, and enhanced aesthetics and access to green space [3–5]. Unlike gray stormwater infrastructure systems that are often large and centralized, green stormwater infrastructure (GSI) can be applied at different spatial scales and decentralized arrangements [6]. GSI like basins [7], bioswales [8], bioretention [9], and constructed wetlands [10] have been adopted and implemented in the transportation infrastructure design. However, such implementation is project-based without analysis at the system level or watershed scale [11]. The individual GSI can mitigate local stormwater runoff but may not lead to performance improvements in the entire stormwater network at the watershed scale [12]. To facilitate a system level analysis for urban stormwater management, a spatial

GSI inventory at a large scale (sub-watershed or watershed) is needed. However, the GSI inventory is currently lacking in many United States (U.S.) cities. This is because the traditional method to create such an inventory is based on survey and inspection data collection [13,14]. It can help build up the GSI inventory accurately, but consumes time and labor meaning that not all cities can afford it. A new framework is needed to construct an inventory of the implemented GSI using the existing geospatial data in a more efficient and economical manner. Such a framework could benefit GSI system-wide assessment and modeling, and future stormwater infrastructure planning.

The previous studies on the topic of urban GSI mapping primarily focused on identifying the potential opportunities for implementing GSI [15–20]. Among the limited number of the studies that mapped implemented GSI, some applied geospatial techniques such as remote sensing to enhance the land use/land cover classification using the remotely sensed images [21–24]. However, most of them focused on GSI detection under the connotation of 'green space.' In other words, they intended to find the GSI footprints without consideration of the unique features of engineered GSI (i.e., GSI types). These studies contributed to the development of the GSI mapping method but lacked actual applications of their methods. Moreover, there is no study focusing on either mapping the implemented GSI or identifying various types of GSI based on their surface features. Hale et al. used topographic data and aerial imagery to identify retention basins; however, this study was limited to the detection of a single GSI type [25]. Only one project focused on creating a comprehensive GSI inventory that was developed by the City of Philadelphia [26]. A GSI database was built for the entire metropolitan area in the project. The GSI mapping was primarily conducted by survey collection (though the mapping method was not explicitly described in the project, the information in the metadata and guidelines matches the survey-based process [26,27]). In addition, errors were found in terms of mapping and GSI type recognition; for example, some sports fields and concrete parking lots were misclassified as GSI, especially in the regions of intensive roads.

To fill the research gap in the GSI mapping, this study aims to develop a framework for creating GSI inventory in a time and labor efficient way. The framework is based on the Geographic Information System (GIS) technique and GSI's visual features. Since it is hard to detect the underground structures from the visual features, e.g., the invisible connections between the inlet and the hybrid GSI nearby [28], the applicability of this framework is limited to surface GSI. All the required data for the framework is available in most municipalities from the government and public organizations. The paper focuses on the transportation-related GSI because the transportation infrastructure planners are key stakeholders for large-scale implementation of GSI. The transportation-related GSI refers to the GSI facilities designed with or serving the road transportation systems including freeways, arterials, collectors, and local roads. The GSI facilities serving only buildings, pedestrian pavements, or parking lots are not included. Therefore, the framework proposed in this paper includes the GSI of bioretention, bioswales (dry or wet swales), basins (dry or wet ponds), infiltration basins, infiltration trenches, and vegetated filter strips (Table 1). Some GSI types are excluded from this study, since they are either rarely applied to transportation planning or commonly applied to pedestrian pavements other than vehicle roads. Table 1 summarizes the type of GSI and their applications to transportation planning. The GSI nomenclature used by the U.S. Environmental Protection Agency (EPA) and the Water Research Foundation [29,30] was adopted in this study, it is worth mentioning that various terms were used interchangeably for some GSI types [31].

Table 1. The green stormwater infrastructures (GSI) and their applications to surface transportation planning.

Green Stormwater Infrastructure	Mechanism Type	Applied to Transportation Planning? If Yes, What Is the Common Place to Use?
Rain barrels/cisterns	Retention/detention	No
Bioretention cells/rain gardens	Filtration/retention	Yes, local roads
Dry/wet swales	Infiltration	Yes, local roads/ highways
Dry/wet ponds	Retention/detention	Yes, highways
Constructed wetlands	Detention	Rarely
Infiltration basin	Infiltration	Yes, highways
Infiltration trenches	Infiltration	Yes, highways
Vegetated filter strips	Filtration	Yes, highways
Sand filters	Filtration	Rarely
Riparian buffers	Filtration	Rarely
Permeable pavements	Infiltration	Yes, pedestrian pavements and parking lots
Downspout disconnection	Site design	No
Urban tree canopy	Site design	Yes, local roads and pedestrian pavements
Green roofs	Site design	No
Land conservation	Site design	No

2. Structure of the GIS-Based Framework

2.1. An Overview of the Framework

The proposed GIS-based framework consists of three steps: Categorizing the roads that may contain GSI nearby, Mapping the existing GSI relevant to transportation, and Identifying GSI types according to their visual features (Figure 1).

All the roads within the area of interest are categorized into major roads and other roads. They are screened by the corresponding criteria and the roads with potential implemented GSI nearby are selected. The land covers of water, grass, tree, and bare soil that fall within the 60-ft buffered areas of the selected roads are identified as the possible GSI footprints, which are confirmed later with the help of Google Earth street view pictures. The types of confirmed GSI sites are identified according to the unique visual characteristics of each GSI type. Eventually the GSI inventory is created with the information collected from the last two steps, including the GSI footprints and types.

The first step of categorizing roads is automated if all the needed data is provided, which helps reduce the workload in the next two steps greatly. For the second step of mapping GSI, the method can automatically find possible GSI footprints, but the confirmation of the GSI footprints requires manual work. The third step of identifying GSI type needs manual work as well. As a result, the framework is half automated.

The framework was tested in Philadelphia, Pennsylvania with accuracy assessment, and then applied in Tampa, Florida. Both the areas of Philadelphia and Tampa adopted gray and green infrastructures for stormwater management during their urban development.

The details in each step are introduced in the following sections.

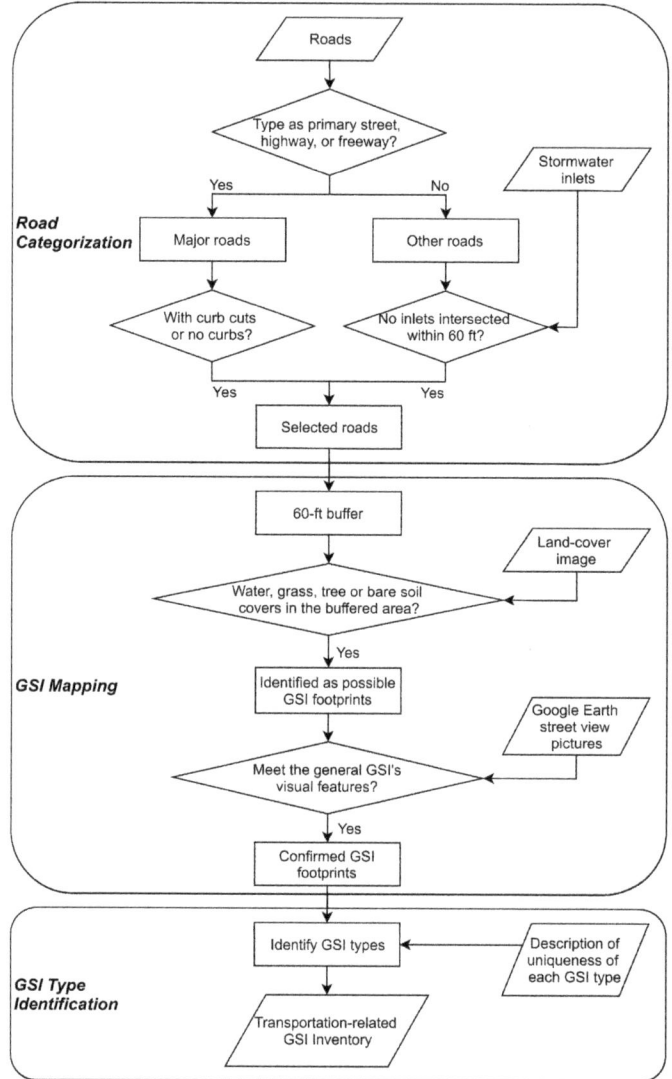

Figure 1. The process diagram to create the GSI inventory.

2.2. Data Requirements

This framework basically requires the GIS data of road centerlines, stormwater management facilities like water inlets, a high-resolution land-cover image, elevation data, and street view pictures as a reference provided by Google Earth.

Specifically, Table 2 lists the collected data and their sources to create the implemented GSI inventory in Tampa, an application of this GIS-based framework method. All the data of road systems and stormwater management facilities were formatted as shapefiles and available to the public through an open data link. The data on stormwater discharge points and open drains are not required but can help select the roads with potential implemented GSI nearby. The non-public raster image of Tampa land cover was created with a rule-based object-orientated classification method utilizing high-resolution imagery, Light Detection and Ranging (LIDAR) data, and ancillary GIS data by the

University of South Florida (USF) Water Institute. It has a 1-ft-by-1-ft resolution, providing extremely high accuracy as a reference map. The one-meter Digital Elevation Models (DEMs) produced by the U.S. Geological Survey (USGS) was used as the elevation layer for identifying GSI types. All the data were adjusted using the "GCS_North_American_1983" ArcGIS file of the coordinate system.

Table 2. Data collected and the sources for creating GSI inventory.

Data	Produced Year	Source
Road centerline	2017	City of Tampa GeoHub [32]
Stormwater inlets	2017	
Stormwater discharge points	2017	
Stormwater open drains	2017	
Tampa land cover	2011	USF Water Institute [33]
Digital Elevation Models	2007, 2010	U.S. Geological Survey [34]

2.3. Categorizing the Roads with Potential Implemented GSI Nearby

In the U.S., the stormwater management is required to be conducted together with surface transportation planning [35]. Both gray and green stormwater infrastructures are considered as options. For instance, community roads usually come with cemented open drains and highways have more water inlets for faster drainage. For the framework developed in this paper, it is critical to find the roads near which GSI may exist, in other words, to exclude the roads that are associated with only gray infrastructure.

In this study, all the roads within the area of interest were categorized into major roads (i.e., interstates, highways, state roads, or county roads) and other roads. The major roads with curb cuts or no curbs and the other roads with no inlets intersected within 60 ft were considered as the ones that may contain GSI nearby and selected for further analysis.

For the major roads, the associated GSI usually exist along with the traditional gray infrastructure to ensure the flood drainage of the major roads under extreme storm events [36,37]. It is common to see GSI and gray water inlets along the same major road. Thus, a better way to determine if the major roads contain GSI nearby is to check if there are curb cuts or even no curbs on the sides of major roads. Those curb cuts or no-curb sides can lead the stormwater runoff to the pervious surface nearby. Some GIS data of road centerlines contain the curb information (e.g., concrete curb, curb cuts, or no curbs) in the attribute table. However, if the curb information is not provided in GIS data, they can be created manually by checking the road pictures (e.g., Google Earth street view pictures) section by section. Each section typically adopts a single curb plan, i.e., full curbs, curb cuts, or no curbs. The manual workload of checking curb information is acceptable because of the limited number of major roads.

For the other roads, usually either green a stormwater solution or gray infrastructure would be implemented. It means GSI would be hardly found along the roads with water inlets. As a result, the other roads with no inlets intersected within 60 ft, as well as the major roads with curb cuts or no curbs, were selected to locate the possible GSI nearby in the next step.

2.4. Mapping GSI Relevant to Transportation

A 60 ft buffer was created for each selected road to determine the search area where the GSI may potentially occur. The 60 ft buffer is the distance from the road centerline to the edge of the road. A single travel lane is usually 10–12 ft wide [38]. For example, the State of Florida adopts 12 ft as the primary travel lane width in the urban area [39]. The roads in the urban area usually consist of one to four lanes in one direction, depending on the type of road, e.g., freeways, arterials, collectors, or local roads. This means a buffer of 48 ft on one side of the road centerline is typically sufficient to cover the road surface. In addition, the setback from the right of way line to the structures (e.g., buildings or parking lots) is required, for instance, Florida requires a minimum distance of 12 ft [40]. The buffer with the selected width should be able to cover the entire road surface in one direction and part of the spacer between the road and the nearby buildings or parking lots, where transportation-related GSI is commonly implemented. After several trials, the 60 ft buffer was chosen as the best fit, which was neither too narrow to cover GSI along some major roads, nor too wide to include the greenspace of non-public properties. Then, the buffer of selected roads was overlapped with the land-cover image. The GSI are usually identified as water, grass, or tree covers, according to the GSI type and their surface covers (e.g., wet ponds would be observed as water, and bioswale as grass or bushes). Therefore, all the water, grass, tree, or bare soil covers in the buffered areas were considered as the possible GSI footprints and converted to vector polygons based on the pixel relativity. The possible GSI polygons were checked manually to determine if they met the general GSI's visual features, with the help of Google Earth street view pictures. Since the possible GSI polygons are limited in amount, the time needed for visual confirmation was reasonable. All the confirmed GSI footprints were stored as GIS datasets for the final GSI inventory with type identification.

2.5. Identifying GSI Types from Visual Features

The framework uses the visual features from the Google Earth street view pictures to identify the GSI types. The visual variables considered include shape, relative elevation, vegetation level, and continuous standing water.

Figure 2 shows the decision-making flowchart that can be used to identify different types of GSI using their visual features. The same shape can be shared by different types of GSI, but it is a useful way to separate them into a couple of groups, namely elongated in shape or not. Swales, infiltration trenches, and vegetated filter strips usually have one of their dimensions being far larger than other dimensions. The aspect ratio of 10:1 was used in this study to determine if the detected GSI was elongated. The value of the aspect ratio is an empirical number and determined from case studies [4,41–43]. Vegetated filter strips in the design of mild slope could then be filtered out of this elongated-shape group because they often do not have a visual elevation difference from the surrounding area [44], while swales and infiltration trenches always do. The elevation difference in the framework refers to the one between the lowest point of the GSI surface and the adjacent point of the road nearby. The Digital Elevation Models (DEMs) produced by USGS were used to show the spatial elevation differences. If the elevation difference is larger than 0.5 m, it can be visually detected in the Google Earth pictures. The elongated-shape GSI with the elevation difference of ≤ 0.5 m can be identified as vegetated filter strips. The level of vegetation can be used to differentiate between swales, infiltration trenches, and the low-lying vegetated filter strips, which all have varying and distinct levels of vegetation. Three categories were developed to represent the vegetation level—tree, grass, and none. "Grass" vegetation level refers to a groundcover with grass as the major vegetation present, while "tree" refers to the vegetation containing other plants as dominant, such as bushes, flowers, and small trees. The vegetation level could be judged from the Google Earth pictures. Another way to classify it is to use the land-cover image that contains the three classes of forest, grass, or bare soil, which can roughly represent the vegetation level of tree, grass, and none, correspondingly [33]. For the group of non-elongated-shape GSI, wet ponds can be simple to sort out, since they are the only element with continuous standing water. The criterion of the vegetation level also helps differentiate between the dry pond/infiltration basin and the bioretention cell/rain garden. The framework does not distinguish infiltration basins from dry ponds, since they share almost the same visual features at the surface.

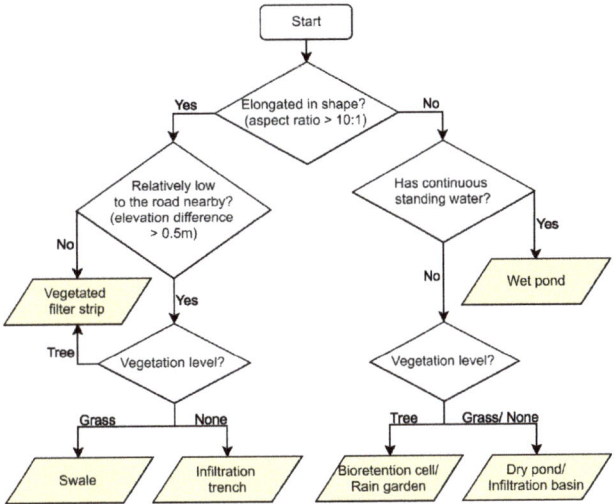

Figure 2. The decision-making flowchart to identify GSI types.

3. Framework Testing

3.1. Test Area and Results

The framework was tested for the GSI inventory in Philadelphia, Pennsylvania, where the dataset of GSI is available to the public [26]. Philadelphia's GSI data were typically collected via survey and the City of Philadelphia claimed no responsibility for the data's accuracy shown in the metadata. A rectangular region in central Philadelphia was selected as the test area, limiting the framework testing at an acceptable scale (Figure 3).

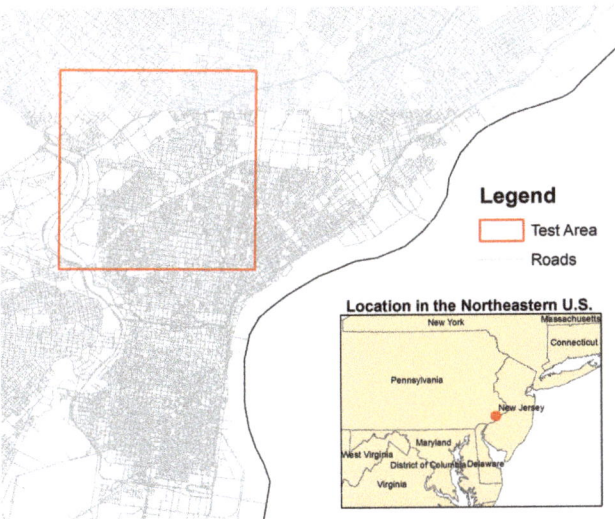

Figure 3. The test area of the framework.

To apply the GIS-based framework developed, the GIS data of roads, water inlets, and land-cover images were acquired within the test area [45]. The GIS-based GSI inventory was created by following the steps mentioned in Section 2. A total of 427 transportation-related GSI elements were detected within the test area, in comparison to the 588 GSI in the same area in the City's inventory. It is important to note that the City's inventory also contains the GSI elements not related to road transportation systems. It took one person 19 h in total to create the GSI inventory, including the whole process of mapping GSI and identifying their types. The time for collecting data listed in Table 2 is not included. There is no record of the time that the City had spent on constructing the GSI inventory, but the challenge of mapping GSI was expressed [46]. The framework is considered an efficient solution for creating GSI inventory with lower time and labor requirement.

3.2. Accuracy Assessment

According to the binomial probability theory and its formula [47],

$$N = z^2pq/E^2, \qquad (1)$$

where z is the number of standard normal deviates (here it is 2, covering 95.4%), p is the expected accuracy in percentage (here it is 90), q is equal to 100-p, and E is the allowable error in percentage (here it is 5). A total of 144 samples were picked randomly from the Google Earth base map. Both the accuracies of GIS inventory created by the city government (Table 3) and this study (Table 4) were assessed. In the accuracy assessment tables, the producer's accuracy means how accurate the GSI in the base map could be identified in the inventory map, and the user's accuracy refers to the one in the opposite way.

According to Tables 3 and 4, the total accuracies of the GSI inventory obtained from the city of Philadelphia are very close to the accuracies of the inventory created using the developed framework, which indicates the GIS-based approach can achieve similar accuracy as the traditional survey-based method. The new GIS-based approach excluded the detection of the GSI types of wetlands and tree trenches identified using the traditional method (Table 3) due to their rare application to the surface transportation planning.

The new method resulted in a slightly lower accuracy for basins compared with the one from the city (86% vs. 89% for producer's accuracy, 89% vs. 91% for user's accuracy), but has higher accuracy for swales and bioretention systems (e.g., 92% vs. 89% for swale in terms of user's accuracy). It implies that the GIS-based framework has a good ability to detect the GSI of small size (e.g., swales and bioretention systems that usually have surface area of 200–10,000 ft^2 [48]), but has the possibility of missing the large-size GSI like basins (requires minimum surface area of 0.25 acres [49]) because they are easily confused with surface waters and grassland landscape from the GIS perspective. In contrast, the traditional method has lower accuracy on mapping the GSI of small size due to the time constraints of surveyors for collecting the information of all small-size GSI. In other words, the GIS-based method scans through the entire studied area and has the advantage of catching the small-size GSI, compared to the labor-intensive survey method.

Overall, the new GIS-based method can achieve similar accuracy as the traditional survey-based method, while saving time and labor on inventory creation. In this case, it took one day to build up the GIS inventory, compared to the survey work that usually takes months.

Table 3. Accuracy assessment of the City of Philadelphia's GSI inventory.

GSI Type	Count	Percent	Accuracy Assessment	
			Producer's	User's
Basin	42	31%	89%	91%
Swale	64	47%	86%	89%
Bioretention cell/rain garden	21	16%	75%	82%
Wetland	6	4%	100%	100%
Tree Trench	2	1%	100%	100%
Total	135		87%	

Table 4. Accuracy assessment of the GSI inventory created in this project.

GSI Type	Count	Percent	Accuracy Assessment	
			Producer's	User's
Basin	46	35%	86%	89%
Swale	60	46%	88%	92%
Bioretention cell/rain garden	24	18%	78%	85%
Total	130		86%	

4. Framework Application

4.1. Study Area

To define the study area for applying the framework to create the GSI inventory, some requirements were taken into consideration:

1. A region in the scale of watershed or subwatershed;
2. A region under flood risk;
3. An area consisting of diverse land uses.

The paper used the Watershed Boundary Dataset from USGS that defines the national hydrological boundary at six different geographical levels from regions to sub-watersheds [50]. To meet the requirements, the Middle Hillsborough River-Spillway 20 subwatershed area (HUC12 code: 031002050503) was selected for this study (see Figure 4). It covers an area of 125 km^2, approximately 30% area of the City of Tampa. Figure 4 also shows the reported street flooding provided by City of Tampa Transportation and Stormwater Services recording the flooding locations during 2015 to 2017. According to the reported street flooding in the last three years, about half of the study area has suffered from the flooding incidents and better stormwater management is needed in the area. Adjacent to downtown Tampa, most of the study area is for urban use, including business, commercial, residential, recreational and some other community mixed uses. The inventory of GSI was created for this study area as a representative of the City of Tampa and the Hillsborough County.

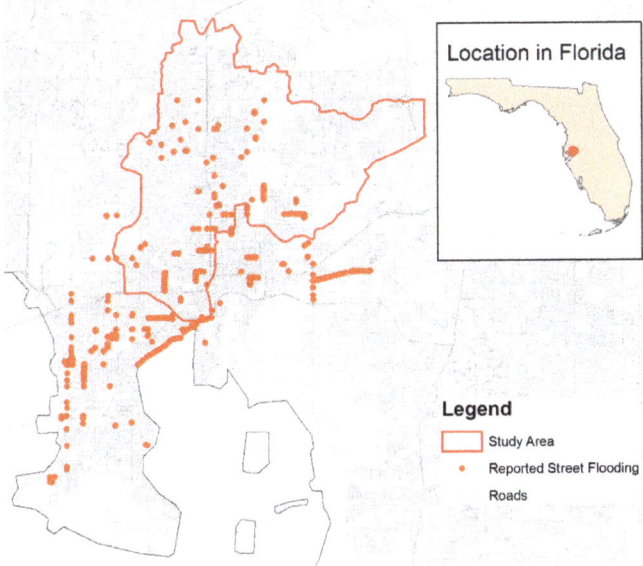

Figure 4. The study area for the framework application.

4.2. Results and Discussion

Using the developed method, a total of 89 GSI were mapped within the study area (see Figure 5). The urban area in Tampa expanded from south to north, indicating the communities in the north were newly built. In line with the characteristics of city development, most of the GSI as new practices of stormwater management were detected in the north of the study area. A limited number of the GSI were implemented along the major roads. This indicates that the gray infrastructure is used as the main stormwater facilities in the major road system in Tampa. For the business districts at the southern corner of the study area, the GSI were rarely detected because gray stormwater features have been preferentially implemented in the downtown and its surrounding areas. For the street flooding, most reported incidents happened in the area with few GSI; less street flooding occurred in the north of the study area where more GSI were implemented. There are many factors that can contribute to the fewer flooding reports in the north area, including the characteristics of the drainage system, the interest of people in reporting issues, as well as the GSI's function of infiltration and storage of stormwater [51].

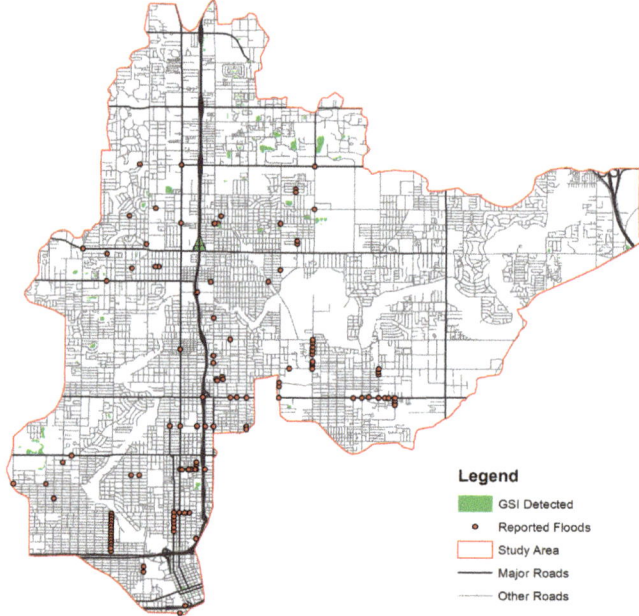

Figure 5. GSI mapped in the study area.

All the GSI detected were identified with their types (see Table 5). Most of the GSI are wet ponds with a relatively larger surface area (43,000–176,000 ft^2 in this case). The GSI with ground vegetation, such as bioretention systems, rain gardens, or vegetated filter strips, were implemented to a very limited extent. Specifically, large-size GSI like dry or wet ponds were easier to be selected by the transportation agency for the stormwater management at the transportation connections, e.g., freeway ramps, or junctions of two major roads. Those regions have a relatively large pervious area without surface cement and asphalt, requiring some GSI type of corresponding surface size. In addition, dry or wet ponds are competitive in costs due to their simpler structure than GSI types with ground vegetation like bioretention systems [30]. On the other hand, small-size GSI like bioretention systems and vegetated filter strips were more often constructed along community roads or near community public areas. This is because bioretention systems, rain gardens, and vegetated filter strips usually have multi-layer designs, performing better in stormwater quality control with the functions of plant uptake, soil adsorption and filtration, and biological treatment. These GSI can benefit the community with better contaminant removal, as well as improving site aesthetics, reducing noise, and providing shade and wind cover [52]. However, their implementation was limited due to the complexity of multi-layer design, relatively small size, and the requirement of active community engagement [53].

Table 5. Different types of GSI identified.

Type	Count	Average Surface Area (1000 ft^2)
Dry pond	5	93.0
Bioretention cell/rain garden	3	1.7
Vegetated filter strip	4	15.4
Wet pond	77	91.8

5. Conclusions

According to the previous studies, GSI as an alternative stormwater management strategy could provide significant benefits such as energy saving and environmental impact reduction, especially when implemented on a large scale (e.g., watersheds) [6,54]. However, to implement GSI on a large scale, an accurate inventory of existing GSI is important for strategic planning for future GSI implementation. Compared with the traditional survey-based method, this study developed an efficient alternative method to map the GSI footprints and identify their types. The newly developed framework was tested with an acceptable accuracy as the traditional survey-based method in the case of Philadelphia. The novelty of the proposed framework lies not in the individual steps but the combination of all steps that can save time and labor to create a relatively accurate GSI inventory. The framework is transferable and can be applied to other locations besides the study area in this research. It can help cities create their own GSI inventory and facilitate the development of GSI relevant to surface transportation planning.

Within the study area in Tampa, the GSI was implemented to a very limited extent for urban transportation stormwater management. Among the GSI mapped, most of them are those with large surface areas (e.g., wet or dry ponds), commonly occurring in the transportation connections. The GSI inventory created for the study area is an example of demonstrating the usefulness of the proposed framework for analyzing the status of GSI implementation and identifying gaps for future planning in terms of potential locations and underrepresented GSI types (e.g., bioretention in this study).

Author Contributions: Conceptualization, Q.Z., Q.L. and X.X.; Methodology, X.X. and Q.Z.; Validation, X.X. and D.S.P.S.; Formal Analysis, X.X., D.S.P.S. and Q.Z.; Investigation, X.X. and D.S.P.S.; Writing—Original Draft Preparation, X.X. and D.S.P.S.; Writing—Review & Editing, Q.Z. and Q.L.; Supervision, Q.Z.; Funding Acquisition, Q.L. and Q.Z.

Funding: This research was funded partially by the U.S. Environmental Protection Agency Grant No. 83556901, and partially by the U.S. Department of Transportation Grant No. 69A3551747119.

Acknowledgments: The contents of this paper reflect the views of the authors, who are responsible for the facts and the accuracy of the information presented herein. This document is disseminated in the interest of information exchange. The study is funded, partially by the Center for Reinventing Aging Infrastructure for Nutrient Management (RAINmgt), supported by US EPA grant 83556901, and partially by the grant from the U.S. Department of Transportation's University Transportation Centers Program under the Agreement ID: 69A3551747119. However, the U.S. Government assumes no liability for the contents or use thereof.

Conflicts of Interest: The authors declare no conflict of interest.

References

1. Forman, R.T.; Alexander, L.E. Roads and their major ecological effects. *Annu. Rev. Ecol. Syst.* **1998**, *29*, 207–231. [CrossRef]
2. Line, D.E.; White, N.M. Effects of development on runoff and pollutant export. *Water Environ. Res.* **2007**, *79*, 185–190. [CrossRef] [PubMed]

3. Wendel, H.E.W.; Downs, J.A.; Mihelcic, J.R. Assessing equitable access to urban green space: The role of engineered water infrastructure. *Environ. Sci. Technol.* **2011**, *45*, 6728–6734. [CrossRef] [PubMed]
4. Bowen, K.J.; Lynch, Y. The public health benefits of green infrastructure: The potential of economic framing for enhanced decision-making. *Curr. Opin. Environ. Sustain.* **2017**, *25*, 90–95. [CrossRef]
5. Demuzere, M.; Orru, K.; Heidrich, O.; Olazabal, E.; Geneletti, D.; Orru, H.; Bhave, A.; Mittal, N.; Feliu, E.; Faehnle, M. Mitigating and adapting to climate change: Multi-functional and multi-scale assessment of green urban infrastructure. *J. Environ. Manag.* **2014**, *146*, 107–115. [CrossRef] [PubMed]
6. Suppakittpaisarn, P.; Jiang, X.; Sullivan, W.C. Green Infrastructure, green stormwater infrastructure, and human health: A review. *Curr. Landsc. Ecol. Rep.* **2017**, *2*, 96–110. [CrossRef]
7. Belizario, P.; Scalize, P.; Albuquerque, A. Heavy metal removal in a detention basin for road runoff. *Open Eng.* **2016**, *6*, 412–417. [CrossRef]
8. Lucas, S.; Clar, M.; Gracie, J.A. Green street retrofit in a Chesapeake Bay community using bioswales. In Proceedings of the 2011 Low Impact Development Conference, Philadelphia, PA, USA, 25–28 September 2011. [CrossRef]
9. Lucke, T.; Nichols, P.W. The pollution removal and stormwater reduction performance of street-side bioretention basins after ten years in operation. *Sci. Total Environ.* **2015**, *536*, 784–792. [CrossRef]
10. Li, J.; Liang, Z.; Gao, Z.; Li, Y. Experiment and simulation of the purification effects of multi-level series constructed wetlands on urban surface runoff. *Ecol. Eng.* **2016**, *91*, 74–84. [CrossRef]
11. Pennino, M.J.; McDonald, R.I.; Jaffe, P.R. Watershed-scale impacts of stormwater green infrastructure on hydrology, nutrient fluxes, and combined sewer overflows in the mid-Atlantic region. *Sci. Total Environ.* **2016**, *565*, 1044–1053. [CrossRef]
12. Roy, A.H.; Wenger, S.J.; Fletcher, T.D.; Walsh, C.J.; Ladson, A.R.; Shuster, W.D.; Thurston, H.W.; Brown, R.R. Impediments and solutions to sustainable, watershed-scale urban stormwater management: Lessons from Australia and the United States. *J. Environ. Manag.* **2008**, *42*, 344–359. [CrossRef] [PubMed]
13. Rowe, A.; Rector, P.; Bakacs, M. Survey results of green infrastructure implementation in New Jersey. *J. Sustain. Water Built. Environ.* **2016**, *2*, 04016001. [CrossRef]
14. Georgetown Climate Center. Green Infrastructure Toolkit. Available online: https://www.georgetownclimate.org/adaptation/toolkits/green-infrastructure-toolkit/introduction.html?full (accessed on 13 November 2018).
15. US EPA. How Do You Select Good Sites for Green Infrastructure? Available online: https://www.epa.gov/sites/production/files/2015-11/documents/how_to_determine_best_opportunites_for_green_infrastructure.pdf (accessed on 15 August 2017).
16. US EPA. *Green Infrastructure Opportunities and Barriers in the Greater Los Angeles Region (EPA 833-R-13-001)*; US Environmental Protection Agency: Los Angeles, CA, USA, 2013.
17. US EPA. *Coastal Stormwater Management Through Green Infrastructure a Handbook for Municipalities (EPA 842-R-14-004)*; US Environmental Protection Agency: Washington, DC, USA, 2014.
18. Huron River Watershed Council. Guide to Green Infrastructure Opportunity Maps for Washtenaw County. Available online: https://www.hrwc.org/wp-content/uploads/2014/03/Guide-to-GI-Opportunities-v2.pdf (accessed on 15 August 2017).
19. Cleland, B.; Karll, K. Identify your green infrastructure priorities. In Proceedings of the Great Lakes and St. Lawrence Green Infrastructure Conference, Detroit, MI, USA, 31 May–2 June 2017.
20. Meerow, S.; Newell, J.P. Spatial planning for multifunctional green infrastructure: Growing resilience in Detroit. *Landsc. Urban Plan.* **2017**, *159*, 62–75. [CrossRef]
21. Dennis, M.; Barlow, D.; Cavan, G.; Cook, P.A.; Gilchrist, A.; Handley, J.; James, P.; Thompson, J.; Tzoulas, K.; Wheater, C.P. Mapping urban green infrastructure: A novel landscape-based approach to incorporating land use and land cover in the mapping of human-dominated systems. *Land* **2018**, *7*, 17. [CrossRef]
22. Labib, S.M.; Harris, A. The potentials of Sentinel-2 and LandSat-8 data in green infrastructure extraction, using object based image analysis (OBIA) method. *Eur. J. Remote Sens.* **2018**, *51*, 231–240. [CrossRef]
23. Liquete, C.; Kleeschulte, S.; Dige, G.; Maes, J.; Grizzetti, B.; Olah, B.; Zulian, G. Mapping green infrastructure based on ecosystem services and ecological networks: A Pan-European case study. *Environ. Sci. Policy* **2015**, *54*, 268–280. [CrossRef]
24. Wang, J.X.; Banzhaf, E. Derive an understanding of Green Infrastructure for the quality of life in cities by means of integrated RS mapping tools. In Proceedings of the 2017 Joint Urban Remote Sensing Event (JURSE), Dubai, UAE, 6–8 March 2017. [CrossRef]

25. Hale, R.L.; Turnbull, L.; Earl, S.R.; Childers, D.L.; Grimm, N.B. Stormwater infrastructure controls runoff and dissolved material export from arid urban watersheds. *Ecosystems* **2015**, *18*, 62–75. [CrossRef]
26. City of Philadelphia. Green Stormwater Infrastructure—SMP Footprints. Available online: http://metadata.phila.gov/#home/datasetdetails/5543864320583086178c4e6f/representationdetails/56eaddf28597748205e975da/ (accessed on 16 August 2017).
27. Philadelphia Water Department. Green Stormwater Infrastructure (GSI) Planning Guidelines. Available online: http://phillywatersheds.org/doc/GSI/150305_PWDGSI_Planning_Guidelines_Version1.0.pdf (accessed on 15 March 2018).
28. Tu, M.-C.; Traver, R. Clogging impacts on distribution pipe delivery of street runoff to an infiltration bed. *Water* **2018**, *10*, 1045. [CrossRef]
29. US EPA. What Is Green Infrastructure? Available online: https://www.epa.gov/green-infrastructure/what-green-infrastructure (accessed on 15 March 2018).
30. WERF. *User's Guide to the BMP and LID Whole Life Cost Models Version 2.0 (Report SW2R08)*; Water Environment & Reuse Foundation: Alexandria, VA, USA, 2009.
31. McPhillips, L.E.; Matsler, A.M. Temporal evolution of green stormwater infrastructure strategies in three US cities. *Front. Built Environ.* **2018**, *4*, 26. [CrossRef]
32. City of Tampa. City of Tampa GeoHub. Available online: http://city-tampa.opendata.arcgis.com/ (accessed on 20 August 2017).
33. USF Water Institute. Tampa Tree Canopy and Land Cover. Available online: http://waterinstitute.maps.arcgis.com/apps/Viewer/index.html?appid=7fa10b957c4d4ee6a4a5bb91f9316118 (accessed on 20 August 2017).
34. U.S. Geological Survey. The National Map—TNM Download. Available online: https://viewer.nationalmap.gov/basic/ (accessed on 30 August 2017).
35. Florida DOT. *Manual of Uniform Minimum Standards for Design, Construction and Maintenance for Streets and Highways (Topic# 625-000-015)*; Florida Department of Transportation: Tallahassee, FL, USA, 2016.
36. Pitt, S.; Pitt, R.; Field, R.; Tafuri, A.N.; Khalid, A.; Reddy, F.; O'Bannon, D. Use of green infrastructure integrated with conventional gray infrastructure for combined sewer overflow control: Kansas City, MO. In Proceedings of the Water Environment Federation, WEFTEC 2011, Los Angeles, CA, USA, 16–19 October 2011.
37. US EPA. *Green Infrastructure Permitting and Enforcement Series Fact Sheet 3: Sanitary Sewer Overflows (EPA 832F12014)*; US Environmental Protection Agency: Washington, DC, USA, 2012.
38. NSCTO. Urban Street Design Guide—Lane Width. Available online: https://nacto.org/publication/urban-street-design-guide/street-design-elements/lane-width/ (accessed on 10 August 2017).
39. Florida DOT. *Roadway Design Bulletin 14-17: Urban Arterial Lane Width and Bicycle Lane Options*; Florida Department of Transportation: Tallahassee, FL, USA, 2014.
40. Florida DOT. *Driveway Information Guide*; Florida Department of Transportation: Tallahassee, FL, USA, 2008.
41. Purvis, R.A.; Winston, R.J.; Hunt, W.F.; Lipscomb, B.; Narayanaswamy, K.; McDaniel, A.; Lauffer, M.S.; Libes, S. Evaluating the Water Quality Benefits of a Bioswale in Brunswick County, North Carolina (NC), USA. *Water* **2018**, *10*, 134. [CrossRef]
42. Vestergren, S. Infiltration Trenches in Scania: A Study of the Hydraulics and the Pollutant Removal Effect. Master Thesis, Lund University, Lund, Sweden, 2010.
43. Bicudo, J.; Stone, R.P.; Pastrik, E.; Patterson, J.; Hazlewood, P.P.; Mlynarz, D. Vegetated Filter Strip Design—Overview and Case Study. In Proceedings of the 2006 ASAE Annual Meeting, St. Joseph, MI, USA, 9–12 July 2006. [CrossRef]
44. Minnesota Pollution Control Agency. Minnesota Stormwater Manual—Vegetated Filter Strips. Available online: https://stormwater.pca.state.mn.us/index.php?title=Vegetated_filter_strips (accessed on 13 September 2017).
45. OpenDataPhilly.com. Open Data in the Philadelphia Region. Available online: https://www.opendataphilly.org/ (accessed on 30 March 2018).
46. CityMart. Philadelphia Green Stormwater Infrastructure Program. Available online: https://www.citymart.com/philadelphia-gsi (accessed on 30 March 2018).
47. USFWS. Accuracy Assessment. Available online: https://www.fws.gov/gisdownloads/R8/individual/Justin/RS_presentations/11%20Accuracy%20assessment.pptx (accessed on 31 March 2018).

48. New Jersey Department of Environmental Protection. *New Jersey Stormwater Best Management Practices Manual Chapter 9.1: Bioretention Systems*; New Jersey Department of Envrionmental Protection: Trenton, NJ, USA, 2009.
49. New Jersey Department of Environmental Protection. *New Jersey Stormwater Management Technical Manual Chapter 6.12: Wet Ponds*; New Jersey Department of Environmental Protection: Trenton, NJ, USA, 2011.
50. U.S. Geological Survey. Watershed Boundary Dataset. Available online: https://www.usgs.gov/core-science-systems/ngp/national-hydrography/watershed-boundary-dataset?qt-science_support_page_related_con=4#qt-science_support_page_related_con (accessed on 20 August 2017).
51. Liu, W.; Chen, W.; Peng, C. Assessing the effectiveness of green infrastructures on urban flooding reduction: A community scale study. *Ecol. Model.* **2014**, *291*, 6–14. [CrossRef]
52. Tzoulas, K.; Korpela, K.; Venn, S.; Yli-Pelkonen, V.; Kaźmierczak, A.; Niemela, J.; James, P. Promoting ecosystem and human health in urban areas using Green Infrastructure: A literature review. *Landsc. Urban Plan.* **2007**, *81*, 167–178. [CrossRef]
53. Allen, W.L. Environmental reviews and case studies: Advancing green infrastructure at all scales: From landscape to site. *J. Environ. Pract.* **2012**, *14*, 17–25. [CrossRef]
54. Pochee, H.; Johnston, I. Understanding design scales for a range of potential green infrastructure benefits in a London Garden City. *Build. Serv. Eng. Res. Technol.* **2017**, *38*, 728–756. [CrossRef]

© 2018 by the authors. Licensee MDPI, Basel, Switzerland. This article is an open access article distributed under the terms and conditions of the Creative Commons Attribution (CC BY) license (http://creativecommons.org/licenses/by/4.0/).

Article

Hydrological Performance of Green Roofs at Building and City Scales under Mediterranean Conditions

Ignacio Andrés-Doménech [1,*], Sara Perales-Momparler [2], Adrián Morales-Torres [3] and Ignacio Escuder-Bueno [1]

1. Instituto Universitario de Investigación de Ingeniería del Agua y Medio Ambiente, Universitat Politècnica de València, Camí de Vera s/n, 46022 Valencia, Spain; iescuder@hma.upv.es
2. Green Blue Management, Avda. del Puerto 180 1B, 46023 Valencia, Spain; sara.perales@greenbluemanagement.com
3. iPresas, Avda. del Puerto 180 1B, 46023 Valencia, Spain; adrian.morales@ipresas.com
* Correspondence: igando@hma.upv.es; Tel.: +34-963-877-610

Received: 23 July 2018; Accepted: 27 August 2018; Published: 31 August 2018

Abstract: Green roofs are one specific type of sustainable urban drainage system (SUDS); they aim to manage runoff at the source by storing water in its different layers, delaying the hydrological response, and restoring evapotranspiration. Evidence of their performance in the Mediterranean is still scarce. The main objective of this paper is to analyse the hydrological performance of green roofs at building and city scales under Mediterranean conditions. A green roof and a conventional roof were monitored over one year in Benaguasil (Valencia, Spain). Rainfall and flow data were recorded and analysed. Hydrological models were calibrated and validated at the building scale to analyse the hydrological long-term efficiency of the green roof and compare it against that obtained for the conventional roof. Results show that green roofs can provide good hydrological performances, even in dry climates such as the Mediterranean. In addition, their influence at the city scale is also significant, given the average runoff coefficient reduction obtained.

Keywords: sustainable urban drainage systems (SUDS); green roof; hydrological efficiency; runoff reduction; city scale

1. Introduction

Sustainable urban drainage systems (SUDS) are an alternative approach to conventional urban stormwater management; they use and promote natural processes to mimic predevelopment hydrology in urbanised areas. SUDS mitigate urban flooding and water pollution [1], save energy, and provide multiple ecosystem services [2]. Hence, SUDS are part of the urban green infrastructure [3] that can be considered as an effective tool to face the challenges that our cities will have to deal with during the next decades regarding the effects of climate change on rainfall regimes and heat waves.

Green roofs are one specific type of SUDS consisting of areas of living vegetation; they are installed on the top of buildings, which promotes the reduction of surface water runoff quantity and pollution, and also provides ecosystem services such as aesthetic benefits, ecological-added value, and an enhancement of building performance [4,5]. Interest in green roofs has increased in recent years, as many more benefits have been reported beyond the hydrological ones: energy savings, thermal benefits, air pollution improvement, carbon sequestration, and psychological benefits [6–8]. Kuronuma et al. [9] demonstrated that green roofs contribute actively to carbon sequestration and energy savings, thus enhancing global warming mitigation. Luo et al. [10] investigated the thermal benefits of green roofs, and also demonstrated the air quality improvement in the green roof surroundings.

From the hydrological perspective, a green roof attenuates runoff hydrographs at the catchment source by providing rainfall retention and runoff detention [11]. Many experimental studies have been reported showing the hydrological performance of green roofs. The literature includes many references on the hydrological performance of both laboratory and full-scale roof installations [12,13]. Results from previous studies show that green roofs can achieve runoff volume reductions ranging from 0% to 100% [13], even though there is a general agreement on the potential of green roofs to effectively manage runoff at the source. Nevertheless, there is still very scarce evidence of their performance under semi-arid climatic conditions such as those in the eastern Mediterranean coastline of Spain [14]. Indeed, many authors have pointed out that the performance of green roofs largely varies with their hydroclimatic exposure [15], especially regarding the rainfall regime (frequency, rainfall volumes) and the soil moisture conditions. Antecedent dry weather periods between storms allow the green roof soil to dry, and consequently have more retention capacity in the next rainfall event. On the other hand, the green roof retention rate will decline with the increase in rainfall event volume, as well as with the increase in rainfall intensity.

Climate change threats on highly urbanised areas represent a great challenge for the coming decades; green infrastructure, including SUDS, represents a promising solution to mitigating these effects [16]. The European Commission recognises green infrastructure as a smart solution for today's challenges [17], providing strong climate change mitigation and adaptation benefits. Managing stormwater in cities through green infrastructure is one of the challenges included in the European Union's green infrastructure strategy, where green roofs are listed as one of the new engineering approaches to provide the above-mentioned benefits. In addition, green roof performance at the building scale is not fully representative of their effect at a higher scale; hence, one must consider the city scale to assess their efficiency within the whole system.

Modelling the hydrological performance of SUDS is a complex issue, given the number of processes involved. Several authors have successfully used the Hydrus-1D software for this purpose. Hilten et al. [18] highlighted the importance of calibrating soil parameters to achieve the good accuracy of the model; Palla et al. [19] achieved good results with Hydrus-1D compared to those obtained with a conceptual model; and Hakimdavar et al. [20] revealed the limitations of Hydrus-1D to reproduce the hydrological performance of a green roof. The Storm Water Management Model (SWMM) has also been widely used to model SUDS through its Low-Impact Development (LID) module [21]. Burszta-Adamiak and Mrowiec [22] reproduced the hydrological response of laboratory green roof pilots in SWMM with unsatisfactory simulation results; Zhang and Guo [23] compared the performance of the SWMM for the continuous simulation of a green roof with an analytical probabilistic model, and highlighted the lack of data for estimating the parameters of the LID module. Finally, Peng and Stovin [11] conducted a critical validation of the potential of the SWMM LID module for representing the hydrological performance of an extensive green roof in response to actual rainfall events. As emphasised by other authors, they revealed the need for calibration to obtain accurate modelling results. In addition, they pointed out the sensitivity of the green roof hydrological model to evapotranspiration.

The objective of this research is to compare the hydrological performance of a green roof and a conventional roof under Mediterranean climatic conditions at two different scales: the plot or building scale, and the city scale. A hydrological model at these two scales is set up using SWMM (v. 5.01.12). Calibration and validation of the model is carried out at the building scale using recorded data from both monitored roofs at the study site in Benaguasil (Valencia, Spain). Hence, the long-term hydrological performance of the green roof and the conventional roof is estimated by simulating a 17-year historical rainfall series. Upscaling at the city scale is analysed through a hypothetical urban area that is representative of the compact and dense cities in the region. Several ratios of green roofs against conventional roofs are analysed to assess the hydrological impact of this type of SUDS at the city scale.

2. Materials and Methods

2.1. Site Description

The study site is located in Benaguasil, which is a Mediterranean city 25 km inland from the city of Valencia, on the eastern Mediterranean Spanish coast. With a mild and semi-arid climate, Benaguasil has an average annual rainfall of 430 mm, with very strong seasonality. Similar to many Mediterranean cities, it experiences very high peak rainfall intensities that are concentrated in short intervals of time, which together with city characteristics such as large impermeability makes urban stormwater management difficult. The average temperature in Benaguasil is around 17 °C, (10 °C in January, and 25 °C in August). This climate regime differs significantly from that of the more northern and temperate climates where SUDS originated. In addition, the lack of experience justifies the need to monitor SUDS in the Mediterranean in order to provide evidence of their performance and tools for engineers and practitioners to properly design and manage the systems.

Nowadays, Benaguasil tackles three major issues in terms of urban water management: frequent pluvial flooding and backup flows from overloaded combined sewers, pollution of watercourses from combined sewer overflows (CSO), and high-energy consumption. Since the solution could not come only from a higher capacity of conveyance and treatment facilities, the Municipality of Benaguasil started to switch the paradigm, understanding that a more nature-based approach to face urban water management might be needed [24].

Within this context, the Municipality of Benaguasil retrofitted a 315 m^2 green roof in 2014 in a public building [25]. Experience in setting up green roofs in Spain is still scarce, especially when focussing on their filtrated water quality performance. To avoid previously observed problems [26], a mineral soil that was poor in nutrients and without brick debris was used. To preserve the drainage capacity of the soil and reduce the runoff colour and turbidity, volcanic gravel (40%) and silica sand (20%) were mixed in the substrate. The remaining 40% corresponded to compost substrate. Organic matter constituted only 13.3%, total nitrogen constituted 0.06%, and phosphorus constituted 0.04%. The nitrogen and phosphorus content fulfils the requirements adopted in the German guidelines for the planning, execution, and upkeep of green roofs [27]. Organic matter is slightly higher than the maximum amount recommended, but it is over half the amount used in previous experiences. With this composition, washing effects were expected to be reduced during the start-up period of the infrastructure. Figure 1 shows the retrofitted green roof and its different layers. The soil layer is 100-mm thick over a storage and drainage plastic layer (25-mm thick). Vegetation consists of a mixed sedum composed of 20% *album*, 18% *acre*, 34% *floriferum*, 17% *spurium*, 3% *rupestre*, 3% *sediform*, and 5% *sexangulare*. Sedums represent one of the most utilised species in green roof surfaces [28]. The previous non-retrofitted roof consists of a conventional inverted flat roof with a standard gravel layer. The inverted roof typology is characterised by a thermal insulation layer over the waterproofing membrane. The thermal insulation layer is protected by a geotextile, and finally, on the top, a 4–5 cm layer of gravel completes the conventional roof section.

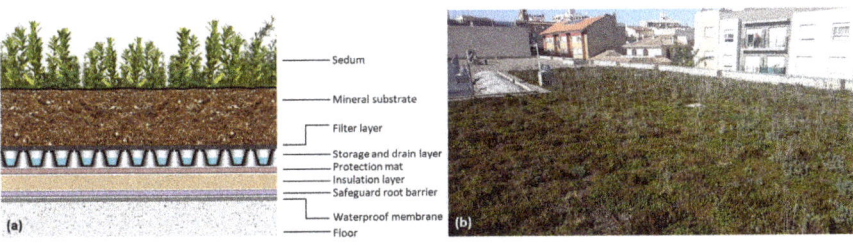

Figure 1. (a) Green roof layer configuration; (b) Green roof vegetative surface composed of a mixed sedum (September 2015).

2.2. Hydrometeorological Data Collection

The monitored period lasted from June 2014 to June 2015. A Detectronic rainfall gauge (0.2-mm accuracy) with a Bühler Montec datalogger was installed near the green roof. A continuous rainfall record was obtained with a two-minute time interval aggregation. For the same period, the flow rate through the downpipes of the green roof was monitored with a tipping bucket flow gauge. To compare the green roof performance against its previous cobbled conventional situation, flows generated by its nearby non-retrofitted conventional roof (with an area of 240 m^2) were also monitored. Hydrographs recorded during this period were used to calibrate and validate the hydrological model.

To analyse the long-term green roof efficiency, a long historical rainfall series is needed. For the analysis developed in this paper, high-resolution rainfall records for the period 1990–2006 in Valencia were available. Data corresponded to a five-minute resolution rain gauge located at the Júcar river basin hydrological service headquarters. Data verification and validation were performed before this study by daily comparison with the Spanish Meteorological Agency observations, as well as monthly and annual accumulation comparisons with nearby rain gauge stations [29].

Finally, the hydrological model requires evapotranspiration data to accurately represent the soil moisture content during the simulation period, and especially at the beginning of a rainfall episode. Evapotranspiration data were obtained from the Spanish Ministry of Agriculture through its agroclimatic information system for irrigation [30]. This system provides data for the potential evapotranspiration (ET_0) and the crop coefficient (K_C), whose product yields the real evapotranspiration (ET). Data corresponds to daily averaged values from data recorded since 1999 at the nearby Moncada station (IVIA, Valencian Institute for Agricultural Research); crop coefficient corresponds to pastures (Table 1). Although some authors have highlighted the sensitivity of green roof hydrological modelling to evapotranspiration [11], the influence of this variable is much more negligible under semi-arid conditions, provided that long dry inter-event periods are expected to occur, so the soil layer is likely to be dry at the beginning of each rainfall episode.

Table 1. Daily average evapotranspiration (ET) rates per month.

Month	ET_0 (mm/day)	K_C	ET (mm/day)
January	1.18	0.5	0.59
February	1.63	0.5	0.82
March	2.56	0.8	2.05
April	3.61	0.9	3.25
May	4.26	0.9	3.83
June	4.96	0.9	4.46
July	5.38	0.9	4.84
August	4.73	0.9	4.26
September	3.27	0.9	2.94
October	2.16	0.9	1.94
November	1.35	0.5	0.68
December	1.03	0.5	0.52

2.3. Hydrological Model

The Storm Water Management Model (SWMM v.5.1.012) [21] was used to model the hydrological response of the green roof and the adjacent conventional roof in Benaguasil. The latest releases of SWMM have implemented the LID (Low-Impact Development) module to simulate the hydrological performance of SUDS such as permeable pavements, bioretention areas, rain gardens, infiltration trenches, vegetative swales, green roofs, and rooftop disconnections. Many authors have shown the potential of the LID module to model SUDS performance [11,31–33].

Nevertheless, many difficulties arise when trying to calibrate and validate a hydrological model using the LID control module. Only focussing on the green roof SUDS type, 14 parameters are needed to characterise the three layers that define the infrastructure: surface (berm height, vegetation fraction, and surface roughness), soil (thickness, porosity, field capacity, wilting point, saturated conductivity, conductivity slope, and suction head) and drainage mat layer (thickness, void ratio, and roughness). This high number of parameters has a direct impact on the results' uncertainty when using this module. As concluded by Peng and Stovin [11], as many parameters are required, the model is not generic, and many uncertainties exist in estimating the values of the parameters. Some processes in the response of real green roofs are not well represented using the SWMM LID module: more robust retention and detention models are required. The authors concluded that the LID module in the SWMM could represent the hydrology of runoff from the green roof only after calibration.

Many factors may influence the detention modelling of the green roof, but the drainage layer parameters have been shown to influence the peak runoff the most, while the conductivity slope influences the smoothness of the runoff profile. Burszta-Adamiak and Mrowiec [22] also proved that the LID module model has limited capabilities in correctly simulating the hydrograph of storm water runoff from green roofs. The model tends to significantly overestimate the maximum flow rates, while its attempts to calibrate the model to the maximum flow rates lead to a significant underestimation of the generated volume of storm water runoff. Rosa et al. [32] demonstrated that uncalibrated SWMM simulations, using parameter values found in the literature and measured in the field, underestimated the runoff from an LID watershed.

These findings represent a barrier to the use of the LID module for planning and designing objectives under uncalibrated circumstances. A more robust model is desirable to minimise uncertainty. Thus, the aim of this work is to set up, calibrate, and validate a hydrological runoff model that is as simple as possible to minimise uncertainty, but robust enough to accurately represent the hydrological performance of the green and the conventional roofs for comparison. Although the Soil Conservation Service–Curve Number (SCS-CN) model [34] was not preliminarily developed for urban areas, it has been successfully used within this context [35,36]. The SWMM allows runoff production to be modelled by using the SCS-CN model. Table 2 shows the data and initial values of the parameters that were used to set up the model.

Table 2. Data and initial values of the hydrological model parameters. CN: curve number.

Parameter	Unit	Conventional Roof	Green Roof	Reference
Subcatchment area	ha	0.0240	0.0315	Building project
Subcatchment width	m	20	35	Building project
Slope	%	0.1	0.1	Building project
Surface roughness	s/m$^{1/3}$	0.10	0.15	[31,33]
Depression storage	mm	1	3	[21,31]
CN	-	98	80	[21,33]

2.4. Calibration and Validation of the Hydrological Model

The hydrological model is calibrated and validated for the green roof and the conventional roof configurations. The calibration and validation procedure consists of comparing the modelled and recorded hydrographs at the outlets of both roof downpipes. The results in SWMM are obtained with a 10-min time interval. The parameters to be calibrated and validated are those shown in Table 2, and initially estimated according to other references: surface roughness, depression storage, and curve number (CN).

To quantify the hydrological model goodness-of-fit, the Nash–Sutcliffe Efficiency index (NSE) is calculated to assess the model accuracy in replicating the recorded outflow hydrographs [37]. In addition, the total runoff volumes that were modelled and recorded per rainfall event are compared to assess the model performance in volumetric terms.

2.5. Continuous Simulation at Building and City Scales

Once the building scale models (green roof and conventional roof) are calibrated and validated, the historical long rainfall series is modelled to assess the long-term hydrological performance of both roofs. Thus, the green roof efficiency in terms of runoff reduction can be compared to the conventional roof. Runoff per rainfall event is extracted from the continuous simulation results, and the cumulative probability functions are inferred. Then, the percentiles of runoff generated by both roofs are finally deducted.

This paper is focussed on assessing the effect of green roofs at the city scale. For this purpose, a hypothetical urban area representative of the study site conditions is considered and modelled in the SWMM with the historical long rainfall series. Four land uses are considered to characterise the subcatchments: roads, conventional roofs, green roofs, and permeable areas (gardens and parks). To assess the effect of green roofs compared to conventional roofs, three scenarios are considered by varying the ratio of the subcatchment area covered by both types of roof. The percentages of roads and permeable areas are constant and representative of the region's dense towns and cities (Table 3). Scenario A represents a conventional land-use distribution in a densely urbanised area of the region. Scenarios B and C are defined to analyse feasible retrofitting strategies, from conventional roofs to green roofs. Scenario C represents the most ambitious but realistic retrofitting strategy. Scenario B is an intermediate stage between scenarios A and C to better assess the impact of moving from conventional roofs to green roofs. Previously calibrated parameters are used for roads and permeable areas [35], while the parameters calibrated and validated within this work are used for conventional and green roofs.

Table 3. Distribution of land uses per scenario in the urban area and hydrological parameters used. The calibrated values are justified in Section 3.

Land Use	Scenario A %	Scenario B %	Scenario C %	CN	Depression Storage (mm)
Road, paved area	40	40	40	98	1
Conventional roof	50	35	20	Calibrated	Calibrated
Green roof	0	15	30	Calibrated	Calibrated
Permeable areas	10	10	10	42	3

3. Results and Discussion

3.1. Hydrometeorological Data

Seventeen rainfall events were recorded during the monitored period (Table 4), totalling 381.7 mm. This annual total amount of rainfall is slightly lower than the annual average rainfall amount for Benaguasil, according to the Spanish meteorological agency. Rainfall volumes per event (V) highlighted the strong seasonality and irregularity of the Mediterranean climate; events from 3.2 mm to 125.2 mm occurred during this period. The average rainfall amount per event was 23.8 mm, and the standard deviation for the monitored period was 34.4 mm. Regarding the maximum 10-min intensities (I_{10}), maximum records were experienced in autumn (45.6 mm/h and 52.8 mm/h), as expected according to the probable convective events occurrence during that season. Nevertheless, a noticeable event (long and intense) occurred in March. The duration of each event (D) is also shown in Table 4.

The runoff volumes (R_{CR} for the conventional roof and R_{GR} for the green roof) and hydrological efficiencies (ratio between the detained volume and the rainfall volume; HE_{CR} for the conventional roof and HE_{GR} for the green roof) are presented for each event. All of the recorded events produced runoff in the conventional roof (R_{CR}). The hydrological efficiency of the conventional roof (HE_{CR}) ranged from 1% for the greater events to 75% for the lower ones. Regarding the green roof, five events out of 17 (29.4%) did not produce runoff (R_{GR} = 0 mm; HE_{GR} = 100% hydrological efficiency). For the events that produced runoff, the hydrological efficiency of the green roof ranged from 53% to 93%.

The average values of hydrological efficiency during the whole monitored year were 13% for the conventional roof and 65% for the green roof.

Table 4. Summary of hydrometeorological data recorded during the monitoring period, runoff volumes, and hydrological efficiencies. Events 03, 11, 12, 13, and 14 did not produce runoff in the green roof. V: volumes per event, D: duration of each event, I_{10}: maximum 10-min intensities, R_{CR}: runoff for the conventional roof, R_{GR}: runoff for the green roof, HE_{CR}: hydrological efficiencies measured as the ratio between the detained volume and the rainfall volume for the conventional roof, HE_{GR}: hydrological efficiencies for the green roof.

Event	Date	Rainfall			Conv. Roof		Green Roof	
		V (mm)	D (hh:mm)	I_{10} (mm/h)	R_{CR} (mm)	HE_{CR} (%)	R_{GR} (mm)	HE_{GR} (%)
01	24 June 2014	14.2	14:50	26.4	7.7	46%	6.0	58%
02	02 July 2014	17.6	27:20	30.0	14.0	20%	8.3	53%
03	07 September 2014	3.2	1:40	6.0	0.8	75%	0	100%
04	22 September 2014	23	1:10	45.6	18.2	21%	7.1	69%
05	29 September 2014	6.2	6:50	2.4	3.4	45%	1.5	76%
06	12 October 2014	6.2	0:30	20.4	3.5	44%	0.6	90%
07	04 November 2014	9.4	1:50	3.6	6.6	30%	0.7	93%
08	11 November 2014	10.8	4:40	4.8	7.8	28%	1.0	91%
09	27 November 2014	89	65:20	52.8	88.2	1%	38.3	57%
10	14 December 2014	27.8	23:20	15.6	26.3	5%	11.4	59%
11	18 January 2015	4.6	13:40	2.4	2.6	43%	0	100%
12	30 January 2015	3.6	11:40	2.4	1.4	61%	0	100%
13	12 February 2015	3.4	7:20	2.4	1.3	62%	0	100%
14	17 February 2015	4.4	5:10	4.8	2.2	50%	0	100%
15	18 March 2015	125.2	174:40	45.6	119.0	5%	55.8	55%
16	19 May 2015	14.4	8:50	7.2	14.1	2%	1.4	90%
17	13 June 2015	32.6	67:50	38.4	26.0	20%	5.5	83%

3.2. Calibration and Validation Results

The calibration and validation of the parameters of the hydrological model in the SWMM have focussed on surface roughness, depression storage, and curve number (CN). The objective functions were the volume error (%) per event (difference between the observed and modelled runoff volume), and the NSE per event, which considers the intra-event progression, especially focussing on the occurrence of peak flows.

Table 5 shows the final values for the calibrated and validated parameters. The conventional roof model finally consisted of an impervious surface (CN = 100) where only depression storage (1.5 mm) affected the rainfall–runoff transformation. Surface propagation kinetics were well performed with a 0.12 roughness coefficient value. The green roof model consists, as expected, of a pervious surface with CN = 85 and no depression storage. This may seem counterintuitive; nevertheless, rainfall losses are accounted for through the infiltration mechanism, as described by the SCS method.

Table 5. Calibrated parameters for the conventional roof and the green roof.

Parameter	Unit	Conventional Roof	Green Roof
Surface roughness	$s/m^{1/3}$	0.12	0.35
Depression storage	mm	1.5	0
CN	-	100	85

The calibration data set was composed of 11 events (65%) whereas the other six events (35%) were used for validation. The criteria to select calibration and validation events consisted of having, in both sets, a representative sample of all of the events recorded according to their duration, rainfall volume, and maximum intensity. Figure 2a,b show the results for the calibration event recorded on 13 June 2015. Figure 2c,d show the results for the validation event recorded on 14 December 2014. The model reproduces the runoff kinetics in both roofs well, although peak flows were not always quantitatively reproduced well.

Sustainability 2018, 10, 3105

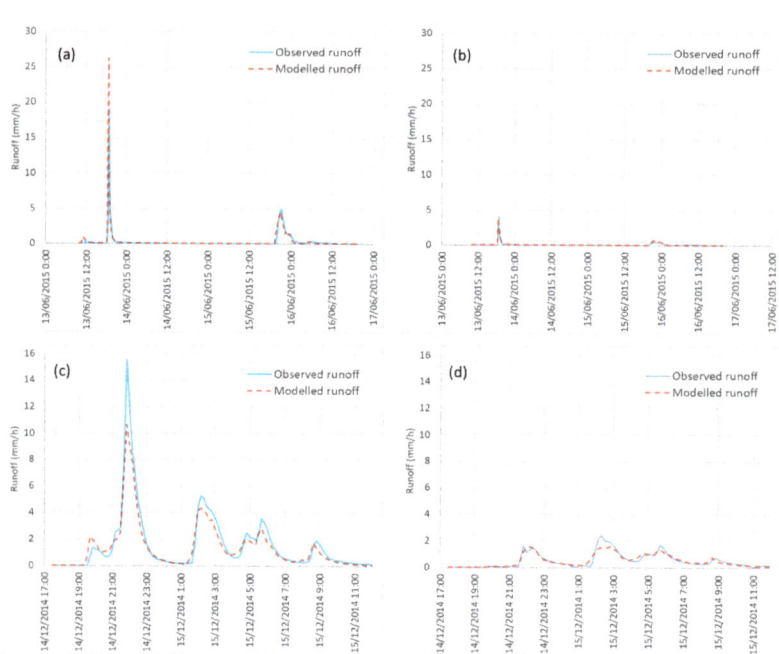

Figure 2. Examples of calibration and validation events. (**a**) Conventional roof calibration event (13 June 2015); (**b**) Green roof calibration event (13 June 2015); (**c**) Conventional roof validation event (14 December 2014); (**d**) Green roof validation event (14 December 2014).

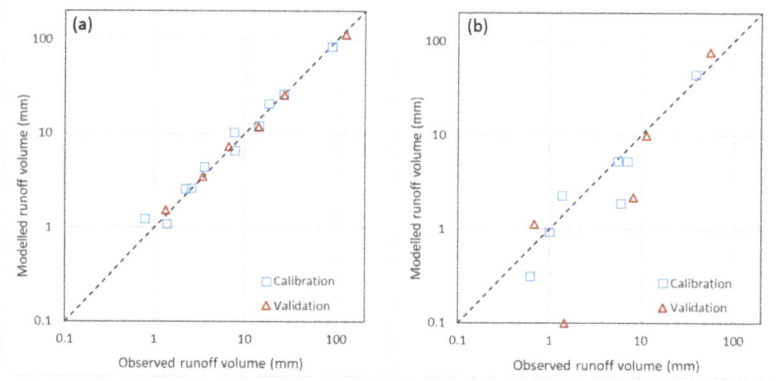

Figure 3. Scatter plots of modelled runoff volumes vs. observed runoff volumes. (**a**) Conventional roof; (**b**) Green roof.

Table 6 shows the complete summary of the model calibration and validation results. Events used for calibration (C) and validation (V) are indicated. Observed (Obs.) and modelled (Mod.) runoff volumes are compared through the volume errors. The Nash–Sutcliffe Efficiency indexes (NSE) are also calculated for each event.

Regarding the conventional roof model, volume errors ranged from −20.1% to 58.1%. The latter corresponds to the smallest event recorded (3.2 mm of rainfall and 0.8 mm of runoff), so a high uncertainty is expected to occur. If we focus on the next event, the maximum volume error falls to 35.4%. The overall volume error for the calibration set is 0.4%, whereas for the validation set, it is −8.0%. For the calibration set, the NSE indexes ranged from negative values (small events) to 0.93 (the average NSE for the calibration set is 0.66); for the validation set, the results were better, and the average NSE is 0.89.

The green roof model performs poorer. Events 3, 11, 12, 13, and 14 are rejected for direct calibration and validation, as they did not produce runoff. They are only considered in the preliminary screening of a feasible range of parameters. Volume errors ranged from −68.5% to 66.4%. The overall volume error for the calibration set was 1.3%, whereas for the validation set, it was −16.5%. For the calibration set, the NSE indexes ranged from negative values (small events) to 0.90. Nevertheless, the average NSE for the calibration set fell to 0.36; for the validation set, better indexes reached 0.91, whereas the average NSE was 0.31. In this case, the differences were due to a weakness in the green roof hydrological model. As highlighted before, evapotranspiration can strongly affect the green roof hydrological response, whereas our ET data comes from the nearby Moncada weather station. Further research must focus on a better estimation of variables affecting evapotranspiration, and consequently soil moisture.

Figure 3 compares the modelled runoff volumes against the observed ones for the conventional roof and the green roof. The Pearson correlation coefficient is 0.99 for the conventional roof (calibration and validation) and 0.98 (calibration) and 0.97 (validation) for the green roof.

Table 6. Calibration (C) and Validation (V) results. NSE: Nash–Sutcliffe Efficiency index. Obs.: observed, Mod.: modelled.

Event	C/V	Conventional Roof				Green Roof			
		Obs. Runoff (mm)	Mod. Runoff (mm)	Vol. Error (%)	NSE	Obs. Runoff (mm)	Mod. Runoff (mm)	Vol. Error (%)	NSE
01	C	7.7	10.4	35.2%	−1.83	6.0	1.9	−68.5%	0.24
02	V	14.0	11.7	−16.8%	0.95	8.3	2.2	−73.6%	0.45
03	C	0.8	1.2	58.1%	−19.46	–	–	–	–
04	C	18.2	21.0	15.1%	0.80	7.1	5.3	−24.9%	0.90
05	V	3.4	3.5	1.2%	0.14	1.5	0.1	−94.0%	−0.34
06	C	3.5	4.4	24.5%	0.78	0.6	0.3	−49.1%	0.65
07	V	6.6	7.3	9.8%	0.87	0.7	1.1	66.5%	0.15
08	C	7.8	6.6	−15.0%	0.58	1.0	0.9	−6.7%	−0.88
09	C	88.2	84.9	−3.8%	0.65	38.3	44.6	16.7%	0.25
10	V	26.3	25.9	−1.6%	0.93	11.4	9.9	−12.7%	0.91
11	C	2.6	2.6	3.0%	0.67	–	–	–	–
12	C	1.4	1.1	−20.1%	0.84	–	–	–	–
13	V	1.3	1.5	13.2%	−5.58	–	–	–	–
14	C	2.2	2.6	16.2%	−0.42	–	–	–	–
15	V	119.0	112.7	−9.8%	0.88	55.8	77.1	38.1%	0.27
16	C	14.1	12.1	−14.7%	0.93	1.4	2.3	66.4%	−0.05
17	C	26.0	26.4	1.5%	0.58	5.5	5.3	−3.9%	0.88
	Calibration set			0.4%	0.66			1.3%	0.36
	Validation set			−8.0%	0.89			16.5%	0.31

3.3. Comparison of Green Roof and Conventional Roof Hydrological Efficiency at the Building Scale

The long-term green roof efficiency is analysed by simulating the 1990–2006 historical rainfall series. According to Andrés-Doménech et al. [35], 464 events occurred within this period, which yields an average of 27.3 events/year. The percentiles of rainfall volumes are represented in Figure 3. The 80% rainfall volume percentile is 19.2 mm, whereas the 90% rainfall volume percentile is 32.8 mm. These percentiles are required in many SUDS design procedures [5,38].

The calibrated and validated SWMM for the conventional and the green roof are simulated with this long rainfall series. In total, 398 events produced runoff in the conventional roof (23.4 events/year), whereas 295 events produced runoff in the green roof (17.4 events/year). In terms of occurrence, at the building scale, 36% of rainfall events were completely detained (no runoff occurs) by the green roof, whereas the conventional roof only fully detained 14% of the registered events. The cumulative rainfall amount during the period 1990–2006 was 6505 mm. Regarding the long-term hydrological efficiency, the cumulative runoff from the conventional roof was 5300 mm (HE_{CR} = 18.5%), whereas the cumulative runoff from the green roof fell to 2136 mm (HE_{CR} = 67.1%). These values are similar to those registered during the monitored period. The average runoff coefficients (total runoff over total rainfall) were 0.81 for the conventional roof and 0.33 for the green roof. These percentages are of paramount interest for design purposes within the region.

Figure 4 shows the cumulative distributions of rainfall volumes, conventional roof runoff volumes, and green roof runoff volumes. The green roof completely smoothens the rainfall regime at the building scale. The 80% percentile of the rainfall regime fell to 3.9 mm (−79.7%), whereas the 90% percentile of the rainfall regime fell to 11.7 mm (−64.3%). These variations for the conventional roof were −11.8% and −15.6%, respectively. The results demonstrate the high potential of green roofs to manage runoff at the source from a hydrological perspective.

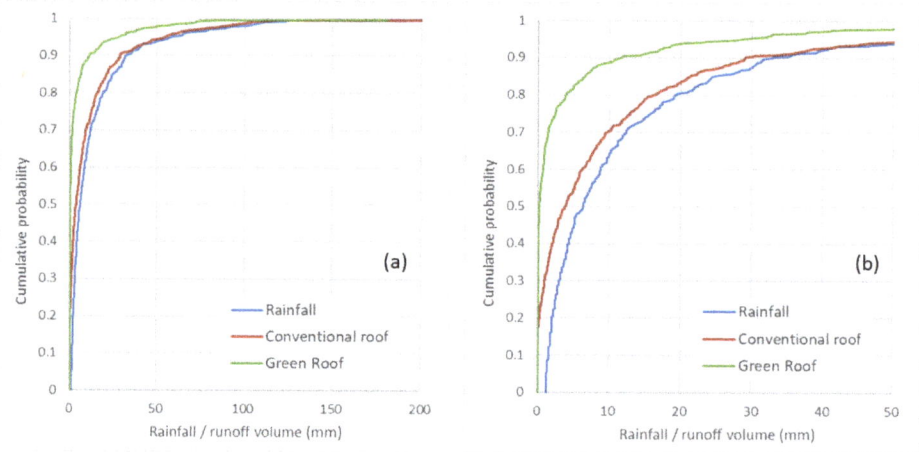

Figure 4. Comparison of runoff cumulative probabilities for the conventional roof and the green roof at the building scale. (**a**) Complete runoff range (0–200 mm); (**b**) Frequent runoff range (0–50 mm).

3.4. Effect of Green Roofs on Hydrological Efficiency at the City Scale

Scenario A represents the baseline to compare the effect of green roofs on the hydrological efficiency at the city scale. Figure 5 shows the cumulative distributions of rainfall volumes and runoffs for scenarios A, B, and C. As expected, the influence of green roofs at the city scale is much more moderate than at the building scale. Nevertheless, their contribution to reducing the magnitude and frequency of runoff volumes is noticeable: the 80% runoff percentile for scenario A is 13.2 mm, whereas it is 11.2 mm (−15%) for scenario B and 9.4 mm (−29%) for scenario C; the 90% runoff percentile for scenario A is 24.7 mm, whereas it is 22.2 mm (−10%) for scenario B and 19.5 mm (−21%) for scenario C. The contribution of green roofs to reducing the 90% runoff percentile is relevant. As stated before, many SUDS manuals recommend managing the 90% rainfall percentile to adequately manage water quality through the SUDS treatment train. As a source technique, green roofs only represent the upstream side of the treatment train (together with permeable pavements, rainwater harvesting, etc.); hence, combination with other types of SUDS to complete the system is needed to achieve the objective of completely managing the 90% rainfall percentile.

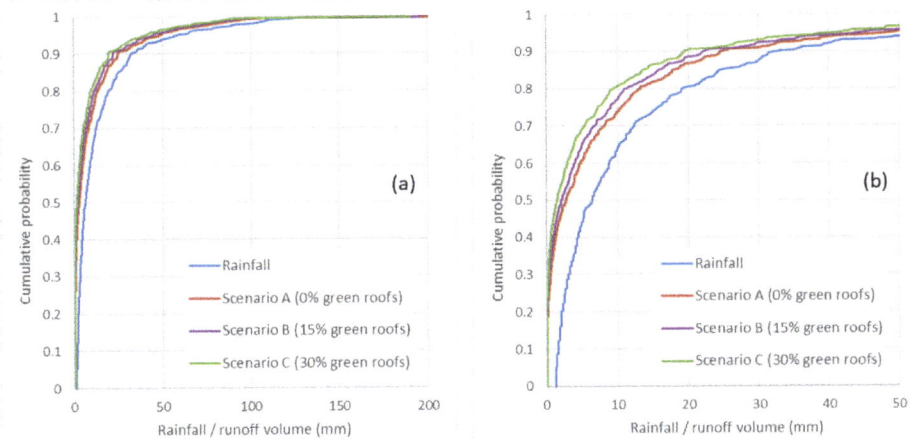

Figure 5. Comparison of runoff non-exceedance probabilities for scenarios A, B, and C at the city scale. (**a**) Complete runoff range (0–200 mm); (**b**) Frequent runoff range (0–50 mm).

In terms of hydrological performance, the cumulative runoff volume within the period 1990–2016 is 4478 mm for scenario A, 4012 mm for scenario B, and 3511 mm for scenario C. Thus, the average runoff coefficients (total runoff over total rainfall) for the city scale are, respectively, 0.69, 0.62 (−10%) and 0.54 (−22%). These results demonstrate the high potential of this type of SUDS to manage runoff at the source.

Figure 6 shows the averaged runoff coefficient for scenarios A, B, and C. The results show that the runoff production reduction is not constant for the complete rainfall range. In addition, the graph shows a change in the trend curves of around 15–20 mm, underlining the effective range of rainfall that can be controlled by the green roof is under this threshold. To better analyse this point, Figure 7 represents the evolution of runoff coefficients for scenarios B and C divided by the corresponding runoff coefficient for scenario A (which is considered the baseline for the comparison). The graph highlights the range where green roofs represent a high contribution for runoff control at the source. For small rainfall events (1–10 mm), runoff coefficients for scenario B are 70–80% of those for scenario A, whereas for scenario C, they are 40–60% of those for scenario A. These differences are the consequence of considering different ratios of green roof surface in each scenario. Nevertheless, the runoff coefficients are more and more similar as rainfall volumes increase. For rainfall events greater than 15–20 mm, the runoff coefficients evolution begins to stabilise, and both scenarios tend show a similar hydrological behaviour. As expected for larger rainfall volumes, runoff coefficients tend to 1, and the differences between scenarios disappear. This evolution was expected to occur as for high rainfall volumes, the hydrological effect of green roofs at the city scale becomes less important: the retention capacity of the green roof is exceeded, and consequently, the hydrological response of the urban catchment increases the runoff rates.

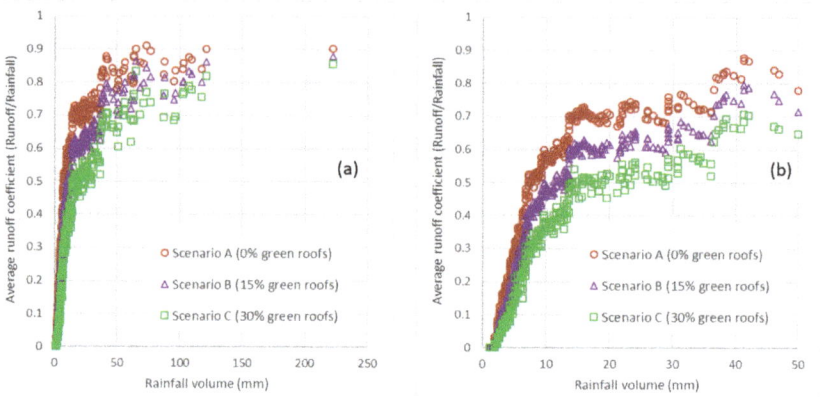

Figure 6. Comparison of average runoff coefficients for scenarios A, B, and C at the city scale. (**a**) Complete rainfall range (0–250 mm); (**b**) Frequent rainfall range (0–50 mm).

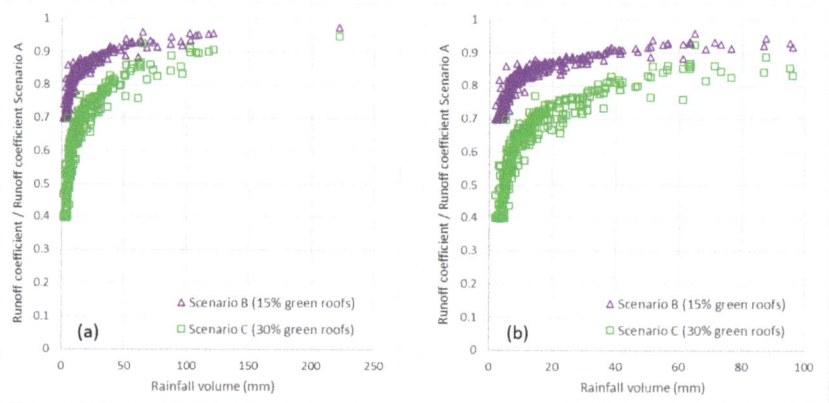

Figure 7. Relative average runoff coefficients for scenarios B and C compared to the average runoff coefficient for scenario A (baseline) at the city scale. (**a**) Complete rainfall range (0–250 mm); (**b**) Frequent rainfall range (0–50 mm).

4. Conclusions

Green roofs have been demonstrated to be an effective type of SUDS for managing runoff at the source. Nevertheless, evidence of their hydrological performance is still scarce in the Mediterranean. The results achieved after one year of monitoring in Benaguasil (Spain) are inspiring, provided that high hydrological efficiencies are also obtained, such as those widely reported for more wet climates.

The hydrological modelling of green roofs is complex; the high number of parameters needed to represent flows and storage through the different layers usually makes these highly physically-based models less robust. The calibration and validation of the SCS-CN model to the recorded data provide good results in terms of volumetric response and long-period performance. Nevertheless, the model does not achieve good performances regarding the peak flow representation. Despite this loss of precision, one of the challenges of this research was to prove whether a simple hydrological model was able to adequately represent the green roof response. The results demonstrate a good performance and consequently, the potential of using simple models when calibration is unfeasible.

The calibrated parameters herein could be used in new green roof projects under similar climatic and design conditions. Further research must focus on a better estimation of variables affecting evapotranspiration, and consequently, soil moisture.

The impact of a green roof at the building scale in the hydrological response is positive. The long-term modelling reinforces the results obtained during the monitored period: the hydrological efficiency of the green roof is high, and the cumulative probabilities of runoff volumes are significantly reduced regarding those produced by a conventional roof. The analysis at the city scale also provides promising results. Results show that the effective range of rainfall that can be controlled by the green roof is around 15–20 mm, which corresponds with the most frequent rainfall events. The average runoff coefficients at the city scale under Mediterranean conditions can be considerably reduced by considering the feasible ratios of green roofs. If half of the current conventional roofs were retrofitted to green roofs, runoff coefficients would be reduced to below 75% of the current ones for the frequent rainfall events. These hydrological benefits, when added to other reported benefits of green roofs, make this type of SUDS a promising solution to face urban challenges caused by climate threats.

Author Contributions: I.A.-D. and S.P.-M. conceived and designed the experiments at the study site; A.M.-T. supervised the study site during the monitoring period and analysed the recorded data; I.A.-D. and S.P.-M. developed the hydrological models. I.E.-B. and I.A.-D. analysed and discussed the results; the four authors wrote the paper.

Funding: This research was funded by the European Regional Development Fund (ERDF) grant number 1C-MED12-14 and by the Spanish Ministry of Economy and Competitiveness with funds from the State General Budget (PGE) and ERDF, grant number BIA2015-65240-C2-2-R. The APC was funded by the Spanish Ministry of Economy and Competitiveness with funds from the State General Budget (PGE) and ERDF, grant number BIA2015-65240-C2-2-R.

Acknowledgments: This research has been conducted as part of the MED program project E^{2}STORMED (Ref. 1C-MED12-14), supported by European Regional Development Fund (ERDF) funding of the European Union and the research project SUPRIS-SUPeI (Ref. BIA2015-65240-C2-2-R MINECO/FEDER, UE), financed by the Spanish Ministry of Economy and Competitiveness with funds from the State General Budget (PGE) and ERDF. Authors also acknowledge the Municipality of Benaguasil for their support during the monitoring period.

Conflicts of Interest: The authors declare no conflict of interest.

References

1. Novotny, V.; Ahern, J.; Brown, P. *Water Centric Sustainable Communities: Planning, Retrofitting, and Building the Next Urban Environment*; John Wiley & Sons, Inc.: Hoboken, NJ, USA, 2010; ISBN 978-0-470-47608-6.
2. Charlesworth, S.M. A Review of the Adaptation and Mitigation of Global Climate Change Using Sustainable Drainage in Cities. *J. Water Clim. Chang.* **2010**, *1*, 165–180. [CrossRef]
3. Xu, M.; Weissburg, M.; Newell, J.P.; Crittenden, J.C. Developing a Science of Infrastructure Ecology for Sustainable Urban Systems. *Environ. Sci. Technol.* **2012**, *46*, 7928–7929. [CrossRef] [PubMed]
4. Rowe, D.B. Green Roofs as a Means of Pollution Abatement. *Environ. Pollut.* **2011**, *159*, 2100–2110. [CrossRef] [PubMed]
5. Woods-Ballard, P.; Wilson, S.; Udale-Clarke, H.; Illman, S.; Scott, T.; Ashley, R.; Kellagher, R. *The Suds Manual*; CIRIA: London, UK, 2015; ISBN 978-0-86017-760-9.
6. Lundholm, J.T.; Williams, N.S.G. Effects of Vegetation on Green Roof Ecosystem Services. In *Green Roof Ecosystems*; Sutton, R.K., Ed.; Springer: Cham, Switzerland, 2015; pp. 211–232. ISBN 978-3-319-14983-7.
7. Morales-Torres, A.; Escuder-Bueno, I.; Andrés-Doménech, I.; Perales-Momparler, S. Decision Support Tool for Energy-efficient, Sustainable and Integrated Urban Stormwater Management. *Environ. Model. Softw.* **2016**, *84*, 518–528. [CrossRef]
8. Mayrand, F.; Clergeau, P. Green Roofs and Green Walls for Biodiversity Conservation: A Contribution to Urban Connectivity? *Sustainability* **2018**, *10*, 985. [CrossRef]
9. Kuronuma, T.; Watanabe, H.; Ishihara, T.; Kou, D.; Toushima, K.; Ando, M.; Shindo, S. CO_2 Payoff of Extensive Green Roofs with Different Vegetation Species. *Sustainability* **2018**, *10*, 2256. [CrossRef]
10. Luo, H.; Wang, N.; Chen, J.; Ye, X.; Sun, Y.F. Study on the Thermal Effects and Air Quality Improvement of Green Roof. *Sustainability* **2015**, *7*, 2804–2817. [CrossRef]

11. Peng, Z.; Stovin, V. Independent Validation of the SWMM Green Roof Module. *J. Hydrol. Eng.* **2017**, *22*. Available online: https://ascelibrary.org/doi/10.1061/%28ASCE%29HE.1943-5584.0001558 (accessed on 29 August 2018). [CrossRef]
12. Palla, A.; Gnecco, I.; Lanza, L.G. Hydrologic Restoration in the Urban Environment Using Green Roofs. *Water* **2010**, *2*, 140–154. [CrossRef]
13. Stovin, V.; Vesuviano, G.; Kasmin, H. The Hydrological Performance of a Green Roof Test Bed under UK Climatic Conditions. *J. Hydrol.* **2012**, *414–415*, 148–161. [CrossRef]
14. Perales-Momparler, S.; Andrés-Domenech, I.; Hernández-Crespo, C.; Vallés Morán, F.; Martín, M.; Escuder-Bueno, I.; Andreu, J. The Role of Monitoring Sustainable Drainage Systems for Promoting Transition towards Regenerative Urban Built Environments: A Case Study in the Valencian Region, Spain. *J. Clean. Prod.* **2017**, *163*, S113–S124. [CrossRef]
15. Akther, M.; He, J.; Chu, A.; Huang, J.; van Duin, B. A Review of Green Roof Applications for Managing Urban Stormwater in Different Climatic Zones. *Sustainability* **2018**, *10*, 2864. [CrossRef]
16. Shafique, M.; Kim, R.; Kyung-Ho, K. Green Roof for Stormwater Management in a Highly Urbanized Area: The Case of Seoul, Korea. *Sustainability* **2018**, *10*, 584. [CrossRef]
17. European Commission. *Building a Green Infrastructure for Europe*; Publications Office of the European Union: Luxembourg, 2013; Volume 24, ISBN 978-92-79-33428-3. [CrossRef]
18. Hilten, R.N.; Lawrence, T.M.; Tollner, E.W. Modeling Stormwater Runoff from Green Roofs with HYDRUS-1D. *J. Hydrol.* **2008**, *358*, 288–293. [CrossRef]
19. Palla, A.; Genaco, I.; Lanza, L. Compared Performance of a Conceptual and a Mechanistic Hydrologic Models of a Green Roof. *Hydrol. Process.* **2012**, *26*, 73–84. [CrossRef]
20. Hakimdavar, R.; Culligan, P.; Finazzi, M.; Barontini, S.; Ranzi, R. Scale Dynamics of Extensive Green Roofs: Quantifying the Effect of Drainage Area and Rainfall Characteristics on Observed and Modeled Green Roof Hydrologic Performance. *Ecol. Eng.* **2014**, *73*, 494–508. [CrossRef]
21. Rossman, L.A. *Storm Water Management Model*, version 5.1.011; National Risk Management Research Laboratory, U.S. Environmental Protection Agency: Cincinnati, OH, USA, 2016.
22. Burszta-Adamiak, E.; Mrowiec, M. Modelling of Green Roofs' Hydrologic Performance Using EPA's SWMM. *Water Sci. Technol.* **2013**, *68*, 36–42. [CrossRef] [PubMed]
23. Zhang, S.; Guo, Y. An Analytical Probabilistic Model for Evaluating the Hydrologic Performance of Green Roofs. *J. Hydrol. Eng.* **2013**, *18*, 19–28. [CrossRef]
24. Perales-Momparler, S.; Andrés-Doménech, I.; Andreu, J.; Escuder-Bueno, I. A Regenerative Urban Stormwater Management Methodology: The Journey of a Mediterranean City. *J. Clean. Prod.* **2015**, *109*, 174–189. [CrossRef]
25. Andrés-Valeri, V.C.; Perales-Momparler, S.; Sañudo-Fontaneda, L.A.; Andrés-Doménech, I.; Castro-Fresno, D.; Escuder-Bueno, I. SuDS in Spain: Case Studies. In *Sustainable Surface Water Management. A Handbook for SUDS*, 1st ed.; Charlesworth, S., Booth, C.A., Eds.; Wiley Blacckwell: Boston, UK, 2016; pp. 355–369. ISBN 9781118897706.
26. Perales-Momparler, S.; Hernández-Crespo, C.; Vallés-Morán, F.; Martín, M.; Andrés-Doménech, I.; Andreu, J.; Jefferies, C. SuDS Efficiency during the Start-Up Period under Mediterranean Climatic Conditions. *Clean-Soil Air Water* **2014**, *42*, 178–186. [CrossRef]
27. Büttner, T.; Rohrbach, J.; Schulze-Ardey, C. *Guidelines for the Planning, Execution and Upkeep of Green-Roof Sites*; FLL, Forschungsgesellschaft Landschaftsentwicklung Landschaftsbau e. V.: Bonn, Germany, 2002; Available online: http://www.greenrooftechnology.com/fll-green-roof-guideline (accessed on 30 August 2018).
28. Morales, J.A.; Cristancho, M.A.; Baquero-Rodríguez, G.A. Trends in the Design, Construction and Operation of Green Roofs to Improve the Rainwater Quality. State of the Art. *Ingeniería del agua* **2017**, *21*, 179–196.
29. Andres-Doménech, I.; Montanari, A.; Marco, J.B. Efficiency of Storm Detention Tanks for Urban Drainage Systems under Climate Variability. *J. Water Resour. Plann. Manag.* **2012**, *138*, 36–46.
30. Sistema de información agroclimática para el riego (Irrigation Advisory Service). Necesidades netas. Available online: Eportal.magrama.gob.es/websiar/NecesidadesHidricas.aspx (accessed on 10 April 2018).
31. Cipolla, S.S.; Maglionico, M.; Stojkov, I. A Long-Term Hydrological Modelling of an Extensive Green Roof by Means of SWMM. *Ecol. Eng.* **2016**, *95*, 876–887. [CrossRef]
32. Rosa, D.J.; Clausen, J.C.; Dietz, M.E. Calibration and Verification of SWMM for Low Impact Development. *J. Am. Water Resour. Assoc.* **2015**, *51*, 746–757. [CrossRef]

33. Palla, A.; Gnecco, I. Hydrologic Modeling of Low Impact Development Systems at the Urban Catchment Scale. *J. Hydrol.* **2015**, *528*, 361–368. [CrossRef]
34. SCS. Hydrology, Soil Conservation Service. In *National Engineering Handbook*; USDA: Washington, DC, USA, 1971; Section 4.
35. Andrés-Doménech, I.; Montanari, A.; Marco, J.B. Stochastic Rainfall Analysis for Storm Tank Performance Evaluation. *Hydrol. Earth Syst. Sci.* **2010**, *14*, 1221–1232. [CrossRef]
36. Andrés-Doménech, I.; Hernández-Crespo, C.; Martín, M.; Andrés-Valeri, V.C. Characterization of Wash-off from Urban Impervious Surfaces and SuDS Design Criteria for Source Control under Semi-arid Conditions. *Sci. Total Environ.* **2018**, *612*, 1320–1328. [CrossRef] [PubMed]
37. Nash, J.E.; Sutcliffe, J.V. River flow forecasting through conceptual models part I—A Discussion of Principles. *J. Hydrol.* **1970**, *10*, 282–290. [CrossRef]
38. Hirschman, D.J.; Kosco, J. *Managing Stormwater in Your Community. A Guide for Building an Effective Post-Construction Programme*; Center for Watershed Protection, EPA: Washington, DC, USA, 2008.

© 2018 by the authors. Licensee MDPI, Basel, Switzerland. This article is an open access article distributed under the terms and conditions of the Creative Commons Attribution (CC BY) license (http://creativecommons.org/licenses/by/4.0/).

Article

Assessing the Runoff Reduction Potential of Highway Swales and WinSLAMM as a Predictive Tool

Bailee N. Young, Jon M. Hathaway *, Whitney A. Lisenbee and Qiang He

Department of Civil and Environmental Engineering, The University of Tennessee, Knoxville, TN 37996, USA; byoung24@vols.utk.edu (B.N.Y.); wlisenbe@vols.utk.edu (W.A.L.); qianghe@utk.edu (Q.H.)
* Correspondence: hathaway@utk.edu; Tel.: +1-865-974-6058

Received: 21 June 2018; Accepted: 8 August 2018; Published: 13 August 2018

Abstract: Across the United States, the impacts of stormwater runoff are being managed through the National Pollutant Discharge Elimination System (NPDES) in an effort to restore and/or maintain the quality of surface waters. State transportation authorities fall within this regulatory framework, being tasked with managing runoff leaving their impervious surfaces. Opportunely, the highway environment also has substantial amounts of green space that may be leveraged for this purpose. However, there are questions as to how much runoff reduction is provided by these spaces, a question that may have a dramatic impact on stormwater management strategies across the country. A highway median swale, located on Asheville Highway, Knoxville, Tennessee, was monitored for hydrology over an 11-month period. The total catchment was 0.64 ha, with 0.26 ha of roadway draining to 0.38 ha of a vegetated median. The results of this study indicated that 87.2% of runoff volume was sequestered by the swale. The Source Loading and Management Model for Windows (WinSLAMM) was used to model the swale runoff reduction performance to determine how well this model may perform in such an application. To calibrate the model, adjustments were made to measured on-site infiltration rates, which was identified as a sensitive parameter in the model that also had substantial measurement uncertainty in the field. The calibrated model performed reasonably with a Nash Sutcliffe Efficiency of 0.46. WinSLAMM proved to be a beneficial resource to assess green space performance; however, the sensitivity of the infiltration parameter suggests that field measurements of this characteristic may be needed to achieve accurate results.

Keywords: stormwater; WinSLAMM; grassed swale; transportation, SCM; runoff, highway

1. Introduction

Increases in imperviousness lead to higher peak flow rates and total runoff volume from watersheds [1], with detrimental effects to stream stability and ecology. One notable source of imperviousness in watersheds is the transportation system. State highway systems are required to operate under municipal separate storm sewer system (MS4) requirements since large amounts of stormwater runoff are transferred from roads to surface waters, carrying with it a range of pollutants associated with vehicle tires, brakes, engine wear, and lubricating fluids [2]. Increasingly, stormwater management techniques have shifted toward green infrastructure applications where runoff reduction is the targeted outcome and ultimately leads to decreased pollutant export.

To achieve post-construction stormwater goals, state transportation authorities are in need of Stormwater Control Measures (SCMs) that can both achieve stormwater management goals and are applicable to the highway environment. Although originally designed primarily as a stormwater conveyance, studies suggest that the vegetated (grassed) swale is one SCM that may have these desired characteristics. Grass swales convey water, yet simultaneously promote infiltration and decreased stormwater velocity [3]. Volume reduction is achieved due to this infiltration, occurring both laterally over the swale side slope and longitudinally along the swale pathway, and through storage in soil pore space (Weiss et al., 2010).

Although a number of field studies have examined the water quality performance of swales, far fewer studies have quantified hydrologic benefits. One example is Lucke et al. [4], which observed the responses of four field swales handling 24 standardized synthetic runoff events. This study found that the swales performed well at attenuating flow, with a mean total flow reduction of 52% in 30 m long swales and a peak flow reduction of 61%. Other authors have reported volume reduction ranging from 30 to 50% and peak flow reductions between 10 and 20% [5–8].

Parameters impacting volume reduction include the duration and depth of rain events and the available storage and length of the swale [5]. During small storm events, complete or high runoff volume reduction is possible; but, during large storm events, soil saturation causes volume reduction to be small and at times, negligible [5,8]. Multiple studies have confirmed this occurrence and suggested the utility of these systems for capturing small events. For instance, Davis et al. [5] defined a complete capture depth for swales ranging from 0.4 to 2.2 cm, with a study from Kaighn and Yu [9] also falling within this range for two 30 m swales (0.5 to 0.7 cm). Yu et al. [10] showed a slightly higher complete capture depth of 1.27 cm for a 274.5 m swale with two check dams. Some of this variability is likely attributed to differences in soil composition, most notably soil infiltration, amongst the sites.

Ahmed et al. [11] took a total of 722 infiltration measurements across six grassed swales and showed high variability in infiltration rates (0.75–15.5 cm/h). High variability in infiltration rates was even apparent in measurements across individual swales, with uncertainties in the geometric mean as high as a factor of 4–7 when a small number of measurements are made at a site (five spatially distributed readings). Studies such as Garcia-Serrana et al. [12] have verified the influence of infiltration rate in controlled field studies of highway sideslope hydrologic function, and field analyses such as Winston et al. [13] show that poorer volume reduction performance can be observed (relative to other studies) for swales constructed in poorly infiltrating soils.

Source Loading and Management Model for Windows (WinSLAMM) was developed to model and analyze stormwater management projects of varying scale, accounting for landscape attributes such as soil infiltration rate [14]. WinSLAMM is an empirical model that is unique in its ability to determine the runoff volume and pollution loading for every source area within a land use for each rainfall event [15]. The analysis accounts for the land use variability and site characteristics, predicts runoff volumes, and evaluates stormwater control effectiveness. After being developed in the mid-1970s, the model started being used in state water quality regulatory agencies in the mid-1980s (Wisconsin Department of Natural Resources for instance). Studies such as Borris et al. [16] and Hurley and Forman [17] have utilized the model to predict the influence of various management and future socio-environmental conditions on urban water systems. Although a common model in the stormwater management field for nearly 40 years, few studies have been performed in literature to assess the performance of this model for SCMs in comparison to field collected data, and no such study was found which focused on grassed swales.

Despite the number of studies performed on vegetated swales, there are still gaps in knowledge regarding their performance. In particular, this is the case for volume reduction, where the influence of local conditions (such as infiltration rate) have been shown in literature, necessitating studies across a range of landscapes. Further, the ability for a common stormwater management analysis tool, WinSLAMM, to predict swale volume reduction is largely unknown. The objectives of this study include: (1) evaluating swale performance for volume reduction; and (2) model the swale in WinSLAMM to determine its ability to provide accurate volume reduction estimates.

2. Materials and Methods

2.1. Study Area

The project site is in Knoxville, TN, in the median of Asheville Highway located near the intersection of Lecil Road (Figure 1). Asheville Highway is a four-lane divided highway with an average annual daily traffic of approximately 27,378 vehicles [18]. Stormwater is drained via two swales connected in series by a pipe for a total of 440 m. The longitudinal slope of the upper swale is 2.5% over a length of 210 m while the longitudinal slope of the lower swale is 1.5% over 230 m. The total catchment area is 0.64 ha, with 0.38 ha of pervious area (including the swale) and 0.26 ha of impervious area, making the contributing area 41% impervious and 59% pervious. The pervious area is made up of loam and silt loam soils [19]. According to the Tennessee Department of Transportation (TDOT) Standard RD01-S-11A, sod ditches are seeded with vegetal retardance classification "C" and are scarified prior to seeding [20,21].

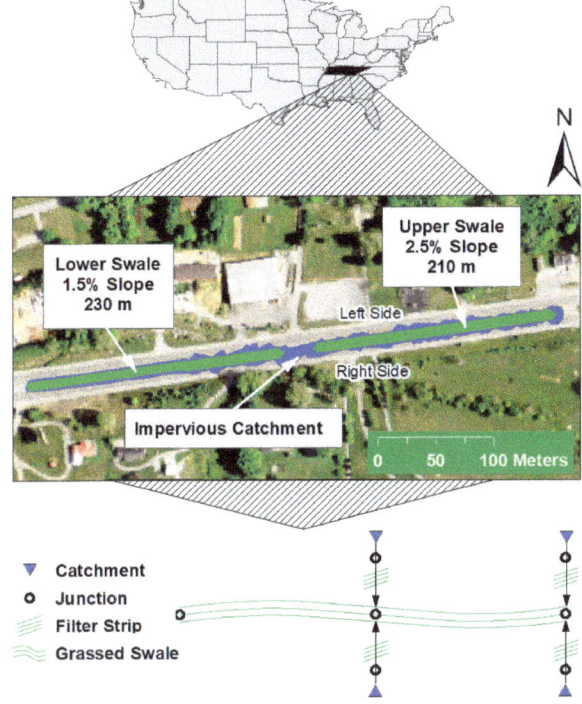

Figure 1. Project site schematic showing aerial view of swale and associated catchment, and the WinSLAMM model representation of the site.

2.2. Swale Outflow Monitoring

Monitoring equipment was installed during the summer of 2016. A 45.75 cm TRACOM fiberglass H-flume was installed at the downslope end of the swale immediately prior to discharge into a storm drain outfall. The presence of the outfall (a drop inlet style structure) allowed free discharge from the flume. Concrete was used to secure the flume approach to prevent flow under the device. Wingwalls were constructed to direct the flow into the flume and to prevent flow from traveling around the sides. An ISCO 6712 equipped with a 730 Bubbler Flow Module was utilized to collect stage data every minute which was converted to flow via standard equations. An ISCO 674 rain gauge was installed on site and connected to an ISCO 4230 flow meter for data collection. The rain data was recorded every 5 min.

2.3. Runoff Estimations

Runoff entered the grassed swale via sheet flow, negating the ability to explicitly monitor inflow. Thus, runoff was estimated following a similar approach to that of Brown and Hunt [22], whereby an initial abstraction of 1.3 mm is assumed (i.e., a curve number of 98) for the impervious area and the remainder of the rainfall produces runoff [23]. The pervious areas of the catchment were all associated with the swale and its sideslopes (i.e., filter strip). The total amount of rainfall that fell on these areas was considered in the runoff volume calculation. To calculate the total runoff volume, the impervious and pervious rainfall volumes were added together. Runoff reduction was calculated per Equation (1):

$$\text{Runoff Reduction (\%)} = \frac{\text{Estimated Inflow} - \text{Measured Outflow}}{\text{Estimated Inflow}} \times 100 \quad (1)$$

2.4. Modeling

WinSLAMM was selected to model the vegetated swale due to its established usage by regulatory agencies, desire to further test its capabilities, and usage in literature [16,17]. For this study, stormwater volume was the focus, being modeled, calibrated, and analyzed for performance using collected site data. Hourly rainfall depths, the finest resolution the model allows, were aggregated from the rain gage data at the site and used to populate the rainfall parameter file. Antecedent moisture content was calculated by the model based on the rainfall file. Other parameter files remained as model suggested values based on the site's location in the southeastern United States.

For the catchment rainfall-runoff modeling, the highway was described in the model as freeway draining to pervious area. The freeway length and average daily traffic were input. All greenspace in the catchment was labeled as a pervious highway traffic urban area. Because runoff entered the greenspace as sheet flow, volume monitoring was not possible at the edge of the pavement, making calibration of these parameters impossible.

To model the swale in WinSLAMM, the contributing area was divided into four catchments. The site was divided between the upper and lower swales and subdivided into northern and southern sections (one on each side of the road). The catchment areas were determined by processing a 1-meter resolution digital elevation model in ArcGIS (Figure 1). Land use calculations were then made. Each catchment was made up of a freeway area (the roadway) and a large turf area (the median). The large turf area consisted of the sideslope/filter strip and the grass swale. To distinguish between the filter strip and swale, the area inundated by the static volume of a five-year frequency storm with a duration of 24 h (not considering slope) was used as the boundary condition. This method produced a depth of 0.21 m in the trapezoidal median, filling the trapezoid to a top width of 3.35 m. The associated area was taken as the extent of the swale, while the remaining area was assumed to represent the filter strip. This depth was found to be conservative (larger) than the depth of flow predicted by the mannings equation for a 10-year, 24-h storm (as estimated from the rational equation), the 10-year storm being the design criteria for TDOT. The characteristics for each control, measured or from literature, were input into WinSLAMM (Table 1). The swale and filter strip lengths, longitudinal slopes, and the swale side slopes were determined using measurement tools within ArcGIS applied to a digital elevation model. The bottom width, grass height, and grass type were determined based on field measurements.

Table 1. Upper and lower grassed swale attributes and model inputs.

Parameter Description	Upper Swale	Lower Swale	Guidance/Source
Total Drainage Area (ha)	0.32	0.47	Measured in ArcGIS
Impervious Drainage Area (ha)	0.13	0.13	Measured in ArcGIS
Length (m)	210	230	Measured in ArcGIS
Bottom Width (m)	0.46	0.46	Measured in ArcGIS
Sideslope (H:V)	6.1	7.3	Measured in ArcGIS
Longitudinal Slope (%)	2.5	1.0	Measured in ArcGIS
Swale Retardance Factor	C	C	WinSLAMM Manual
Typical Grass Height (cm)	20	20	Field Measurement
Infiltration Rate (cm/h) *	3.4	5.5	Field Measurement

* Further detail provided in "Infiltration Measurements" Section.

Due to the filter strips' steep slope (12 to 17%), WinSLAMM automatically removes 3 m from their length per standard procedure (performed when slopes are >5%: PV & Associates, 2015). For the Asheville Highway site, this is the entire length of the filter strips which were measured as having lengths from 2.1 to 3 m. Thus, the lack of filter strip representation in the model as a control practice is likely a source of some error. It should be noted that the filter strip still was included in the runoff model as a pervious land use. Thus, Table 1 focuses on the attributes of the grassed swales.

2.5. Infiltration Measurements

Infiltration rates were determined by conducting field tests using double-ring infiltrometers (DRI) on the northern filter strip, southern filter strip, and grass swale [24]. Graphs of the results from the DRI tests were used to determine the point at which the infiltration rates reached an equilibrium. Field tests were performed in triplicate for each grassed swale and filter strip on two occasions (a total of six readings at each location—Table 2). WinSLAMM requires dynamic infiltration rate as an input, which is equivalent to the measured static infiltration rates divided by two [25]. High variability was noted for the site as has been shown in other studies of highway green space. The infiltration rates of the side slopes varied from those at the center of the swale, and the measured infiltration rates were higher than WinSLAMM's defined infiltration rates for loam and silt loam soil types (the predominate soil type in the surrounding area). Ahmed et al. [11] obtained similar results from a roadside swale study. Large differences were observed between the geometric mean infiltration rate of the side slopes and that of the center of the swale. Ahmed et al. [11] also observed that soil texture class did not have a statistically significant effect on the mean field-saturated hydraulic conductivity of a swale, which supports the observation of higher measured infiltration rates than implied by the soil type. It should be noted that soil type is a typical input to WinSLAMM from which infiltration rate is estimated.

Table 2. Measured infiltration rates (cm/h).

Location	Right FS			Swale			Left FS		
	Min	Max	Avg	Min	Max	Avg	Min	Max	Avg
Upper Section	11.9	16.2	13.7	3.0	4.2	3.4	4.7	6.4	5.3
Lower Section	8.7	12.2	10.1	2.5	9.2	5.5	1.0	6.2	3.7

3. Results and Discussion

3.1. Data Summary

Data was collected for 11 months from 18 August 2016 until 18 July 2017, with 65 rainfall events monitored. The events ranged from a minimum rainfall of 2.8 mm to a maximum rainfall of 138.9 mm (see Figure 2). The rainfall events were distributed over the four seasons with the most (40%) occurring during spring and the least (6%) occurring during autumn when abnormally low rainfall totals occurred. The other two seasons, summer and winter, constituted 32% and 22% of events, respectively.

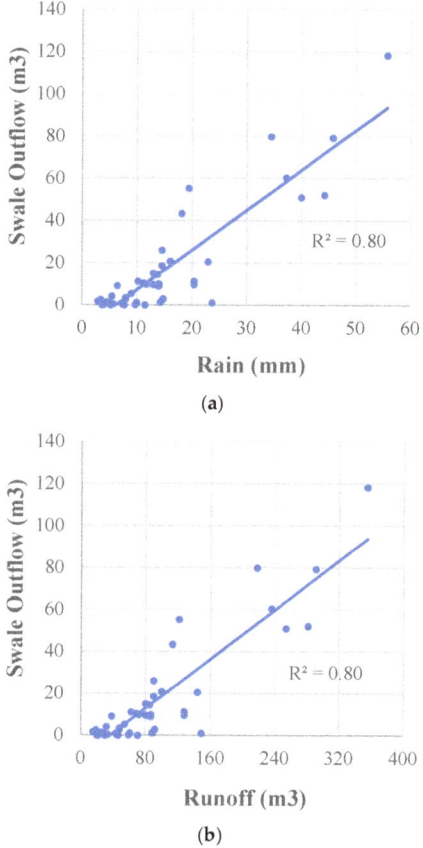

Figure 2. (a) Rainfall-Outflow and (b) Runoff-Outflow trends with Outliers Removed.

3.2. Hydrology Results

Rainfall-outflow data was plotted and resulted in a linear relationship with a correlation coefficient of 0.8 (Figure 2a). Two potential outliers in the data lacked agreement with the rainfall-outflow pattern. These were the largest two events monitored, a 139-mm storm showed substantially less outflow than expected (119 m^3), while the 98-mm storm showed substantially more (418 m^3). The runoff volumes from both events were removed from further analysis as there appeared to be monitoring error for these events.

The swale hydrologic performance exceeded what has been seen in previous literature. The swale runoff reduction ranged from 54.1% to 100% with a mean (average of all individual event reductions) of 87.2% (Figure 2b). Percent runoff reduction for swales in literature ranges from 20–52% [4,8,13,26–29]. Davis et al. [5], Deletic [8], and Yu et al. [10] also observed complete capture for small storm events, ranging from 4 to 22 mm. Similarly, in this study, rain events up to 22 mm approached complete capture with an average runoff reduction of 89%. Rainfall events below 12.5 mm varied between complete capture and producing a small runoff volume, relative to rainfall, with runoff reduction varying from 75 to nearly 100% (average of 93%). This performance variability could be a result of the soil's antecedent moisture content at the time of the event.

3.3. WinSLAMM Output

Due to the model structure, whereby the impervious roadway drains onto the median green space (filter strip and swale), nearly no runoff was produced by the model for any events unless soil compaction was set to "severe" within the rainfall-runoff variables. As runoff was observed and monitored for many events conflicting with model results (as seen in Figure 2), this was the first point of calibration in the modeling procedure. After compaction severity was adjusted, the model was found to underestimate swale outflow values (NSE of 0.35) when the field measured dynamic infiltration rates of 1.7 and 2.7 cm/h were input for the upper swale and lower swale, respectively (Figure 3a).

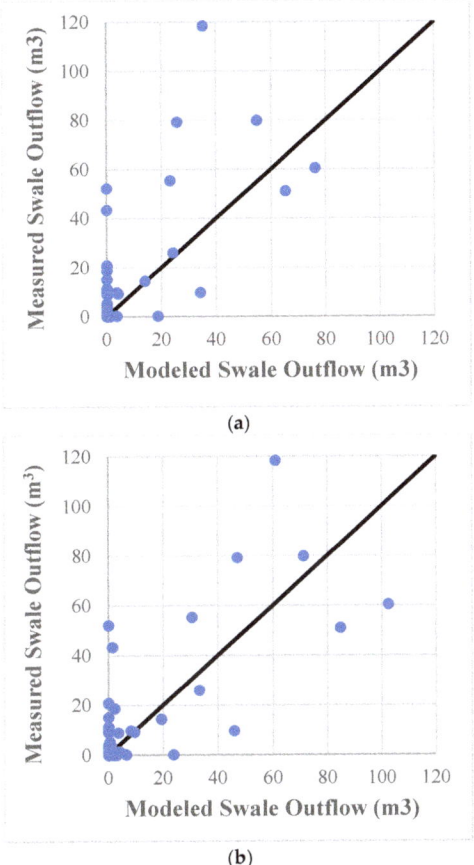

Figure 3. Modeled vs. Measured Swale Outflow for (**a**) Measured Infiltration Rates and (**b**) 80% of Measured Infiltration Rates.

Underestimation of swale outflow suggested that either the catchment was providing more flow to the system than the model predicted, or that the swale was retaining more water than the model predicted (i.e., the infiltration rate was too high). Errors due to the temporal resolution of rainfall data (1-h precipitation) may have also contributed due to how rapidly many processes occur in small, urban watersheds. Since runoff was only measured at the outfall and not quantified at the edge of pavement, the runoff coefficients could not be calibrated to observed data. Further, it was anticipated that the runoff coefficients in WinSLAMM are generally reasonable, given their determination through extensive field monitoring, calibration, and verification [30]. Most input variables for the model were default values or taken from field measurements. However, infiltration measurements within the swale were noted to be highly variable, from 2.5 to 9.2 cm/h for the lower swale, providing substantial error to that parameter and making it the most likely to need calibration. This is supported by recent studies by Ahmed et al. [11] who also showed high variability in the information rate of highway green space. A sensitivity analysis was performed to understand the effect of this parameter on the model performance.

The measured dynamic infiltration rates were multiplied by factors from 0.5 to 1.2, a range of 0.9 to 2.1 cm/h and 1.4 to 3.3 cm/h for the upper and lower swales, respectively. The model was run with each adjusted infiltration rate, and the Nash-Sutcliffe model efficiency coefficient (NSE) was generated for each model iteration [31]. NSE values and modeled infiltration rates were plotted to determine under which multiplier the maximum NSE value occurs. The optimum dynamic infiltration rates were found at 80% of measured values (Figure 4).

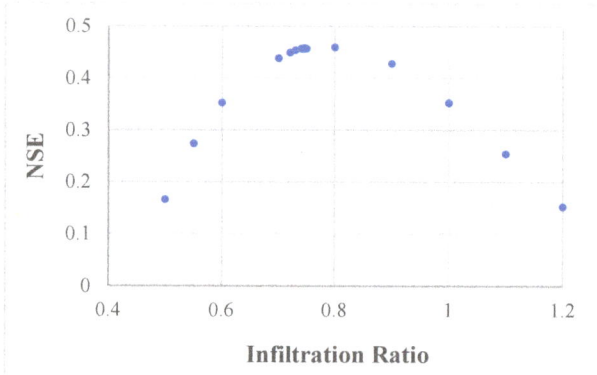

Figure 4. NSE vs. Infiltration Ratio.

As an additional observation, the calibrated infiltration rates fall within the range of sandy loam and loamy sand per the WinSLAMM manual. Given that the soils surrounding the site are made up of loam and silt loam per the USDA soil survey of the area, the native soils do not correspond with the calibrated infiltration rate, which is higher than expected based on soil type [32]. Similarly, the saturated hydraulic conductivity field tests performed by Garcia-Serrana et al. [12] also showed higher infiltration rates than the native soil texture classes indicated. It is possible that the dense stand of grass provided improved permeability over time due to root action, that fill soils were used for the roadway (thus making the soil map inaccurate), and/or that an organic layer developed over time and provided additional water storage. Regardless, it is apparent that infiltration tests should be performed instead of assuming that infiltration rates in highway medians will correspond with those assumed based on soil texture. As suggested by Ahmed et al. [11], this may require a large number of infiltration tests (between 10–40 per swale, depending on desired uncertainty factor) to be performed for a given location, likely exceeding the number of tests performed herein. This is particularly important in light of how sensitive this variable was shown to be during calibration.

Figure 3b shows the measured vs. modeled runoff volumes for the final calibrated model. The measured runoff volume for each rain event during the study period was totaled and every modeled runoff volume was totaled; the percent difference was calculated to be 28.4% over the entire study. Percent differences for other catchments modeled by WinSLAMM have ranged from 0 to

27%, with the site size ranging from 1.6 to 390 ha of varying land use [14]. Although this project from literature was focused on catchment modeling, these values do provide some context for past WinSLAMM performance. The max NSE was approximately 0.46 using only one calibration parameter which approaches the value suggested as acceptable model performance (>0.5) in such studies as Dongquan et al. [33] and Santhi et al. [34]. The Root Mean Square Error was 19.0 m^3, corroborating the fair, but not flawless, performance of the model suggested by the NSE. WinSLAMM appears to be a reasonable planning model for highway managers, but further study from other locations is needed to both verify the results herein and to understand which parameters can be estimated based on literature and which require field measurement. In particular, site specific infiltration measurements may be required to achieve reasonable performance.

4. Conclusions

This study investigated the potential for highway grassed swales to contribute to the stormwater management goals of entities such as state transportation agencies to meet MS4 requirements. The results were favorable for volume control, as the swale reduced runoff volume by a median 88.2%. One explanation for the high reduction percentage is the elevated infiltration rates measured for the site. Despite soil maps of the area identifying soils as primarily loam and silt loam, on-site infiltration tests showed relatively high infiltration rates compared to literature (2.5 cm/h to 9.2 cm/h). This parameter became critical in modeling the system, showing high sensitivity during the calibration process. The final, calibrated WinSLAMM model showed a percent difference of 28.4% between observed and modeled for the entire study period with an NSE of 0.46 and RMSE of 19.0 m^3. The modeling process reiterated the importance of collecting localized infiltration data when modeling these systems and confirmed the findings of other studies [11,12] that infiltration rates can be highly variable in highway environments. In addition, these results suggest the value of using WinSLAMM for estimating the performance of highway green space for stormwater management.

Although there are a number of studies examining the performance of swales as stormwater management features, the performance variability in literature suggests further study is needed to allow them to be properly credited by regulators. In particular, there is a need to better understand how infiltration rates vary in the highway environment and if tools to estimate these rates are feasible. Examining additional sites to see if infiltration rates are more elevated than the native soil texture class suggests would be beneficial for scaling estimates of highway swale performance from the local to regional level. In addition, WinSLAMM was shown to be a reasonable tool for modeling swale performance, but further study is needed to determine if the observed performance can be replicated in other sites and if improvements such as allowing a 15-min resolution rainfall input would improve performance. Using this tool, highway stormwater managers may also be able to determine how swale performance would vary given a range of infiltration rates, catchment sizes, and swale geometries.

Author Contributions: Conceptualization, J.M.H. and Q.H.; Methodology, B.N.Y., J.M.H. and Q.H.; Formal Analysis, B.N.Y., W.A.L., J.M.H. and Q.H.; Investigation, B.N.Y.; Writing-Original Draft Preparation, B.N.Y.; Writing-Review & Editing, B.N.Y., W.A.L. and J.M.H.

Funding: Funding for this research was provided by the Tennessee Department of Transportation Grant No. RES 2016-21.

Conflicts of Interest: The authors declare no conflict of interest.

References

1. Weis, P.T.; Gulliver, J.S.; Erickson, A.J. *The Performance of Grassed Swales as Infiltration and Pollution Prevention Practices: A Literature Review*; University of Minnesota: Minneapolis, MN, USA, 2010; Available online: http://stormwater.safl.umn.edu/publications/reports (accessed on 21 June 2018).
2. USEPA. *Stormwater Discharges from Transportation Sources*; USEPA: Washington, DC, USA, 2015. Available online: https://www.epa.gov/npdes/stormwater-discharges-transportation-sources#overview (accessed on 27 July 2016).
3. USEPA. *Storm Water Technology Fact Sheet: Vegetated Swales*; EPA Reference No. 832-F-99-006; USEPA: Washington, DC, USA, 1999. Available online: http://nepis.epa.gov/Exe/ZyPDF.cgi/200044A8.PDF?Dockey=200044A8.PDF (accessed on 20 July 2016).
4. Lucke, T.; Mohamed, M.; Tindale, N. Pollutant Removal and Hydraulic Reduction Performance of Field Grassed Swales during Runoff Simulation Experiments. *Water* **2014**, *6*, 1887–1904. [CrossRef]
5. Davis, A.P.; Stagge, J.H.; Jamil, E.; Kim, H. Hydraulic performance of grass swales for managing highway runoff. *Water Res.* **2011**, *46*, 6775–6786. [CrossRef] [PubMed]
6. Barrett, M.E. Comparison of BMP Performance Using the International BMP Database. *J. Irrig. Drain. Eng.* **2008**, *134*, 556–561. [CrossRef]
7. Shafique, M.; Kim, R.; Kyung-Ho, K. Evaluating the capability of grass swale for the rainfall runoff reduction from an urban parking lot, Seoul, Korea. *Int. J. Environ. Res. Public Health* **2018**, *15*, 537. [CrossRef] [PubMed]
8. Deletic, A. Modelling of water and sediment transport over grassed areas. *J. Hydrol.* **2001**, *248*, 168–182. [CrossRef]
9. Kaighn, R.J.; Yu, S.L. Testing of Roadside Vegetation for Highway Runoff Pollutant Removal. *Transp. Res. Rec. J. Transp. Res. Board* **1996**, *1523*, 116–123. [CrossRef]
10. Yu, S.L.; Kuo, J.T.; Fassman, E.A.; Pan, H. Field Test of Grassed-Swale Performance in Removing Runoff Pollution. *J. Water Resour. Plan. Manag.* **2001**, *127*, 168–171. [CrossRef]
11. Ahmed, F.; Gulliver, J.S.; Nieber, J.L. Field infiltration measurements in grassed roadside drainage ditches: Spatial and temporal variability. *J. Hydrol.* **2015**, *530*, 604–611. [CrossRef]
12. Garcia-Serrana, M.; Gulliver, J.S.; Nieber, J.L. Infiltration capacity of roadside filter strips with non-uniform overland flow. *J. Hydrol.* **2017**, *545*, 451–462. [CrossRef]
13. Winston, R.J.; Powell, J.T.; Hunt, W.F. Retrofitting a grass swale with rock check dams: Hydrologic impacts. *Urban Water J.* **2018**. [CrossRef]
14. PV & Associates (PVA). *Using WinSLAMM: Meeting Urban Stormwater Management Goals*; Workshop Presentations; PVA: Madison, WI, USA, 20–21 April 2017.
15. Pitt, R. WINSLAMM Documentation. 2013. Available online: http://rpitt.eng.ua.edu/SLAMMDETPOND/WinSlamm/MainWINSLAMM_book.html (accessed on 27 August 2017).
16. Borris, M.; Leonhardt, G.; Marsalek, J.; Osterlund, H.; Viklander, M. Source-Based Modeling of Urban Stormwater Quality Response to the Selected Scenarios Combining Future Changes in Climate and Socio-Economic Factors. *Environ. Manag.* **2016**, *58*, 223–237. [CrossRef] [PubMed]
17. Hurley, S.E.; Forman, R.T.T. Stormwater ponds and biofilters for large urban sites: Modeled arrangements that achieve the phosphorus reduction target for Boston's Charles River, USA. *Ecol. Eng.* **2011**, *36*, 850–863. [CrossRef]
18. Knoxville Geographic Information System (KGIS). Traffic Counts. KGIS Maps. 2017. Available online: http://www.kgis.org/KGISMaps/Map.htm (accessed on 7 July 2016).
19. United States Department of Agriculture (USDA). Natural Resources Conservation Service. Web Soil Survey. Available online: https://websoilsurvey.sc.egov.usda.gov/App/WebSoilSurvey.aspx (accessed on 18 March 2017).
20. TDOT. Roadside Ditch Details for Design and Construction (RD01-S-11A). Standard Drawings Library. 2002. Available online: http://www.tn.gov/assets/entities/tdot/attachments/RD01S11A_101502.pdf (accessed on 10 November 2017).
21. TDOT. Section 801—Seeding. TDOT Spec Book. 2015. Available online: https://www.tn.gov/assets/entities/tdot/attachments/2015_TDOT_Spec_Book.pdf (accessed on 10 November 2017).
22. Brown, R.A.; Hunt, W.F. Underdrain configuration to enhance bioretention exfiltration to reduce pollutant loads. *J. Environ. Eng.* **2011**, *137*, 1082–1091. [CrossRef]

23. Pandit, A.; Heck, H.H. Estimations of soil conservation service curve numbers for concrete and Asphalt. *J. Hydrol. Eng.* **2009**, *14*, 335–345. [CrossRef]
24. American Society for Testing and Materials International (ASTM). *Standard Test Method for Infiltration Rate of Soils in Field Using Double-Ring Infiltrometer*; ASTM International: West Conshohocken, PA, USA, 2009.
25. PV & Associates (PVA). WinSLAMM Model Algorithms. 2015. Available online: http://winslamm.com/Select_documentation.html (accessed on 27 August 2017).
26. Backstrom, M. Grassed swales for stormwater pollution control during rain and snowmelt. *Water Sci. Technol.* **2003**, *48*, 123–132. [CrossRef] [PubMed]
27. Barrett, M.E.; Irish, L.B.; Malina, J.F.; Charbeneau, R.J. Characterization of Highway Runoff in Austin, Texas, Area. *J. Environ. Eng.* **1998**, *124*, 131–137. [CrossRef]
28. Rushton, B.T. Low-impact parking Lot design reduces runoff and pollutants loads. *J. Water Res. Plan. Mgnag.* **2001**, *127*, 172–179. [CrossRef]
29. Knight, E.M.P.; Hunt, W.F.; Winston, R.J. Side-by-side evaluation of four level spreader-vegetated filter strips and a swale in eastern North Carolina. *J. Soil Water Conserv.* **2013**, *7*, 5330–5346. [CrossRef]
30. Pitt, R. Calibration of WinSLAMM. 2008. Available online: http://winslamm.com/Select_documentation.html (accessed on 27 August 2017).
31. Nash, J.E.; Sutcliffe, J.V. River flow forecasting through conceptual models part I—A discussion of principles. *Elsevier* **1970**, *10*, 282–290. [CrossRef]
32. Pitt, R.; Lantrip, J.; Harrison, R.; Henry, C.; Hue, D. *Infiltration through Disturbed Urban Soils and Compost-Amended Soil Effects on Runoff Quality and Quantity*; EPA 600/R-00/016; U.S. Environmental Protection Agency, Water Supply and Water Resources Division, National Risk Management Research Laboratory: Cincinnati, OH, USA, 1999.
33. Zhao, D.; Chen, J.; Wang, H.; Tong, Q.; Cao, S.; Sheng, Z. GIS-based urban rainfall-runoff modeling using an automatic catchment-discretization approach: A case study in Macau. *Environ. Earth Sci.* **2009**, *59*, 465–472.
34. Santhi, C.; Arnold, J.G.; Williams, J.R.; Dugas, W.A.; Srinivasan, R.; Hauck, L.M. Validation of the SWAT model on a large river basin with point and nonpoint sources. *J. Am. Water Resour. Assoc.* **2001**, *37*, 1169–1188. [CrossRef]

© 2018 by the authors. Licensee MDPI, Basel, Switzerland. This article is an open access article distributed under the terms and conditions of the Creative Commons Attribution (CC BY) license (http://creativecommons.org/licenses/by/4.0/).

Article

A Retrospective Comparison of Water Quality Treatment in a Bioretention Cell 16 Years Following Initial Analysis

Jeffrey P. Johnson * and William F. Hunt

Department of Biological and Agricultural Engineering, North Carolina State University, Campus Box 7625, Raleigh, NC 27695, USA; bill_hunt@ncsu.edu
* Correspondence: jeffrey_johnson@ncsu.edu; Tel.: +1-919-515-7475

Received: 18 March 2019; Accepted: 29 March 2019; Published: 2 April 2019

Abstract: One of the most popular stormwater practices in (sub-)urban North Carolina is bioretention. While bioretention has been researched intensively to determine the most efficient designs, few long-term studies have attempted to assess the performance of older bioretention. However, previous research and design guidance for bioretention has predicted long-term water quality treatment. This study compared discharged concentrations and loads of nitrogen and phosphorus from a bioretention cell (1) post-construction and (2) following 17 years of treatment. A conventionally-drained bioretention cell with lateral underdrains in Chapel Hill, North Carolina, USA, was first monitored post-construction for 10-months from 2002–2003 and, again following continuous use, for 14 months from 2017–2018. Estimated mass load reductions during the initial monitoring period were 40% for total nitrogen (TN) and 65% for total phosphorus (TP). Mass load reductions were increased 17 years after construction, with reductions of 72% and 79% for TN and TP, respectively. Plant growth, death, and decay over the 17-year life of the bioretention cell are hypothesized to have contributed additional nitrogen assimilation and carbon to the fill media, serving as a catalyst for nitrogen treatment. Phosphorus removal remained relatively unchanged between the two monitoring periods. Filter media samples indicated the top 20 cm of filter media were nearing phosphorus saturation, but with 1.2 m of filter media, lower depths would most likely continue to provide treatment. If designed, built, and maintained correctly, bioretention appears to provide sustained treatment of stormwater runoff for nitrogen and phosphorus for nearly two decades, and likely longer.

Keywords: stormwater management; green infrastructure; bioretention; biofilter; sustainable drainage systems; water quality; low impact development; nitrogen; phosphorus

1. Introduction

To ameliorate deleterious environmental impacts of urbanization, developers utilize low impact development (LID) practices to reduce stormwater runoff and treat stormwater on-site to improve downstream water quality [1]. By employing decentralized treatment via detention and infiltration, LID practices have been shown to reduce stormwater runoff volumes, nutrient loading, and sediment loading compared to traditional stormwater practices [2–4].

First developed in the early 1990's, bioretention is now one of the most popular LID stormwater control measures (SCMs) in the United States and Australasia as research has demonstrated success in meeting both hydrologic and water quality goals in laboratory and field settings [5–10]. Bioretention cells (BRCs) are a (depressed) landscape feature; underlying the landscape is engineered filter media, and in many cases, an underdrainage system [9]. Stormwater runoff fills the BRC bowl while simultaneously infiltrating into the filter media. Runoff stored within the BRC either exits via exfiltration to in-situ soils, is discharged through an underdrainage system, or evapotranspires. BRC filter media guidance varies by state and country. However, a typical filter media consists of a mixture of predominantly sand with native soil, gravel, and an electron donor (e.g., organic matter) [9]. Per its design, bioretention employs adsorption, filtration, sedimentation, volatilization, ion exchange, and biological decomposition [9].

While bioretention has been researched intensively to determine the most efficient design with respect to media depth, media selection, vegetative cover, drainage configuration, ponding depth, and capture volume [9], few long-term studies have assessed the function of older BRCs. Previous research and design guidance for BRCs, however, does predict long-term water quality treatment. Komlos and Traver [11] reported sustained orthophosphate (OP) removal over a nine year monitoring period from a rain garden in Philadelphia, PA. The authors observed P saturation within the top 10 cm of filter media after nine years, but estimated that saturation of deeper depths would not occur for 20 years [11]. Similarly, Johnson and Hunt [12] described elevated concentrations of phosphorus, zinc, and copper in the filter media of an 11-year-old BRC, but predicted sustained removal with routine maintenance. Willard et al. [13] compared post-construction pollutant removal in a BRC with that which occurred following seven years of service and noted removal of nutrients and sediment during both monitoring periods. This limited research suggests BRCs could still be performing as originally designed, if not better.

The research presented herein explores changes in the performance of a BRC with age. Discharged concentrations and loads of nitrogen and phosphorus are compared for one BRC (1) post-construction as previously reported by Hunt et al. [14] and then (2) following 17 years of service.

2. Methodology

2.1. Site Description

A bioretention cell was constructed in Chapel Hill, NC, USA, at the University Mall shopping center (35°55′39.0″ N, 79°01′29.6″ W), in Fall 2001 to treat parking lot stormwater runoff (Figure 1; Table 1). The original drainage area consisted entirely of a 0.06-hectare asphalt parking lot, but was resurfaced between the initial and second (post-17 years) monitoring periods, resulting in an enlarged watershed of 0.11 hectares for the second monitoring period. While the drainage area nearly doubled, the BRC was initially oversized, designed to capture and store 95 mm of precipitation within the surface storage bowl. Conventional design suggests runoff from 25 mm of rainfall be captured [9]. The BRC was constructed over low permeability hydrologic soil group (HSG) D, white store-urban land complex, soils [15].

Figure 1. Bioretention cell during initial monitoring period (**left**) and return monitoring period (**right**).

The surface area of the BRC was 90 m² with a surface storage depth of 10 cm. The BRC was excavated to a depth of 1.2 m and filled with a sandy filter media sourced from a local quarry. Shortly after construction, filter media saturated hydraulic conductivity was 3.3–7.6 cm/h, and the media was low in bioavailable P (3.7–11.1 mg Mehlich-3 P/kg filter media). The BRC was drained using two lateral 10 cm underdrains installed beneath the filter media. The surface was mulched and planted with perennial grasses, shrubs, and trees (Figure 1).

Table 1. Chapel Hill bioretention cell (BRC) characteristics.

Characteristic	Chapel Hill BRC
Year constructed	2001
Underlying soil	Clay, clay loam, and silty clay
2002–2003 drainage area (m²)	607
2017–2018 drainage area (m²)	1133
Imperviousness	100%
BRC surface area (m²)	90
Bowl storage depth (cm)	10
Media depth (m)	1.2
Median infiltration rate (cm/h)	3.3–7.6
Original media P-index	4–12 (3.7–11.1 mg Mehlich-3 P/kg)
Underdrain type	Conventional (no IWS)
Vegetative cover	Perennial grasses, trees, shrubs

2.2. Monitoring

The Chapel Hill (CH) BRC was monitored for hydrology and water quality during both monitoring periods. Hydrologic data were used to calculate pollutant loads. As Hunt et al. [14] monitored the CH BRC 16 years prior to the second period, monitoring technology differed, but techniques remained constant between monitoring periods.

During the initial monitoring period (June 2002–April 2003), rainfall data were collected using a tipping bucket RG600™ rain gauge. American Sigma900 Max™ automatic samplers were coupled with pressure transducers to calculate flow over v-notch weirs at both the inlet to the BRC (120°) and from a weir box (30°) attached to the outlet of the underdrain system. Influent and effluent samples were collected on a flow-weighted basis and composited following each storm event.

During the second monitoring period (February 2017–March 2018), rainfall data were collected with a ISCO model 674 tipping bucket rain gauge and checked for accuracy with a manual plastic rain gauge. Inflow and underdrain flow were measured with ISCO 730 bubbler modules and a 90° sharp crested v-notch weir at the inlet to the BRC and a 45° sharp crested v-notch weir on a weir box attached to the outlet of the underdrain system. Onset HOBO U20 water level loggers were installed from February–March 2018 to monitor internal water levels and calculate media infiltration rates. While overflow was not directly measured during either study, overflow volumes for load calculations were estimated using routing methods described by Malcom [16]. Composite water quality samples were taken on a flow proportional volumetric basis using ISCO 6712 portable samplers.

For both monitoring periods, rainfall events were defined as a minimum of 5 mm of rainfall, having a minimum antecedent dry period of 6 h. Water quality samples were collected within 36 h of rainfall cessation. Water quality samples were placed on ice and immediately transported to a nearby laboratory for analysis. During the second monitoring period, samples were analyzed at the Environmental Analysis Laboratory at NC State University for total ammoniacal nitrogen (TAN), nitrate/nitrite nitrogen (NO_3-N), total Kjeldahl nitrogen (TKN), total phosphorus (TP), and orthophosphate (OP). Total nitrogen (TN), organic nitrogen (ON), and particulate bound phosphorus (PBP) were calculated (Table 2).

Table 2. Water quality analysis methods of the second monitoring period (2017–2018).

Analyte	Method
NO_3-N	EPA Method 353.2
TAN	Standard Method 4500-NH3 G
TKN	EPA Method 351.2
ON	TKN–TAN
TN	NO_3-N + TKN
TP	Standard Method 4500-P F
OP	EPA Method 365.1
PBP	TP–OP

In February 2018, seven soil samples were collected from the filter media following the standard method for sampling with a scoop [17]. The BRC surface was cleared of debris, and samples were taken from the top 20 cm of filter media following guidance by Li and Davis [18]. Samples were placed in individual sealable plastic bags and labeled with site location. Samples were analyzed at the Environmental Analysis Laboratory at NC State University for bulk density, pH, TP, and carbon:nitrogen (C:N) ratio.

2.3. Data Analysis

Influent and effluent event mean concentrations (EMCs) were compared to assess pollutant removal during each sampled storm event using the efficiency ratio (ER), calculated as,

$$ER\ (\%) = \left(\frac{EMC_{in} - EMC_{out}}{EMC_{in}}\right) \times 100, \quad (1)$$

where EMC_{in} is the influent event mean concentration and EMC_{out} is the effluent event mean concentration.

Flow data were analyzed in ISCO Flowlink® software to calculate flow volumes. Individual storm loads were then calculated using Equation (2).

$$Event\ Load\ (g) = EMC_{i,j} \cdot V_{i,j}, \quad (2)$$

where $EMC_{i,j}$ is the observed EMC (mg/L) for a particular storm at either the inlet or outlet sampling location and $V_{i,j}$ is storm-associated flow volume (L).

Cumulative loads were normalized by annual rainfall using Equation (3).

$$Cumulative\ Load\ (kg) = \frac{\sum_{i=1}^{n} c_i V_i}{DA \cdot 1 \times 10^6} \cdot \frac{P_{annual}}{P_{observed}}, \quad (3)$$

where c_i is the observed or event median concentration (mg/L), V_i is the runoff or outflow volume (L), DA is the drainage area of the BRC (ha), P_{annual} is the normal annual rainfall in Chapel Hill, NC (1129 mm [19]), and $P_{observed}$ is the measured rainfall during each monitoring period (mm).

Statistical analyses were performed using R statistical software [20]. Data were inspected for normality and log-normality visually and by using the Shapiro–Wilk, Anderson–Darling, and Cramer–von Mises tests for normality. Data were uniformly non-normal, and statistical comparisons utilized Wilcoxon signed rank non-parametric statistical analyses. Differences were considered significant at α = 0.05.

3. Results and Discussion

During the initial monitoring period (June 2002–April 2003), Hunt et al. [14] collected water quality samples from ten events with precipitation ranging from 17.3 to 58.4 mm. Median EMCs were reduced for TAN and OP, while median EMCs increased for TKN, NO_3-N, TN, ON, and TP

(Figures 2 and 3; Table 3). Median PBP EMCs were unchanged. Median TN and TP EMCs increased by 38% and 21%, respectively. Using observed EMCs and volumes for runoff and outflow samples, Hunt et al. [14] were able to estimate 10 month loads for nitrogen and phosphorus species. Estimated annual mass loads exported during the initial monitoring period were calculated as 3.2 kg/ha/year and 0.4 kg/ha/year for TN and TP, respectively. The BRC provided 40% and 65% reductions in TN and TP loads, respectively. Mass load of nitrate was reduced by only 13.2%. These observations were in line with other research at the time which demonstrated variable removal of TN from BRCs, yet suggested that BRCs could reduce loads via volume reduction [21–24].

During the second monitoring period (February 2017–March 2018), 18 separate storm events were sampled for water quality, with individual storm precipitation ranging from 5.6 to 50.8 mm. Statistically significant reductions ($p < 0.05$) in EMCs were observed for TAN, NO_3-N, TN, and TP (Figures 2 and 3; Table 3). Annual TN and TP loads exported from the BRC during the return period were 5.0 kg/ha/year and 0.4 kg/ha/year, respectively, and represented percent reductions of 72% and 79%. NO_3-N loads during the return period were reduced by 84%.

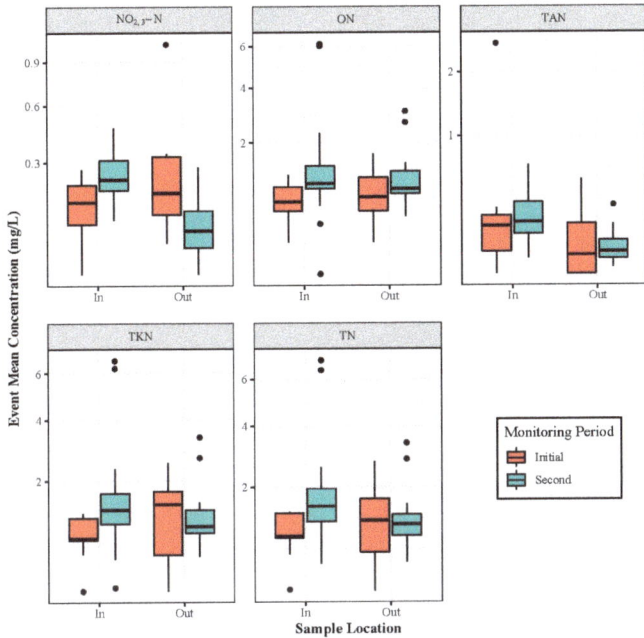

Figure 2. Event mean nitrogen species concentrations for sampled storm events at the inlet (in) and outlet (out) of the Chapel Hill bioretention cells (BRC) during each monitoring period.

Based upon these two monitoring periods, N and P removal improved over time (Table 3). Although annual TN loads exported from the BRC increased between the initial and second monitoring periods, it should be noted that TN loads into the BRC increased from an annualized 5.4 kg/ha/year during the initial monitoring period to 17.8 kg/ha/year during the second monitoring period, mainly because the watershed size was much larger for a similar amount of rain (1063.1 mm initial vs. 924.6 mm second). Regardless, mass removal rates for both TN and TP were higher during the 2017–2018 monitoring period. This BRC might typify sustainable treatment for extended periods of time, with the caveat of regular maintenance [11,12,25].

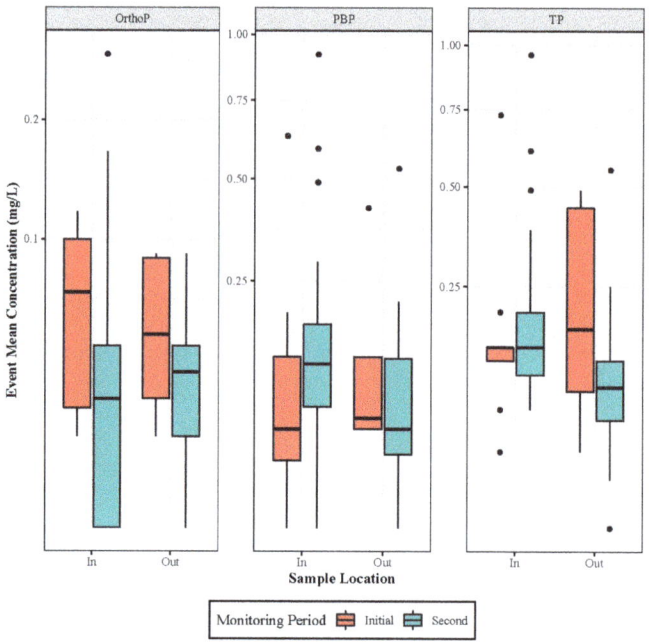

Figure 3. Event mean phosphorus species concentrations for sampled storm events at the inlet (in) and outlet (out) of the Chapel Hill BRC during each monitoring period.

Table 3. Median EMCs and efficiency ratio (ER) for sampled analytes during each monitoring period.

Pollutant	Initial Monitoring Period			Second Monitoring Period		
	EMC In	EMC Out	Change	EMC In	EMC Out	Change
	(mg/L)		(%)	(mg/L)		(%)
TN	0.89	1.23	+37.6 *	1.51	1.12	−25.8 *
TKN	0.74	1.41	+90.5 *	1.29	0.95	−26.4
TAN	0.17	0.05	−70.6	0.19	0.06	−68.4 *
NO_3-N	0.15	0.18	+20.0 *	0.23	0.08	−67.4 *
ON	0.56	0.70	+25.0 *	0.95	0.84	−12.1
TP	0.14	0.17	+21.4	0.14	0.09	−39.3 *
OP	0.07	0.05	−28.6	0.02	0.03	+50.0
PBP	0.04	0.04	0.0	0.11	0.04	−63.6

* denotes statistical significance ($p < 0.05$).

When compared to target thresholds for ambient water quality vis-à-vis benthos species in the Piedmont of NC [26], the Chapel Hill BRC performed better for TP and the same for TN after 17 years of maturation compared to when it was initially monitored (Figures 4 and 5). For TN, effluent EMCs exceeded the "good" threshold for approximately 55–57% of storm events for both monitoring periods (Figure 4). Effluent TP EMCs met target thresholds for 36% initially, but increased to 57% after 17 years. Results from the latter period compare favorably to studies from newer BRCs [21,27,28].

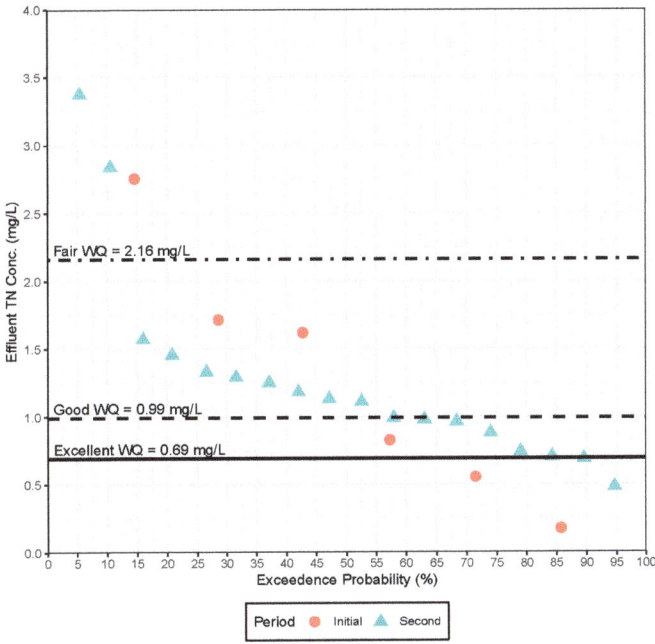

Figure 4. Chapel Hill BRC effluent total nitrogen (TN) exceedance probability compared with McNett et al. [26] threshold for water quality.

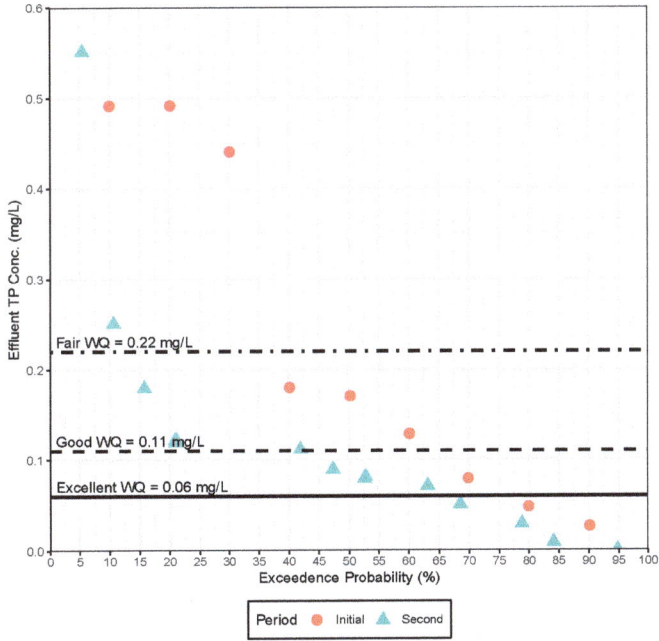

Figure 5. Chapel Hill BRC effluent total phosporous (TP) exceedance probability compared with McNett et al. [26] threshold for water quality.

3.1. Nitrate

A noteworthy improvement between the two monitoring periods was observed for NO_3-N. During the initial monitoring period NO_3-N mass removal was 13%; this jumped to 86% 17 years post-construction. The median effluent NO_3-N concentration in 2002–2003 was 0.18 mg/L; while that of the second period was 0.08 mg/L. During the initial monitoring period, effluent nitrate concentrations exceeded influent concentrations for four of seven observed events, yet in 2017–2018, effluent concentrations exceeded influent concentrations for only one of 18 events (Figure 6). It is postulated that nitrate conversion is attributed to (1) increased N uptake following maturation of vegetation in the BRC and/or (2) cycling of plant material through the fill media of the BRC which increased media carbon content (a necessary component for denitrification).

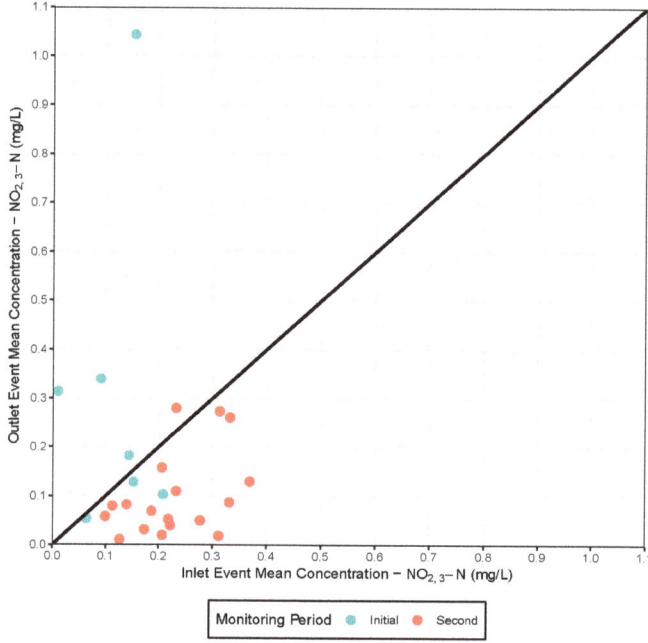

Figure 6. Influent and effluent nitrate concentrations observed at the Chapel Hill BRC during each monitoring period.

The importance of maturation vis-à-vis denitrification has been demonstrated for another vegetated SCM [29]. Assimilation has been shown to be a key component in BRC nitrogen removal [10,30–32], representing up to 88% of nitrate removal [30]. The steady growth of vegetation, as evidenced in Figure 1, provides a greater opportunity for the uptake of nitrogen. A key consideration for sustained removal of nitrogen via vegetative uptake is maintenance of vegetation. Vegetation of the CH BRC received regular pruning and landscaping, following the guidance of Payne et al. [30]; regular vegetation maintenance prevents the release of nitrogen via senescence and extends plant nitrogen demand by removing sources of internal N cycling. Further, mature vegetation will have a greater root mass than when first planted and may provide greater N removal due to a larger surface area for nutrient uptake and microbial communities [31]. Seasonal cycling of decaying plant matter into the soil media will also provide an optimal environment for nitrogen-fixing microbes [33]. Microbial communities provide rapid sequestration of nitrogen and can outcompete vegetation for nitrogen [34].

Increasing vegetation and root density may also provide additional opportunities for denitrification within the rhizosphere of BRC plants [35].

Plant growth, death, and decay over the 17-year life of the BRC likely contributed carbon to the fill media, a catalyst for denitrification [36]. While the CH BRC is conventionally drained, denitrification still occurs within microsites of saturated fill media [10,37].

Seven soil samples were taken from the fill media to quantify carbon content. The carbon to nitrogen ratio (C:N) ranged from 6.4 to 11.6 with a mean C:N of 8.8 ± 2.0, both of which are less than a maximum C:N of 44 observed in a seven year-old BRC in Virginia [13], but similar to those found in cultivated soils [38]. The average carbon content of the media was 0.67%. Initial C:N values for the fill media during installation were not available, but the media was a sandy media mined from a local quarry [14]. Considering that Peterson et al. [39] suggest an optimal soil media carbon content of 4.5% by weight to promote denitrification, nitrate removal is poised to continue at the Chapel Hill BRC as additional carbon is introduced to the filter media over time.

Willard et al. [13] examined denitrifying bacteria populations within a seven-year-old BRC in Virginia, finding that despite the presence of a dedicated anoxic zone for denitrification, the greatest populations of denitrifying bacteria were located in the surface layers of the BRC. The authors hypothesized that the abundance of carbon and organic material at the surface would result in preferential denitrification there. As a 10–15 cm layer of organic matter had accumulated at the surface of the CH BRC, it is possible that conditions within the upper layers of the CH BRC surface became more favorable to denitrification due to maturation.

3.2. Phosphorus

TP concentrations were significantly reduced after the site was 17 years old. The young site (1–2 years old) did not have a significant difference between inflow and outflow concentrations. During the initial monitoring period median influent and effluent TP concentrations were 0.14 mg/L and 0.17 mg/L, respectively (an ER of −21.4%). Fifteen years later, median influent and effluent TP concentrations were 0.14 mg/L and 0.09 mg/L, respectively (an ER of 39.3%). Median effluent OP concentrations were 0.05 mg/L and 0.03 mg/L during the initial and second monitoring periods, respectively, and remained proportional (approximately one third) to TP concentrations. Annual export loads of TP were 0.4 kg/ha/year during both monitoring periods, suggesting that TP export remained relatively constant over the life of the BRC.

That phosphorus removal has persisted for 17 years in this BRC and is supported by previous research [11,40]. A BRC in Philadelphia, PA, had no loss of function with respect to OP removal over a seven-year monitoring period [11]. The BRC investigated by Willard et al. [13] continued to significantly remove TP loads for seven years.

To assess the rate of phosphorus accumulation within the CH BRC filter media, soil media samples were collected post-construction during the initial monitoring period and again during the second monitoring period. Post-construction samples were analyzed for Mehlich-3 phosphorus (M3-P, an estimation of bioavailable P), while 15 years later, the samples were analyzed for total phosphorus using K_2SO_4-$CuSO_4$ digestion and colorimetry. Soil media samples collected during the initial monitoring period had M3-P concentrations of 3.7 and 11.1 mg/kg, which was expected for a sand-based media [14]. Soil media samples collected during the second monitoring period had TP concentrations ranging from 133.0–302.6 mg P/kg filter media. While return period samples did not represent bioavailable P, it can be estimated using research by Lammers and Bledsoe [41] who related various TP measurement methods to M3-P in streambank sediments from 662 soil samples. Their investigation reported that M3-P concentrations were, on average, 11.7% of those of TP.

Following the results of Lammers and Bledsoe [41], M3-P concentrations were estimated from those of TP (by multiplying TP by 0.117). This allowed comparison between both monitoring periods. After 17 years, M3-P concentrations ranged from 15.6–35.4 mg/kg; the mean M3-P concentration was

24.3 ± 7.1 mg/kg. Using both periods' M3-P concentrations, the mass of M3-P within the filter media was calculated using Equation (4).

$$Mass_{M3\text{-}P} = \frac{A \times D \times \rho_b \times c_{M3\text{-}P}}{1000} \quad (4)$$

where $Mass_{M3\text{-}P}$ is the mass of Mehlich-3 P in the filter media (g), A is the surface area of the BRC (m^2), D is the depth of media (m), ρ_b is the bulk density of the filter media (kg/m^3), and $c_{M3\text{-}P}$ is the mean concentration of Mehlich-3 P in the filter media (mg M3-P/kg filter media).

As soil media samples were collected from the top 20 cm, Equation (4) was applied using a depth of 0.2 m rather than the total depth of the filter media. Since bulk density values from the first monitoring period were not reported, mean bulk density from the later monitoring period samples were used for calculations.

M3-P within the filter media was calculated as 136.4 g and 446.9 g for the first and second periods, respectively. An increase of 310.6 g represents an annualized accumulation rate of 19.4 g/year. While sorption capacity of the BRC filter media will vary based on the chemical composition of media, previous research by Hsieh and Davis [40] estimated the sorption capacity of two sand based media mixes (92–95% sand, similar to that at CH BRC) to be 28 mg M3-P/kg media. With an average M3-P concentration of 24.3 mg/kg, the top 20 cm of BRC media appears to be nearing sorption capacity. At an annual M3-P accumulation rate of 19.4 g/year, the CH BRC filter media would reach sorption capacity in approximately 3.5 more years.

That the top 20 cm of filter media are at, or nearing, sorption capacity for phosphorus following 17 years of treatment, is expected. There appears to remain multiple decades of capacity at this BRC because of the 1.2 m media depth. Komlos and Traver [11] found the top 10 cm of BRC filter media to be saturated with phosphorus while estimating 20+ years to saturation at lower depths. However, should media replacement be necessary, simply replacing the top 20 cm of saturated filter media might prove less costly and is supported by previous research [18,42].

3.3. Regulatory Implications

As regulation of nitrogen and phosphorus export in nutrient sensitive watersheds continues, the sustained treatment performance of the CH BRC is promising. Estimates of performance are usually predicated on early succession research [7,9]. Were BRC nutrient reduction ability to improve with time, then BRCs may be undervalued by regulators and models. For example in NC, BRCs are assigned specific TN (1.20 mg/L) and TP (0.12 mg/L) effluent concentrations for nutrient regulation calculations [43]. Post-construction median effluent TN and TP concentrations at the CH BRC were 1.23 and 0.17 mg/L, respectively. Fifteen years later, median TN and TP concentrations had decreased to 1.12 and 0.09 mg/L, respectively. With maturation, the CH BRC is performing better than the NCDEQ-assigned credit. As BRCs are a biologically-derived treatment practice, their performance will change over time and will be dependent on maturation and maintenance, much like constructed stormwater wetlands [29]. This BRC appears to be a very sustainable SCM with regard to nutrient removal.

4. Conclusions

The preponderance of bioretention research is concentrated on the treatment capabilities of "young" BRCs, post-construction. As bioretention is a biological filtration system, physical and biogeochemical processes that drive stormwater treatment in BRCs are subject to temporal changes [44], yet little research has addressed exactly how BRC treatment changes with time. The research presented herein illustrates that BRC nutrient removal can be sustained. If designed, built, and maintained following research-based guidance [9,25,45], bioretention provides excellent treatment of stormwater runoff for nitrogen and phosphorus for prolonged periods of time.

Author Contributions: conceptualization, J.P.J. and W.F.H.; data curation, J.P.J.; formal analysis, J.P.J.; funding acquisition, W.F.H.; investigation, J.P.J.; project administration, J.P.J. and W.F.H.; writing—original draft, J.P.J.; writing—review and editing, J.P.J. and W.F.H.

Funding: This research was funded by the North Carolina Policy Collaboratory.

Acknowledgments: The authors would like to thank Shawn Kennedy for his assistance with the installation and maintenance of monitoring equipment at the research site.

Conflicts of Interest: The authors declare no conflict of interest. The funders had no role in the design of the study; in the collection, analyses, or interpretation of data; in the writing of the manuscript, or in the decision to publish the results.

References

1. Davis, A.P. Green Engineering Principles Promote Low-impact Development. *Environ. Sci. Technol.* **2005**, *39*, 338A–344A. [CrossRef]
2. Dietz, M.E. Low Impact Development Practices: A Review of Current Research and Recommendations for Future Directions. *Water Air Soil Pollut.* **2007**, *186*, 351–363. [CrossRef]
3. Line, D.E.; Brown, R.A.; Hunt, W.F.; Lord, W.G. Effectiveness of LID for Commercial Development in North Carolina. *J. Environ. Eng.* **2012**, *138*, 680–688. [CrossRef]
4. Wilson, C.E.; Hunt, W.F.; Winston, R.J.; Smith, P. Comparison of Runoff Quality and Quantity from a Commercial Low-Impact and Conventional Development in Raleigh, North Carolina. *J. Environ. Eng.* **2015**, *141*, 05014005. [CrossRef]
5. Bratieres, K.; Fletcher, T.; Deletic, A.; Zinger, Y. Nutrient and sediment removal by stormwater biofilters: A large-scale design optimisation study. *Water Res.* **2008**, *42*, 3930–3940. [CrossRef]
6. Coffman, L.S.; Green, R.; Clar, M.; Bitter, S. Development of Bioretention Practices for Stormwater Management. *J. Water Manag. Model.* **1994**, *6062*, 23–42. [CrossRef]
7. Davis, A.P.; Hunt, W.F.; Traver, R.G.; Clar, M. Bioretention Technology: Overview of Current Practice and Future Needs. *J. Environ. Eng.* **2009**, *135*, 109–117. [CrossRef]
8. Davis, A.P.; Traver, R.G.; Hunt, W.F.; Lee, R.; Brown, R.A.; Olszewski, J.M. Hydrologic Performance of Bioretention Storm-Water Control Measures. *J. Hydrol. Eng.* **2012**, *17*, 604–614. [CrossRef]
9. Hunt, W.F.; Davis, A.P.; Traver, R.G. Meeting Hydrologic and Water Quality Goals through Targeted Bioretention Design. *J. Environ. Eng.* **2012**, *138*, 698–707. [CrossRef]
10. Lucas, W.C.; Greenway, M. Nutrient Retention in Vegetated and Nonvegetated Bioretention Mesocosms. *J. Irrig. Drain. Eng.* **2008**, *134*, 613–623. [CrossRef]
11. Komlos, J.; Traver, R.G. Long-Term Orthophosphate Removal in a Field-Scale Storm-Water Bioinfiltration Rain Garden. *J. Environ. Eng.* **2012**, *138*, 991–998. [CrossRef]
12. Johnson, J.P.; Hunt, W.F. Evaluating the spatial distribution of pollutants and associated maintenance requirements in an 11 year-old bioretention cell in urban Charlotte, NC. *J. Environ. Manag.* **2016**, *184*, 363–370. [CrossRef]
13. Willard, L.L.; Wynn-Thompson, T.; Krometis, L.H.; Neher, T.P.; Badgley, B.D. Does It Pay to be Mature? Evaluation of Bioretention Cell Performance Seven Years Postconstruction. *J. Environ. Eng.* **2017**, *143*, 04017041. [CrossRef]
14. Hunt, W.F.; Jarrett, A.R.; Smith, J.T.; Sharkey, L.J. Evaluating Bioretention Hydrology and Nutrient Removal at Three Field Sites in North Carolina. *J. Irrig. Drain. Eng.* **2006**, *132*, 600–608. [CrossRef]
15. Soil Survey Staff. Web Soil Survey. 2018. Available online: https://websoilsurvey.sc.egov.usda.gov/ (accessed on 18 April 2018).
16. Malcom, H.R. *Elements of Urban Stormwater Design*; North Carolina State University: Raleigh, NC, USA, 1989.
17. *Standard Practice for Sampling with a Scoop*; ASTM Standard D5633; Technical Report; ASTM: West Conshohocken, PA, USA, 2012. [CrossRef]
18. Li, H.; Davis, A.P. Urban Particle Capture in Bioretention Media. I: Laboratory and Field Studies. *J. Environ. Eng.* **2008**, *134*, 409–418. [CrossRef]
19. NOAA NCEI. *NCDC Climate Normals*; NOAA NCEI: Washington, DC, USA, 2018.
20. R Core Team. *R: A Language and Environment for Statistical Computing*; R Core Team: Vienna, Austria, 2017.
21. Davis, A.P. Field Performance of Bioretention: Water Quality. *Environ. Eng. Sci.* **2007**, *24*, 1048–1064. [CrossRef]

22. Davis, A.P.; Shokouhian, M.; Sharma, H.; Minami, C. Water Quality Improvement through Bioretention Media: Nitrogen and Phosphorus Removal. *Water Environ. Res.* **2006**, *78*, 284–293. [CrossRef]
23. Dietz, M.E.; Clausen, J.C. A Field Evaluation of Rain Garden Flow and Pollutant Treatment. *Water Air Soil Pollut.* **2005**, *167*, 123–138. [CrossRef]
24. Li, H.; Davis, A.P. Water Quality Improvement through Reductions of Pollutant Loads Using Bioretention. *J. Environ. Eng.* **2009**, *135*, 567–576. [CrossRef]
25. Brown, R.A.; Hunt, W.F. Improving bioretention/biofiltration performance with restorative maintenance. *Water Sci. Technol.* **2012**, *65*, 361. [CrossRef]
26. McNett, J.K.; Hunt, W.F.; Osborne, J.A. Establishing Storm-Water BMP Evaluation Metrics Based upon Ambient Water Quality Associated with Benthic Macroinvertebrate Populations. *J. Environ. Eng.* **2010**, *136*, 535–541. [CrossRef]
27. Hunt, W.F.; Smith, J.T.; Jadlocki, S.J.; Hathaway, J.M.; Eubanks, P.R. Pollutant Removal and Peak Flow Mitigation by a Bioretention Cell in Urban Charlotte, N.C. *J. Environ. Eng.* **2008**, *134*, 403–408. [CrossRef]
28. Passeport, E.; Hunt, W.F.; Line, D.E.; Smith, R.A.; Brown, R.A. Field Study of the Ability of Two Grassed Bioretention Cells to Reduce Storm-Water Runoff Pollution. *J. Irrig. Drain. Eng.* **2009**, *135*, 505–510. [CrossRef]
29. Merriman, L.S.; Hunt, W.F. Maintenance versus Maturation: Constructed Storm-Water Wetland's Fifth-Year Water Quality and Hydrologic Assessment. *J. Environ. Eng.* **2014**, *140*, 05014003. [CrossRef]
30. Payne, E.G.I.; Fletcher, T.D.; Russell, D.G.; Grace, M.R.; Cavagnaro, T.R.; Evrard, V.; Deletic, A.; Hatt, B.E.; Cook, P.L.M. Temporary Storage or Permanent Removal? The Division of Nitrogen between Biotic Assimilation and Denitrification in Stormwater Biofiltration Systems. *PLoS ONE* **2014**, *9*, e90890. [CrossRef]
31. Read, J.; Wevill, T.; Fletcher, T.; Deletic, A. Variation among plant species in pollutant removal from stormwater in biofiltration systems. *Water Res.* **2008**, *42*, 893–902. [CrossRef]
32. Turk, R.P.; Kraus, H.T.; Hunt, W.F.; Carmen, N.B.; Bilderback, T.E. Nutrient Sequestration by Vegetation in Bioretention Cells Receiving High Nutrient Loads. *J. Environ. Eng.* **2017**, *143*, 06016009. [CrossRef]
33. Chen, G.; Zhu, H.; Zhang, Y. Soil microbial activities and carbon and nitrogen fixation. *Res. Microbiol.* **2003**, *154*, 393–398. [CrossRef]
34. Dunn, R.M.; Mikola, J.; Bol, R.; Bardgett, R.D. Influence of microbial activity on plant–microbial competition for organic and inorganic nitrogen. *Plant Soil* **2006**, *289*, 321–334. [CrossRef]
35. Birgand, F.; Skaggs, R.W.; Chescheir, G.M.; Gilliam, J.W. Nitrogen Removal in Streams of Agricultural Catchments—A Literature Review. *Crit. Rev. Environ. Sci. Technol.* **2007**, *37*, 381–487. [CrossRef]
36. Kim, H.; Seagren, E.A.; Davis, A.P. Engineered bioretention for removal of nitrate from stormwater runoff. *Water Environ. Res.* **2003**, *75*, 355–367. [CrossRef]
37. Seitzinger, S.; Harrison, J.A.; Böhlke, J.K.; Bouwman, A.F.; Lowrance, R.; Peterson, B.; Tobias, C.; Drecht, G.V. Denitrification Across Landscapes and Waterscapes: A Synthesis. *Ecol. Appl.* **2006**, *16*, 2064–2090. [CrossRef]
38. Brady, N.C.; Weil, R.R. *The Nature and Properties of Soils*; Pearson Prentice Hall: Columbus, OH, USA, 2008.
39. Peterson, I.J.; Igielski, S.; Davis, A.P. Enhanced Denitrification in Bioretention Using Woodchips as an Organic Carbon Source. *J. Sustain. Water Built Environ.* **2015**, *1*, 04015004. [CrossRef]
40. Hsieh, C.h.; Davis, A.P.; Needelman, B.A. Bioretention Column Studies of Phosphorus Removal from Urban Stormwater Runoff. *Water Environ. Res.* **2007**, *79*, 177–184. [CrossRef]
41. Lammers, R.W.; Bledsoe, B.P. What role does stream restoration play in nutrient management? *Crit. Rev. Environ. Sci. Technol.* **2017**, *47*, 335–371. [CrossRef]
42. Hatt, B.E.; Fletcher, T.D.; Deletic, A. Hydraulic and Pollutant Removal Performance of Fine Media Stormwater Filtration Systems. *Environ. Sci. Technol.* **2008**, *42*, 2535–2541. [CrossRef]
43. N.C. DEQ. *North Carolina Stormwater Control Measure Credit Document*; North Carolina Department of Environmental Quality: Raleigh, NC, USA, 2017.
44. Davis, A.P.; Traver, R.G.; Hunt, W.F. Improving Urban Stormwater Quality: Applying Fundamental Principles. *J. Contemp. Water Res. Educ.* **2010**, *146*, 3–10. [CrossRef]
45. Blecken, G.T.; Hunt, W.F.; Al-Rubaei, A.M.; Viklander, M.; Lord, W.G. Stormwater control measure (SCM) maintenance considerations to ensure designed functionality. *Urban Water J.* **2017**, *14*, 278–290. [CrossRef]

© 2019 by the authors. Licensee MDPI, Basel, Switzerland. This article is an open access article distributed under the terms and conditions of the Creative Commons Attribution (CC BY) license (http://creativecommons.org/licenses/by/4.0/).

Article

Design and Performance Characterization of Roadside Bioretention Systems

Rajendra Prasad Singh [1,2], Fei Zhao [1,2], Qian Ji [1,2], Jothivel Saravanan [1,2] and Dafang Fu [1,2,*]

[1] School of Civil Engineering, Southeast University, Nanjing 211189, China; rajupsc@seu.edu.cn (R.P.S.); 108209024@seu.edu.cn (F.Z.); 220174166@seu.edu.cn (Q.J.); nobelsaravanan@gmail.com (J.S.)
[2] Southeast University-Monash University Joint Research Centre for Future Cities, Nanjing 211189, China
* Correspondence: fdf@seu.edu.cn

Received: 28 February 2019; Accepted: 2 April 2019; Published: 5 April 2019

Abstract: In the current study, three roadside bioretention systems with different configurations were constructed to investigate their pollutant removal efficiency in different rainfall recurrence intervals. The bioretention systems (referred as units) (unit A: 700 mm height material without submerged zone; unit B: 400 mm height material with 300 mm submerged zone; unit C: 400 mm height material without submerged zone) were used to conduct the rainfall events with uniform 120 min rainfall duration for 2-, 5-, 10-, 15-, and 30-year recurrence intervals. Results reveal that the gradual increase of rainfall return period would have negative effects on TN and NH_4^+-N removal. The higher filler layer may increase pollutant removal efficiency. Setting a submerged zone could improve the COD_{Mn} and TN removal compared to TP and NH_4^+-N removal. The values for comprehensive reduction rate of pollutant load in the three bioretention systems were recorded as follows: 64% in SS, 50%~80% in TP, 69% in NH_4^+-N, and 28%~53% in NO_3-N separately. These results provide greater understanding of the design and treatment performance of bioretention systems.

Keywords: bioretention; pollutant removal; urban road runoff; stormwater management

---

1. Introduction

Since the beginning of the 21st century, the urbanization process has developed rapidly in China. By the end of 2015, the level of urbanization in China had exceeded 54%, and there is still much room for urbanization. Although urbanization can promote the rapid development of regional economies, at the same time, hardened pavement has been expanded on a large scale, gradually replacing the soil-rock pavement composed of vegetation, which has greatly changed the original permeability of the pavement, triggering the generation of urban rainwater runoff [1]. The nature of urban underlying surfaces is the most important factor affecting the form of runoff, while pavement hardening will lead to the growth of urban rainwater runoff and the pollution of surface water, which will lead to various water environmental problems, especially rainstorms and floods, and seriously affect peoples' daily lives [2,3]. At present, facing the double problems of frequent urban waterlogging disasters and shortage of freshwater resources, improving urban drainage system, increasing rainwater recycling efficiency, and balancing water resources have become the most important issues in the development of urbanization [4]. Many researchers are engaged in research related to flood control and retention facilities. One of the essential needs for retention reservoirs is to reduce the volume of wastewater flow in sewer systems. Their main advantage is the potential to increase retention in the system, which in turn improves hydraulic safety by reducing the risk of node flooding and the emergence of the phenomenon of "urban flooding" [5]. The most effective method of adapting sewer systems suffering from hydraulic overloads to the new hydraulic conditions is to apply rainwater retention at the various stages of rainwater handling and disposal [6–10]. Therefore, the development of alternative, more sustainable, and cost-effective urban drainage systems has been growing in size [11].

A new urban water resource management strategy called the Sponge City Program (SCP) was proposed by China in 2013, which totally deviated from the traditional rapid-drainage techniques [12]. The Sponge City concept is an effective urban water management plan that aims to absorb rainwater for reuse in a proper scientific manner [13]. Meanwhile, the construction of the sponge urban road facility emphasizes the imitation of natural ecological rainwater treatment, as well as the application of low impact development techniques (LID) [14]. Much attention is also paid to the importance of slow release and distributed rainwater in order to address urban road flooding rationally, and make full use of rainwater resources [15]. Adoption of the sponge urban facilities (such as pumps, storage tanks, wetlands, and a bioretention facility) is aimed at alleviating the urban flooding problem, which has always been a key issue in China [16].

Bioretention is a key element of LID practices, which is implemented for hydrological and water treatment purposes, flood control, and water resource management [17,18]. Bioretention systems are extremely important due to their ability to be used as seawater in cities, reducing stormwater runoff, eliminating pollutants, as well as reusing water resources [19]. In addition, bioretention systems also have an aesthetic function valued by the public [20]. As an important part of the urban ecosystem, most of the urban green space does not meet its design expectations. Attention is only paid to the aesthetics, causing poor performance regarding reduction of peak flow and pollutant load [21]. Therefore, the transformation of green space into a roadside bioretention system is necessary to resolve this problem.

The objectives of this study are: (1) to evaluate the comprehensive reduction rates of pollutant load, including suspended solids (SS), chemical oxygen demand (COD_{Mn}), total phosphorus (TP), ammonium-nitrogen (NH_4^+-N), nitrate-nitrogen (NO_3^--N), and total nitrogen (TN) of a rainfall event of 120 min duration for recurrence intervals of 2, 5, 10, 15, and 30 years in three roadside bioretention units; (2) investigate the relationship between the height of the filler layer and pollutant removal efficiency; and (3) investigate the effects of setting of a submerged area on pollutant removal efficiency.

2. Methodology

2.1. Experimental Site

In this study, an urban road green space along the arterial road was selected to be transformed into bioretention systems, which was located in the Hefei High-tech Zone Park, Hefei, China. The catchment area of this green space was approximately 60 m^2.

Three types of bioretention units with different configurations were designed and constructed according to the design idea of a rain garden. Design of the experimental bioretention units utilized in the current work is presented in Figure 1. Detailed configuration of the bioretention units is provided in Supplementary Materials (Table S1). All the experimental units had a uniform surface area: 2 × 1.5 m. The first bioretention unit (unit A) had a 700 mm height without a submerged zone. The artificial filler layer was uniformly mixed material of coarse sand, original soil, and wood chips in a ratio of 1.5:1:0.5. The original soil was undisturbed natural soil and the wood chips were added to the artificial filler layer mainly because they served as organic matter [12]. The second bioretention unit (unit B) had a 400 mm substrate with a 300 mm submerged zone, meanwhile, a vertical PVC tube of 300 mm height was laid at the bottom of the perforated pipe. A gravel layer with a height of 300 mm was laid on the upper part of the perforated pipe, and the gravel particle size was 6.35~12.7 mm. The third bioretention unit (unit C) had a 400 mm height without a submerged zone. The upper part of the filter material layer was covered with 50 mm planting soil. The construction details of the system are supposed to adapt to local environmental conditions [22], and *Photinia fraseri* were planted in the unit, which is a commonly available plant species (Hefei, China) with a planting density of 15 units per square meter. At the bottom of each unit, perforated water collection pipes with 75 mm diameter were arranged, and a perforated collector outlet was located at the original rain water storage tank located in the park. The overflow port in the pool was directly connected with the municipal pipe network.

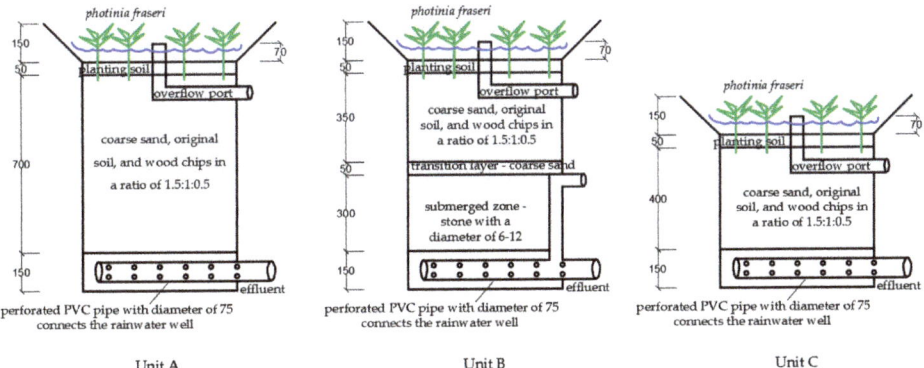

Figure 1. Bioretention units: structural details (Unit: mm).

2.2. Rainfall Event Selection and Hydrologic Parameters

In this study, the Chicago Design Storm (CDS) was selected for rainfall event simulation and analysis, which was proposed by Keifer and Chu in 1957 [23]. It is a non-uniformly designed rainfall storm pattern based on the relationships of intensity-duration-frequency (IDF) [4]. The selected rainfall patterns were 120 min rainfall duration separately under 2-, 5-, 10-, 15-, and 30-year rainfall recurrence intervals. The Chicago Design Storm determination process (Figure 2) includes the determination of the integrated rainfall peak position coefficient (r) and the rainfall hydrograph model [4]. Statistical results of the rainfall peak position coefficient (r) in different countries and regions, including China, are presented in the Supplementary Materials (Table S2) [24].

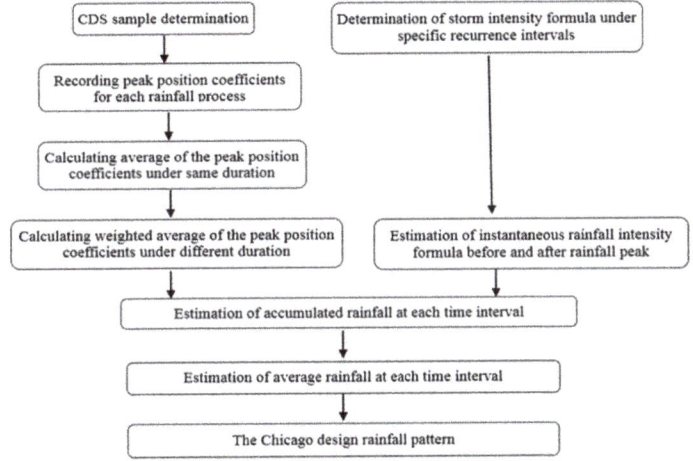

Figure 2. Flow chart of the Chicago Design Storm (CDS) Model [4].

The instantaneous rainfall intensity before and after the rain peak can be calculated by the following formula [4]:

$$i(t_b) = \frac{A\left[\frac{(1-n)t_b}{r} + b\right]}{\left[\left(\frac{t_b}{r}\right) + b\right]^{n+1}} \quad (1)$$

$$i(t_a) = \frac{A\left[\frac{(1-n)t_a}{1-r} + b\right]}{\left[\left(\frac{t_a}{1-r}\right) + b\right]^{n+1}} \quad (2)$$

where i(t$_b$) is the instantaneous intensity before rain peak, i(t$_a$) is the instantaneous intensity after rain peak, t$_a$ and t$_b$ are the corresponding rainfall duration, r is the rainfall peak position coefficient. Taking a certain recurrence interval, the storm intensity formula is $i = \frac{A}{(t+b)^n}$, noting that A, b, and n are constant parameters for a certain return period, which are calculated and determined by statistical methods. Specifically, A is the rainfall intensity parameter, indicating the 1 min design rainfall (mm) at the return period of 1 a; b is the rainfall duration correction parameter, which represents a time constant (min) added to the logarithm of the two sides of the storm intensity formula to draw the curve into a straight line; n is the rain attenuation index, related to the return period.

The Chicago synthetic rainstorm process line can be calculated by using formula (1) and (2). Afterwards, the Chicago design rainfall pattern corresponding to a certain recurrence interval and rainfall duration can be determined. The integrated rainfall peak position coefficient of Hefei in the current study was determined to be 0.4. The other parameters in the Hefei storm intensity formula are shown in Table 1, noting that parameter C is the rain force variation parameter (dimensionless). The CDS simulation rainfall patterns with 120 min duration in Hefei for return periods of 2, 5, 10, 15, and 30 years were calculated by EXCEL (Supplementary Materials (Table S3)). Rainfall data in different periods in Hefei during the 10-year recurrence interval and the average influent flow rate per 5 min under different recurrence intervals are provided in Supplementary Materials (Tables S3 and S4).

Table 1. Chicago design storm parameters [4].

Integrated rainfall peak position coefficient (r)	0.4
Total rainfall duration (T) min	120
Rain force parameter (A) mm	36.61
Rain force variation parameter (C)	0.77
Rainfall duration correction parameter (b) min	23.15
Rain attenuation index (n)	0.93

2.3. Influent Water Volume and Quality

Based on the simulation results of the rainfall intensity at each time interval in Hefei, the inflow flow rate per minute and the total water inflow of 120 min in the bioretention system under different rainfall recurrence intervals could be obtained (Supplementary Materials (Table S3)). Since the rainfall intensity was not changed much in a short period of time in the case of the experiment condition, the inflow flow rate was adjusted every 5 min in order to make the bioretention operation easier.

Xie et al. analyzed the SS, COD, TN, and TP water quality indicators of different rainfall runoffs and natural rainfall samples from 4 different types of underlying surfaces in Hefei city, and the scouring effect was compared between the rainfall runoff of different types of underlying surfaces and the concentration distribution of pollutants in natural rainfall samples, and the average concentration of incidents of rainfall runoff pollutants was calculated [25]. According to the water quality characteristics of rainfall runoff on different underlying surfaces in Hefei city, the synthetic runoff in the current study was prepared to mimic local highly polluted runoff characteristics of COD, TP, TN, NH_4^+-N, and NO_3^--N (Table 2).

Table 2. Mean inflow concentrations of pollutants in the synthetic runoff.

Pollutant	Mean Inflow Concentration (mg/L)	Source
COD	120	Glucose ($C_6H_{12}O_6$)
TP	1.0	Monopotassium phosphate
TN	8.0	Alanine
NH_4^+-N	4.0	Ammonium chloride
NO_3^--N	2.0	Potassium nitrate

2.4. Experimental Procedure

Three mesocosm bioretention units were installed and operated with synthetic runoff for two hours at one-week intervals for five rainfall events (five rainfall recurrence intervals) during the month of May to September, 2016 (at 23~38 °C). According to the runoff pollution load shown in Table 2, the synthetic water was prepared in the laboratory and added to the storage unit (bucket) (Figure 3), and the amount of sample water was the total rainfall under a specific return period. In the experimental process, the water inflow of 5-min intervals is shown in Supplementary Materials (Table S5). The total water inflows under 2-, 5-, 10-, 15-, and 30-year recurrence intervals were 3.22 m^3, 4.02 m^3, 4.62 m^3, 4.97 m^3, and 5.58 m^3, respectively. The turbine flow meter was used to control the outlet flow and total water output and the valves were adjusted according to the outlet flow rate until the experimental requirements were met. Bioretention units A, B, and C were sequentially conducted to the above experimental steps, while the interval of rainfall was supposed to be one week. The above test steps were repeated with different pollution loads under 2-, 5-, 10-, 15-, and 30-year recurrence intervals.

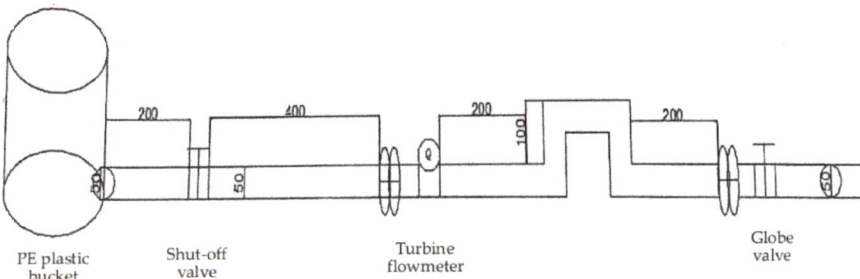

Figure 3. Experimental device and process flow chart (unit: mm).

2.5. Sampling and Data Analysis

Discrete samples were collected to investigate the variation in water quality during a runoff event. The inlet and outlet water samples were collected at the 30-min intervals after the unit began to have stable water discharge. The total volume of each sample was 2.0L. The samples collected and analyzed for SS, COD_{Mn}, TP, NH_4^+-N, NO_3—N, and TN, which are the primary stormwater pollutants [26].

In this study, the event mean concentration (EMC) removal method, recommended by the American Society of Civil Engineers and the Environmental Protection Agency (ASCE-EPA) [27], was used to calculate the removal efficiency:

$$\text{Removal efficiency (\%)} = \left[\frac{EMC_i - EMC_o}{EMC_i}\right] \times 100 \qquad (3)$$

where EMC_i and EMC_o were the event mean concentration of various pollutants in the inflow and outflow, respectively.

All the box-and-whisker plots were drawn using OriginPro 2015 (OriginLab Corp., Northampton, MA, USA). Significant statistical difference analyses were performed by the statistical software package PASW Statistics 19.0 (SPSS Inc., Chicago, IL, USA). Meanwhile, ANOVA was used to test the significant difference in the removal efficiencies of various pollutants.

3. Results and Discussion

To evaluate the pollutant removal efficiency of the three bioretention systems, the concentration of contaminants in the water at the outlet under the given five rainfall scenarios was detected and compared to the inlet water. Statistical analysis was conducted on pollutant removal and load reduction rate for the bioretention facilities with the consideration that the pollutant removal and load reduction rate are 100% if no outflow and overflow occur. The pollutant removal efficiency and load reduction of the bioretention facilities are presented in Table 3.

Table 3. Mean removal efficiency of pollutants for each bioretention system.

Unit	Recurrence Interval (Years)	Mean Removal Efficiency (%)					
		SS	COD_{Mn}	TP	NH_4^+-N	NO_3^--N	TN
A	2	52.4	51.4	65.8	82.9	32.1	68.1
	5	73.5	60.3	30.4	77.6	38.2	59.3
	10	66.7	78.7	51.0	68.3	50.5	62.4
	15	66.7	69.2	75.8	66.6	42.9	57.1
	30	55.2	47.5	84.5	51.7	23.5	48.3
B	2	71.3	12.7	51.4	74.4	47.9	63.6
	5	68.5	38.5	17.0	68.1	49.5	57.1
	10	62.2	63.4	54.4	70.1	67.3	71.4
	15	57.1	83.4	71.9	61.7	58.3	68.3
	30	62.1	79.9	65.0	68.4	42.3	56.2
C	2	49.9	30.4	30.1	55.4	23.1	41.1
	5	68.2	54.2	24.7	48.6	27.8	37.4
	10	57.7	67.8	30.7	50.8	37.4	43.9
	15	48.2	58.7	35.1	39.3	28.6	32.6
	30	41.7	50.4	31.2	28.0	25.2	25.1

3.1. SS Removal

Earlier studies revealed that bioretention has the ability to capture SS and other solids due to its capacity for sedimentation and filtration [28]. Figure 4 reveals that the removal rate of SS in all 3 units reached the highest level at the 5-year return period, and decreased slightly afterwards. The removal efficiency of SS fluctuated greatly in the single inlet process, but the average removal efficiency varied slightly in different structures of the bioretention units and in different recurrence intervals. However, the impact of climate-change-induced precipitation increase on SS retention has not been clearly assessed [29]. The concentration of SS in effluent was generally stable at 30 mg/L, and some effluent SS could even be optimized below 20 mg/L. In the current study, under the five rainfall scenarios, the average removal efficiency in unit A and B was around 64%, comparatively lower than other studies. Brown and Hunt demonstrated that removal performance for the SS was less variable than other nutrient parameters, the value of which was between 70 to 80% [28], whereas the SS removal rate was consistently over 95% in the work of Bratieres et al. [30]. Hatt et al. presented that loads of SS were effectively reduced by all filter types, despite variability in inflow pollutant concentrations [31]. Only the initial runoff SS concentration reached 150 mg/L, and then the influent SS concentration was maintained at 60 mg/L in the current experiment. When compared to the removal rate of unit A and unit B, it indicated that a higher filter layer would bring better results, which is consistent with the conclusion obtained by Brown and Hunt [28].

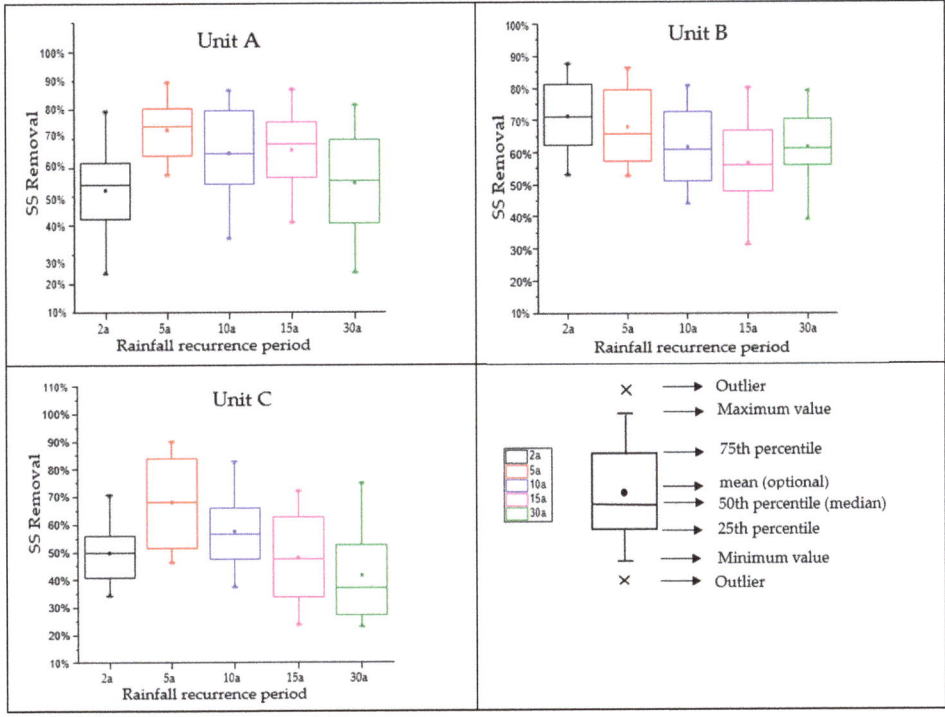

Figure 4. Removal efficiency of SS in bioretention units (A–C).

3.2. COD_{Mn} Removal

Figure 5 presents the mean concentration removal efficiency for COD_{Mn}. It revealed that in the early stage of the systems, due to the decay of the wood chips added by the artificial filler, the number of microbes in the system was too small to decompose large quantities of pollutants, resulting in the lower removal efficiency ($\geq 12.7\%$ in unit B). After the systems were stabilized, the degradation of organic matter was promoted with the increase of oxygen content and the microbial population of the systems, leading to the rapid increase of the COD_{Mn} removal in the three bioretention units (the highest removal rate of unit B in the 15 year recurrence interval reached up to 94.2%). Li et al. found that the COD_{Mn} removal efficiency with different thicknesses of filter varied between 65%~75% [32]. However, as the experiment continued, with the further increase of the rainfall recurrence interval, the COD_{Mn} adsorption capacity of the filler was gradually saturated. Under scouring action of the runoff, the COD_{Mn} leaching from the filler led to a gradual decline of the removal effect. The same phenomenon appeared in the experiment of Jiang et al. and Shrestha et al. [19,33]. Among the three bioretention units, unit A showed the best performance on COD_{Mn} removal. It could be deduced that COD_{Mn} removal was least affected by the rainfall intensity in the case of the higher filler layer, which was similar to the findings of Liu [34]. Unit B showed better performance than unit C, indicating that setting a submerged zone could improve COD_{Mn} removal efficiency in some extent.

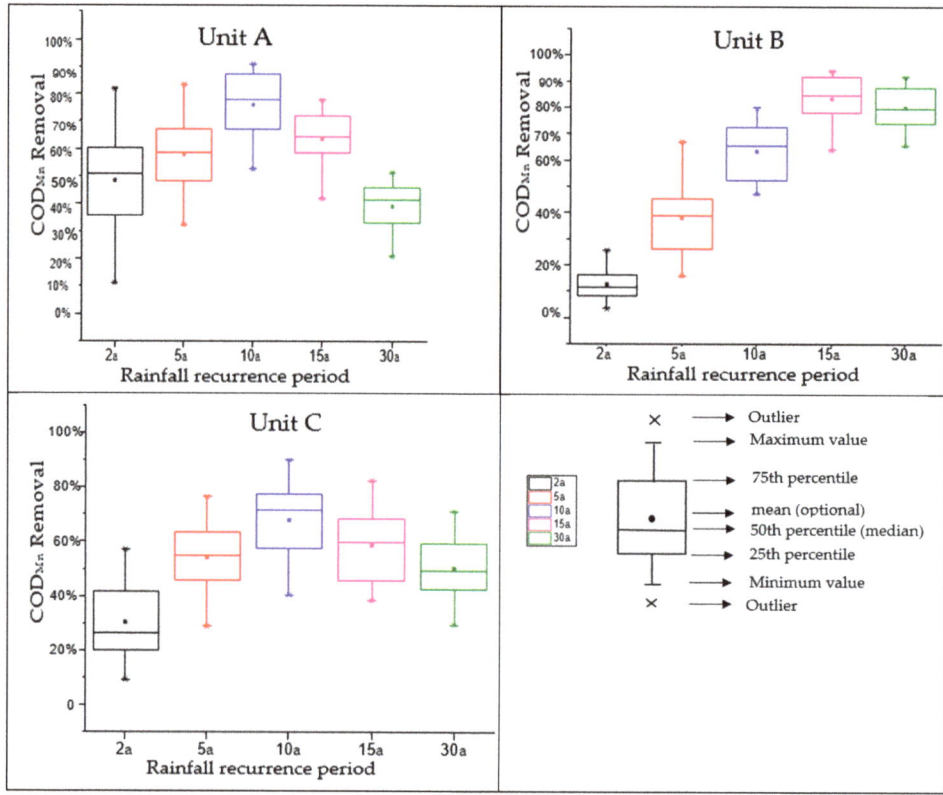

Figure 5. Removal efficiency of COD$_{Mn}$ bioretention units (A–C).

3.3. TP Removal

The mean TP removal rates in this study are presented in Figure 6, which shows unit A fluctuated between the range of 50%~85%, unit B was slightly lower than unit A, and unit C had the least ideal performance (at around 30%). Several studies have measured the TP removal efficiency in bioretention systems. For example, reports of labile P removal by bioretention to date have been extremely variable: from 78 to 98% for soluble reactive phosphorus (SRP) [35]. Li et al. found that soluble reactive phosphorus removal rate in bioretention systems was in the range of 65%~85% [36]. Shrestha et al. noted that the removal rate of TP in bioretention systems with different configurations was 50%~70% [33]. In contrast to these studies, TP removal efficiency in this study was generally lower. This could be because the original soil around the bioretention units was buried in the park building and domestic garbage, which caused the removal efficiency to fluctuate greatly and be relatively low. The relatively higher TP reduction happened in unit A, which demonstrated that the height of the filler layer was critical for the removal of TP, while the setting of the submerged area results in a reduction in TP removal. Earlier studies also reveal that setting a submerged area exerted minimal influence on the TP removal in different filler combinations [33,36].

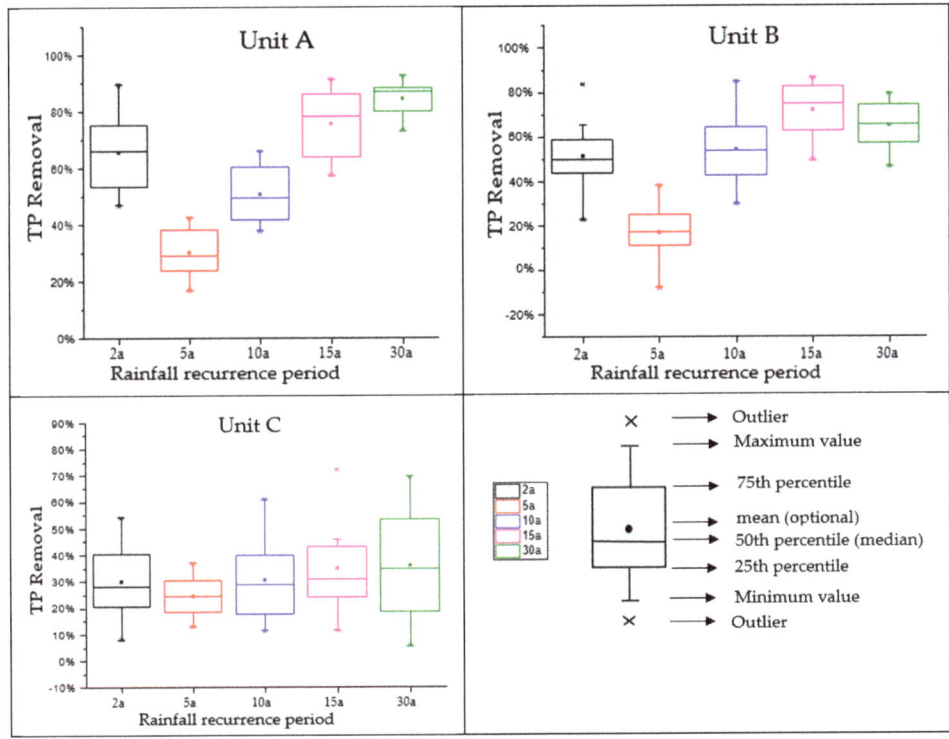

Figure 6. Removal efficiency of TP in bioretention units (A–C).

3.4. Nitrogen Removal

In contrast with other synthetic stormwater pollutant parameters, nitrogen species are more difficult to remove due to being highly soluble and strongly affected by the variable wetting and drying regime inherent in bioretention system [37]. Removal efficiency of various nitrogen species in bioretention units is presented in Figure 7. In the current study, both unit A and B were effective on NH_4^+-N with a removal rate of 69%. Results reveal that three units were not effective for removing NO_3—N where the mean removal was in the range of 28%~53%. For TN removal, unit A and unit B showed better removal effect, and the effluent water quality could reach the surface V water standard, which is stipulated by the Standard for Surface Water Environmental Quality (GB3838-2002), representing water qualities that are suitable for agricultural water use, as well as general landscape requirements, and water bodies that exceed V water quality standards are useless. Meanwhile, the unit C removal effect was poor, which was more affected by the rainfall return period. With the gradual increase of rainfall return period, the effects of NH_4^+-N and TP removal were significantly reduced, similar to the results of Li et al. [32].

Comparing the performance of the three units, results indicated that the height of the filler layer had a greater influence on the removal effect of NH_4^+-N and TN under different rainfall recurrence intervals, which is the same conclusion as that obtained by Liu [34]. The NH_4^+-N removal efficiency in unit B was lower than that of unit A under most recurrence intervals. The reason may be that the existence of submerged area affected the aerobic environment of the whole system, resulting in lower removal efficiency [37,38]. However, unit B with a submerged area had the best removal effect on TN, and was least affected by the change of rainfall intensity, which indicated the same conclusion as that obtained by Afrooz and Boehm [39]. All the three units had a large fluctuation in the removal effect of NO_3^--N, and the removal effect was not ideal. Nitrate removal in bioretention systems equipped with drains is usually worse than that in NH_4^+-N due to its poor adsorption capacity on most of the soils or filtration media [40]. The results from the experiment of Brown et al. revealed another reason, which reported that the residual ammonia and organic nitrogen could be nitrated to nitrate during aerobic circumstances in the filler layer during the rainfall events, resulting in low nitrate removal or even excessive leaching during subsequent events [37].

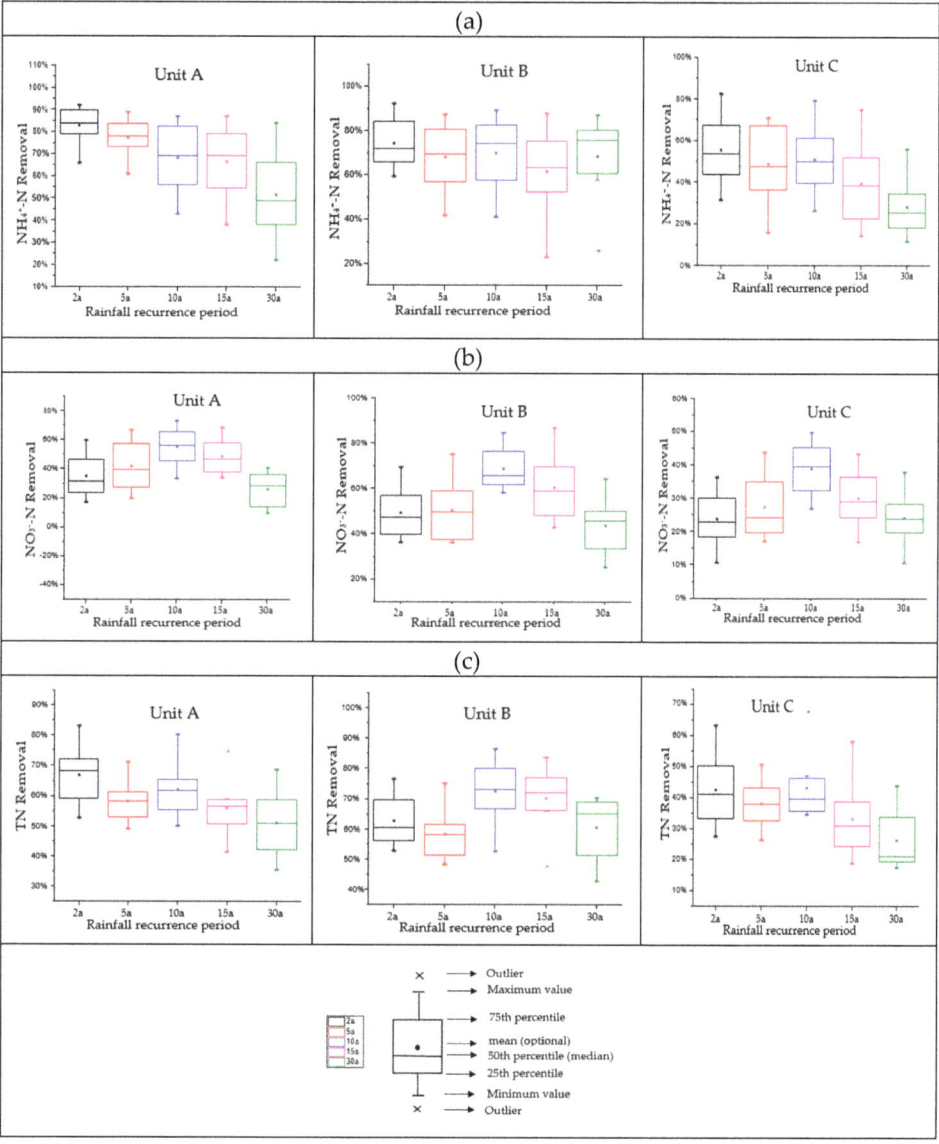

Figure 7. Removal efficiency of nitrogen species in bioretention units (A–C); (**a**) NH_4^+-N; (**b**) NH_3—N; and (**c**) TN.

The performance of bioretention systems is dynamic as a function of media depth, rainfall pattern, and retention time. The configuration of the saturated zone and optimized depth of filler layers can provide a favorable condition for pollutant removal under random inflow conditions. The factors, including climate, rainfall pattern, and runoff pollutant load, should be taken into consideration when designing a roadside bioretention system. Given the innovated pollutant removal ability of the bioretention system under different rainfall recurrence intervals in the case of 120 min rainfall duration, further studies should evaluate the design and removal efficiency under different rainfall duration.

4. Conclusions

The results of the simulated rainfall events under five different rainfall intervals in three different bioretention systems reveal that: (1) with the gradual increase of rainfall recurrence interval, TN and NH_4^+-N concentration reduction showed a significant decrease, while COD_{Mn} removal rate also declined slightly. SS and TP showed no obvious change under different intervals. (2) In most of the cases, the pollutant parameters presented higher removal efficiency with the higher filler layer in the current study. (3) A submerged zone improved COD_{Mn} and TN removal rates, but resulted in a slight reduction of TP and NH_4^+-N removal. (4) For the comprehensive reduction rate of the pollutant load, suspended solids (SS) were effectively removed, with a 64% mean concentration removal in unit A and unit B. The COD_{Mn} removal rates of the three bioretention units fluctuated heavily, reaching 85% in the case of the stabilized system. However, the phenomenon of COD_{Mn} leaching from the filler layer was common, which caused the higher concentration of the outflow. TP removal efficiency in the current study was 50%~80%, which might be affected by the surrounding soil conditions. For nitrogen species removal, NH_4^+-N concentration removal was about 69%; the TN concentration of the outlet in unit A and unit B could reach the surface V water standard. The NH_3^--N removal was the least (28%~53%) and it requires further study for a clearer explanation.

Supplementary Materials: The following are available online at http://www.mdpi.com/2071-1050/11/7/2040/s1, Tables S1~S5.

Author Contributions: D.F. and R.P.S. conceived and designed the experiments. D.F. supervised the experiments and managed the funding. F.Z., J.S., and Q.J. performed the experiments, and along with R.P.S. analyzed the data. R.P.S. wrote the paper.

Acknowledgments: This work was financially supported by the National Key Research and Development Program of China (Grant No. 2018YFC0809900).

Conflicts of Interest: The authors declare no conflict of interest.

References

1. Liu, Z.; Li, J.; Li, P.; Li, Y.; Li, W. Study of bioretention system on heavy-metal removal effect. *Pol. J. Environ. Stud.* **2017**, *27*, 163–173. [CrossRef]
2. Brezonik, P.L.; Stadelmann, T.H. Analysis and predictive models of stormwater runoff volumes, loads, and pollutant concentrations from watersheds in the twin cities metropolitan area, Minnesota, USA. *Water Res.* **2002**, *36*, 1743–1757. [CrossRef]
3. Mohammed, A.; Babatunde, A.O. Modelling heavy metals transformation in vertical flow constructed wetlands. *Ecol. Model.* **2017**, *354*, 62–71. [CrossRef]
4. Uruya, W.; Winai, C.; Muhammad Mudassar, R.; Sutat, W. Modification of a design storm pattern for urban drainage systems considering the impact of climate change. *Eng. Appl. Sci. Res.* **2017**, *44*, 161–169.
5. Pochwat, K.B.; Słyś, D. Application of artificial neural networks in the dimensioning of retention reservoirs. *Ecol. Chem. Eng. S* **2018**, *25*, 605–617. [CrossRef]
6. Mazurkiewicz, K.; Skotnicki, M.; Cimochowicz-Rybicka, M. The influence of synthetic hyetograph parameters on simulation results of runoff from urban catchment. *E3S Web Conf.* **2018**, *30*, 01018. [CrossRef]
7. Davydova, Y.; Volkova, Y.; Nikonorov, A.; Aleksandrovskiy, M. Drainage of small volume reservoirs on the technogenic territories. *MATEC Web Conf.* **2018**, *170*, 02025. [CrossRef]

8. Douglas, N.I. On-site stormwater detention: Improved implementation techniques for runoff quantity and quality management in Sydney. *Water Sci. Technol.* **1995**, *32*, 85–91.
9. Guo, X.C.; Zhao, D.Q.; Du, P.F.; Li, M. Automatic setting of urban drainage pipe monitoring points based on scenario simulation and fuzzy clustering. *Urban Water J.* **2018**, *15*, 700–712. [CrossRef]
10. Liang, J.; Melching, C.S. Experimental evaluation of the effect of storm movement on peak discharge. *Int. J. Sediment Res.* **2015**, *30*, 167–177. [CrossRef]
11. De, M.M.; Rosa, A.; Do, C.L.; Mendiondo, E.M.; De, V.S. Learning from the operation, pathology and maintenance of a bioretention system to optimize urban drainage practices. *J. Environ. Manag.* **2017**, *204*, 454–466.
12. Wang, C.; Wang, F.; Qin, H.; Zeng, X.; Li, X.; Yu, S.-L. Effect of saturated zone on nitrogen removal processes in stormwater bioretention systems. *Water* **2018**, *10*, 162. [CrossRef]
13. Hao, W.; Chao, M.; Liu, J.H.; Shao, W.W. A new strategy for integrated urban water management in China: Sponge city. *China Technol. Sci.* **2018**, *61*, 317–329.
14. Zhang, L.; Lu, Q.; Ding, Y.; Peng, P.; Yao, Y. Design and performance simulation of road bioretention media for sponge cities. *Perform. Constr. Facil.* **2018**, *32*, 04018061. [CrossRef]
15. Davis, A.P. Green engineering principles promote low impact development. *Environ. Sci. Technol.* **2005**, *39*, 338–344. [CrossRef]
16. Mei, C.; Liu, J.; Wang, H.; Yang, Z.; Ding, X.; Shao, W. Integrated assessments of green infrastructure for flood mitigation to support robust decision making for sponge city construction in an urbanized watershed. *Sci. Total Environ.* **2018**, *639*, 1394–1407. [CrossRef]
17. Chin, D.A. Designing bioretention areas for stormwater management. *Environ. Proc.* **2017**, *4*, 1–13. [CrossRef]
18. Gülbaz, S.; Kazezyılmaz-Alhan, C.M. Hydrological model of LID with rainfall-watershed-bioretention system. *Water Resour. Manag.* **2017**, *31*, 1931–1946. [CrossRef]
19. Jiang, C.; Li, J.; Li, H.; Li, Y.; Chen, L. Field performance of bioretention systems for runoff quantity regulation and pollutant removal. *Water Air Soil Pollut.* **2017**, *228*, 468. [CrossRef]
20. Kim, S.; An, K. Exploring psychological and aesthetic approaches of bioretention facilities in the urban open space. *Sustainability* **2017**, *9*, 2067. [CrossRef]
21. Guo, J.; Yu, B.; Zhang, Y.; Che, S. Predicted models for potential canopy rainfall interception capacity of landscape trees in Shanghai, China. *Eur. J. For. Res.* **2017**, *136*, 387–400. [CrossRef]
22. Tahvonen, O. Adapting bioretention construction details to local practices in Finland. *Sustainability* **2018**, *10*, 276. [CrossRef]
23. Keifer, C.J.; Chu, H.H. Synthetic storm pattern for drainage design. *J. Hydraul. Div.* **1957**, *83*, 1–25.
24. Taylor, S.R.; Mclennan, S.M. The geochemical evolution of the continental crust. *Rev. Geophys.* **1995**, *33*, 241–265. [CrossRef]
25. Xie, J.F.; Hu, Z.X.; Xu, T.; Han, H.Y.; Yin, D.Q. Analysis on characteristics of rainfall runoff water quality of different underlying surfaces in Hefei City. *J. China Environ. Sci.* **2012**, *32*, 1018–1025. (In Chinese)
26. Mangangka, I.R.; Liu, A.; Egodawatta, P.; Goonetilleke, A. Performance characterisation of a stormwater treatment bioretention basin. *J. Environ. Manag.* **2015**, *150*, 173–178. [CrossRef]
27. American Public Health Association; American Water Works Association; Water Environment Federation. *Standard Methods for the Examination of Water and Wastewater*, 22nd ed.; American Public Health Association, American Water Works Association, Water Environment Federation: Washington, DC, USA, 2012.
28. Brown, R.A.; Hunt, W.F. Impacts of media depth on effluent water quality and hydrologic performance of undersized bioretention cells. *J. Irrig. Drain. Eng.* **2011**, *137*, 132–143. [CrossRef]
29. Cording, A.; Hurley, S.; Adair, C. Influence of critical bioretention design factors and projected increases in precipitation due to climate change on roadside bioretention performance. *J. Environ. Eng.* **2018**, *144*, 04018082. [CrossRef]
30. Bratieres, K.; Fletcher, T.; Deletic, A.; Zinger, Y. Nutrient and sediment removal by stormwater biofilters: A large-scale design optimisation study. *Water Res.* **2008**, *423*, 930–3940. [CrossRef] [PubMed]
31. Hatt, B.E.; Fletcher, T.D.; Deletic, A. Hydraulic and pollutant removal performance of fine media stormwater filtration systems. *Environ. Sci. Technol.* **2008**, *42*, 2535–2541. [CrossRef]
32. Li, J.; Zhao, R.; Li, Y.; Chen, L. Modeling the effects of parameter optimization on three bioretention tanks using the Hydrus-1d model. *J. Environ. Manag.* **2018**, *217*, 38–46. [CrossRef]

33. Shrestha, P.; Hurley, S.E.; Wemple, B.C. Effects of different soil media, vegetation, and hydrologic treatments on nutrient and sediment removal in roadside bioretention systems. *Ecol. Eng.* **2018**, *112*, 116–131. [CrossRef]
34. Liu, J. The Design and Operation of the Bioretention with Submerge Area Zone. Ph.D. Thesis, Southeast University, Nanjing, China, 2015. (In Chinese)
35. Lefevre, G.H.; Paus, K.H.; Natarajan, P.; Gulliver, J.S.; Novak, P.J.; Hozalski, R.M. Review of dissolved pollutants in urban storm water and their removal and fate in bioretention cells. *J. Environ. Eng.* **2015**, *141*, 04014050. [CrossRef]
36. Li, J.; Liang, Z.; Li, Y.; Li, P.; Jiang, C. Experimental study and simulation of phosphorus purification effects of bioretention systems on urban surface runoff. *PLoS ONE* **2018**, *13*, e0196339. [CrossRef] [PubMed]
37. Brown, R.A.; Birgand, F.; Hunt, W.F. Analysis of consecutive events for nutrient and sediment treatment in field-monitored bioretention cells. *Water Air Soil Pollut.* **2013**, *224*, 1581. [CrossRef]
38. Wu, J.; Cao, X.; Zhao, J.; Dai, Y.; Cui, N.; Li, Z.; Cheng, S. Performance of biofilter with a saturated zone for urban stormwater runoff pollution control: Influence of vegetation type and saturation time. *Ecol. Eng.* **2017**, *105*, 355–361. [CrossRef]
39. Afrooz, A.R.M.N.; Boehm, A.B. Effects of submerged zone, media aging, and antecedent dry period on the performance of biochar-amended biofilters in removing fecal indicators and nutrients from natural stormwater. *Ecol. Eng.* **2017**, *102*, 320–330. [CrossRef]
40. Li, L.; Davis, A.P. Urban stormwater runoff nitrogen composition and fate in bioretention systems. *Environ. Sci. Technol.* **2014**, *48*, 3403–3410. [CrossRef]

© 2019 by the authors. Licensee MDPI, Basel, Switzerland. This article is an open access article distributed under the terms and conditions of the Creative Commons Attribution (CC BY) license (http://creativecommons.org/licenses/by/4.0/).

Article

Rainfall Runoff Mitigation by Retrofitted Permeable Pavement in an Urban Area

Muhammad Shafique [1,2], Reeho Kim [1,2,*] and Kwon Kyung-Ho [3]

1. Department of Smart City and Construction Engineering, Korea Institute of Civil Engineering and Building Technology, University of Science and Technology (UST), 217, Gajeong-ro, Yuseong-gu, Daejeon 34113, Korea; shafique@ust.ac.kr
2. Environmental & Plant Engineering Research Institute, Korea Institute of Civil Engineering and Building Technology, 83, Goyangdae-ro, Ilsanseo-gu, Goyang-si, Gyeonggi-do 10223, Korea
3. Urban Water Cycle Research Center, Korea Institute of Safe Drinking Water Research, Anyang si, Gyeonggi-do 14059, Korea; kwonkh@kisd.re.kr
* Correspondence: rhkim@kict.re.kr; Tel.: +82-31-9100-291

Received: 21 March 2018; Accepted: 12 April 2018; Published: 18 April 2018

Abstract: Permeable pavement is an effective low impact development (LID) practice that can play an important role in reducing rainfall runoff amount in urban areas. Permeable interlocking concrete pavement (PICP) was retrofitted in a tremendously developed area of Seoul, Korea and the data was monitored to evaluate its effect on the hydrology and stormwater quality performance for four months. Rainfall runoff was first absorbed by different layers of the PICP system and then contributed to the sewage system. This not only helps to reduce the runoff volume, but also increase the time of concentration. In this experiment, different real rain events were observed and the field results were investigated to check the effectiveness of the PICP system for controlling the rainfall runoff in Songpa, Korea. From the analysis of data, results showed that the PCIP system was very effective in controlling rainfall runoff. Overall runoff reduction performance from the PCIP was found to be around 30–65% during various storm events. In addition, PICP significantly reduced peak flows in different storm events which is very helpful in reducing the chances of water-logging in an urbanized area. Research results also allow us to sum up that retrofitted PICP is a very effective approach for rainfall runoff management in urban areas.

Keywords: permeable interlocking concrete pavement; runoff mitigation; storm events; rainwater management; urban area

1. Introduction

Climate change and urbanization are two dominant factors that are altering the natural hydrological cycle as well as boosting the flash flooding in cities [1,2]. High speed urban growth increases the peak runoff, which involves the rapid discharge of rainfall to the conventional drainage system. The effect of climate change alters the rainfall intensity which results in raising the peak discharge and runoff volumes that might exceed the capacity of existing drainage infrastructure such as sewer systems. However, urbanization has replaced the natural surface with hard infrastructure such as roads, buildings and parking lots, which have decreased the natural infiltration rate of the soil [3–5]. These adverse impacts of urbanization have created multiple problems, such as flash flooding, stream bank erosion and water quality degradation [4,5]. Under these circumstances, our traditional storm water management approaches need to be redesigned to perform well under the extreme climate conditions [6]. To mitigate these adverse impacts of urbanization, there is an urgent need for development of new sustainable urban water management approaches around the globe. Permeable pavement has been applied globally for the past years and it can reduce flooding and water quality degradation problems [7–9]. Permeable pavement is generally used to collect, treat, and absorb rainfall runoff; allow groundwater recharge; and prevent pollution [9,10]. Previous studies have shown that permeable pavements have an ability to reduce surface runoff by permitting the infiltration into the underground soil [11–13] (even though the infiltration rate is low) and enhancing the infiltrated water quality through different layers [14,15].

Typically, permeable pavement consists of permeable block pavers with open joint spaces that allow the infiltration of surface water. Underneath the pavers lies the storage layers (bedding aggregate + base course aggregate), which generally consists of open graded aggregate size varies from 2 mm to 20 mm [4]. The rainwater infiltrates through the bedding layers and infiltrates to the ground surface. Sometimes, geotextiles are used which helps for separation and strengthening layers under new roads and car parking lots. The lowest layer is the subgrade, which is a generally native soil. If the infiltration rate of permeable pavements is lower than the perforated drainage pipe, it can be applied to the subgrade layer in the pavement design [4,15]. Figure 1 below shows details of cross section of a permeable block pavement. It also shows the different layers of the permeable pavement system.

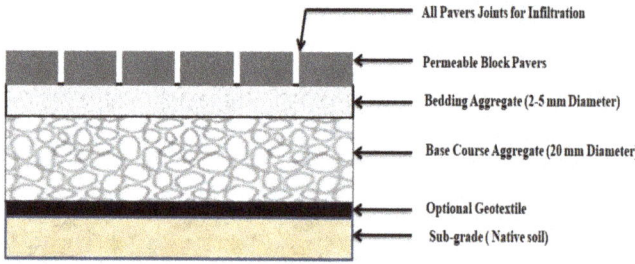

Figure 1. Shows the cross section of permeable block pavement.

Permeable pavement is referred to as permeable concrete (PC), permeable interlocking concrete pavers (PICP), porous asphalt (PA), or concrete grid pavers (CGP). Numerous studies have suggested permeable pavement as a sustainable stormwater management practice because it helps manage stormwater quantity as well as the quality [14,15]. Permeable pavement allows the rainfall runoff to infiltrate through the pores which help to reduce the small flash flood in developed areas. Different studies showed that the permeable pavement reduces the runoff volume and peak flow and also increases the time of concentration [8,15]. Brattebo and Booth [9] evaluated the four permeable pavements and found that if the rainfall is less than 5 mm, permeable pavements can reduce peak runoff by 95% and total runoff volume by 90%.

Researchers have indicated that rainfall runoff from permeable pavement is notably less as compared to traditional pavement (asphalt pavement) [13,16,17]. Collins et al. [16] investigated the hydrological performance of four different permeable pavements. Results indicated that the all four permeable pavements reduce runoff volume by 36–67% and peak flow by 60–70% as compared to the asphalt pavements. Palla et al. [17] studied the hydrological properties of permeable pavements in a laboratory. This study utilized two types of permeable pavements: concrete cell (CC) and deep pervious brick (PB). Results showed that there was no runoff observed from both permeable pavements in the first 15 min of constant rainfall intensity. In addition, the study indicated that the higher slope and bigger aggregate size enhanced the drainage [17]. Similarly, Drake [18] investigated permeable pavement performance and showed it can reduce runoff volumes by up to 43% in urban areas. Infiltration rates of the permeable pavement system greatly enhance its performance for controlling runoff in an urban area. Research shows that the newly constructed permeable pavement has higher infiltration rates and hence is more effective in managing stormwater in the urban area [19].

Some studies showed the efficiency of the permeable pavements in case of flood reduction [20,21]. For instance, Huang et al. [20] indicated that the application of permeable pavements decreased

35.6% of total rainfall runoff and 28.7% of peak flow in Tianjin University, China. The peak flow of designed five-year recurrence storm events was reduced by approximately 24.7% with the application of permeable pavements in the small village of Jurong, China [21]. Additionally, the rainfall runoff reduction performance of permeable pavements counts on the types of materials, the life of pavement, usage, operation and maintenance [16,22]. Collins et al. [16] indicated that the concrete grid pavers produced the more rainfall runoff volumes than the other kinds of permeable pavements in a field test of a parking lot in Carolina, USA. On the other hand, Kumar et al. [22] found that the infiltration rates of the pavement system depend on the service life of the pavement. From the field experiment, he found that the infiltration rates of three kinds of permeable pavements were decreased significantly from the second year because of clogging effect. Few studies also have discussed the material's effects on the efficiency of permeable pavements in controlling runoff quantity and quality at the small scale [23]. In real world applications, clogging of the pavement system is one the serious issue which needs to be investigated further for coming years [24]. In this way, we can develop a sustainable design for permeable pavements that can solve urban water related problems in the future.

Previous studies have shown that permeable pavements are very effective in controlling a large amount of rainfall runoff in urban areas. From a water quantity perspective, permeable pavements cited above mostly showed the promising results in capturing larger rain events in their sub-base layer. This process not only reduces the surface runoff but also decreases the risks of flash flooding. This study investigates the effectiveness of retrofitted permeable interlocking concrete pavement (PICP) in controlling the runoff in a populated area of Seoul, Korea. The main objectives of the paper are: (1) to investigate the performance of permeable interlocking concrete pavement in retaining runoff in different storm events; and (2) to estimate the capability of the permeable interlocking concrete pavement system to decrease the chances of flash flooding in an urban area of Seoul, Korea.

2. Materials and Methods

2.1. Site Detail Description

Permeable pavements were retrofitted in series in a highly developed area of Songpa, (Seoul, Korea, 37°49′11.33″ N 127°13′29.05″ E). Seoul perceives an average of 1300 mm of precipitation every year [25]. The research area is near the residential apartments in Songpa, Seoul, Korea (Figure 2). Because of the highly developed urban area, there are more chances of flash flooding in the rainy season. The non-busy residential road is selected for this study. This site was studied and it was found that it is composed of sandy loam soils. Eight different permeable pavement sections (permeable pavement, reinforced permeable pavement, asphalt pavement, block pavement) were retrofitted with the total length of 180 m and width of about 7 m. Lengths of each section of pavement are about 20 m and about 7 m wide. The infiltration rate of the PICP system was found to be around 0.7 mm/s during this study. The main purpose of these permeable pavements was to reduce the stormwater runoff as well as to improve the water quality performance in that area. In this study, we had selected the permeable interlocking concrete pavement (PICP) section to check the rainfall runoff reduction and water quality performance in Seoul, Korea. The PICP system consisted of different aggregate size (3–20 mm) in the sub-base with a native sandy loam soil as the sub-grade. Permeable block pavers were used on the top surface with the slope of 3%. Perforated pipe was used in the sub-grade layer to capture the infiltrate stormwater and move it to the sewer system. This PICP was designed to capture rainfalls ranging in intensity from about 20 to 50 mm/h as occurs with various storm events.

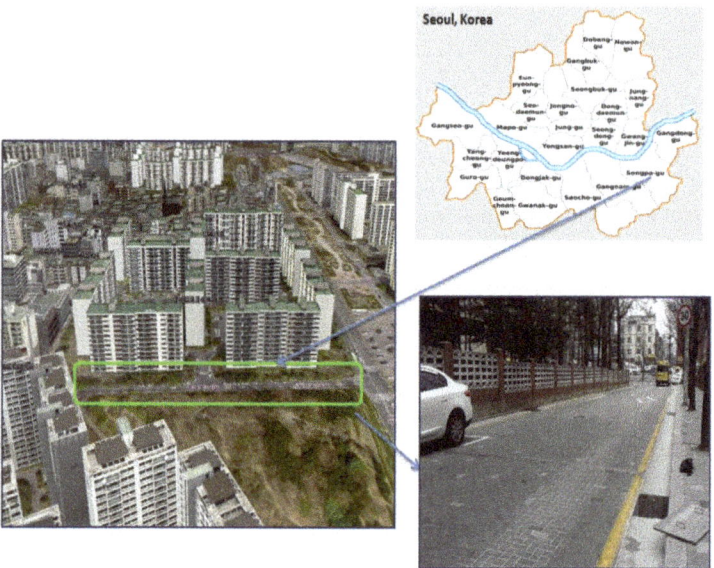

Figure 2. Study area of the permeable interlocking concrete pavement (PICP) in Seoul, Korea.

Figure 3, below shows the rainfall of Seoul, Korea throughout the year. It also shows that the maximum rainfall occurred in July of every year.

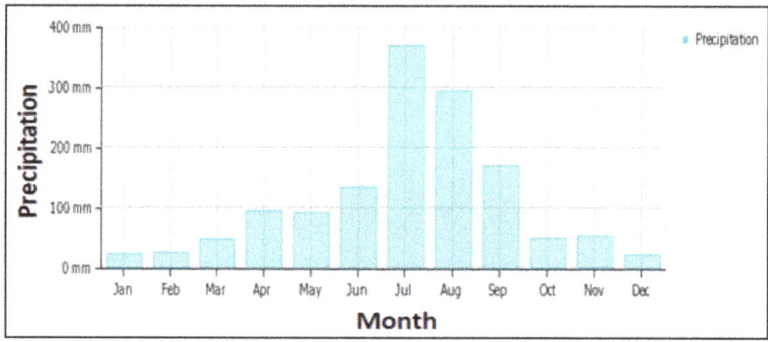

Figure 3. Monthly precipitation of Seoul, Korea in 2016 [25].

Figure 4, below, shows the permeable pavement site before the permeable pavement when it was an asphalt pavement road. This also shows the construction stage and retrofitted permeable pavement in Songpa permeable pavement systems. The construction of this site was completed in December 2017. Overall parameters of the PICP system are explained in Table 1 given below.

Figure 4. Shows the site of permeable pavement before and after construction; (**a**) Site before permeable pavement (asphalt pavement); (**b**) Site under the construction of permeable pavement; (**c**) Retrofitted permeable pavement (PICP).

Table 1. Parameter values used for the study of PICP system.

Parameter	PICP System
PICP total drainage area	Around 200 (m^2)
Hydraulic conductivity of sub-grade soil	6.5 (mm/s)
Watershed land use	Residential and asphalt pavement
Drainage area: PICP area	3:1
Underlying soil classification	Sandy loam
Rainfall intensity	30–120 (mm/h)
PICP system length	20 m
PICP system width	7 m
Monitoring period	22 April 2017–16 July 2017

2.2. Monitoring and Data Calculation

The Hydrological performance of the PICP system was investigated from 22 April 2017–16 July 2017. Regular data readings were collected to evaluate the PICP system in mitigating the rainfall runoff during various storm events in the highly urbanized area of Seoul, Korea. Measured precipitation readings were calculated from the Korea Meteorological Administration (KMA) online site by though the nearest rain gauge [25]. The uncertainty in the collected storm data from the KMA site was estimated to be less than 13% during the study investigation. Continuous rainfall runoff water flow measurements were taken during the different rain events. In this field study, a Stingray 2.0 (Greyline instruments Inc., Largo, FL, USA) portable level velocity logger was placed at the PICP site to investigate the water flows of the PICP system during the analysis period. PICP system rainfall runoff was calculated at the side of the vertical infiltration trench. Figure 5 below shows the point of PICP system where the rainfall runoff was calculated. After rainfall, rainwater first infiltrates in the ground soil through the PICP system, and the extra water becomes surface runoff and collects in the vertical infiltration trench where the surface rainfall runoff is calculated. Field tests carry out during real rain conditions for a period of about four months to evaluate the runoff mitigation performance of the PICP system in Korean geographical conditions.

Figure 5, shows (a) the mechanism of rainfall runoff distribution within the PICP system; (b) indicated the location of the instrument where water flow is measured. Stingary 2.0 portable level velocity logger was installed in the PICP system to check the water flow of the PICP system during different storm events.

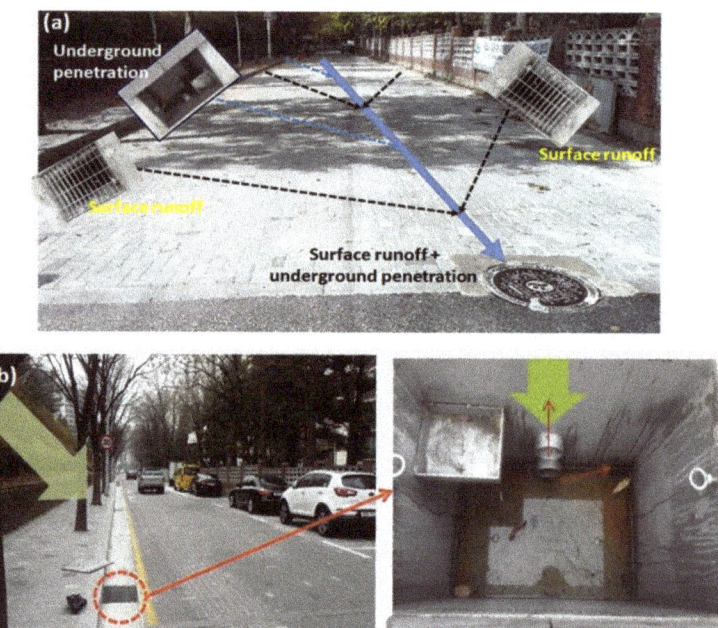

Figure 5. (a) Permeable block runoff collecting mechanism; (b) location of data collection and permeable infiltration trench.

Figure 6 below, shows the stormwater management mechanism on the site. It also shows that the permeable pavements have a vertical underground infiltration trench on the left side of the pavement system that is collected the infiltrated water through the underground porous pipe which further infiltrates water to the ground surface. In the middle of a permeable pavement system, the perforated porous pipe installed, which is used to gather all the infiltrated rainwater from all permeable pavements and throw extra water into the sewer system. This system helps to absorb more rainfall runoff into the ground surface in many different ways and is very effective in controlling flooding in urban areas.

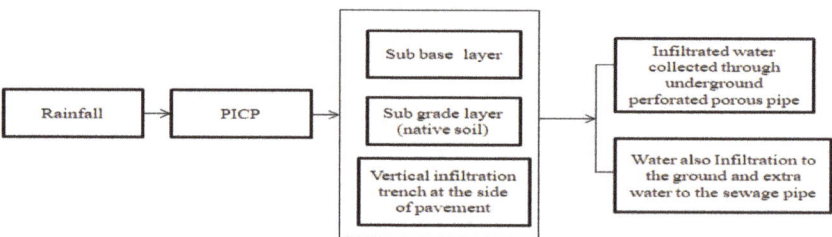

Figure 6. Shows the mechanism of stormwater management of the retrofitted PICP system.

3. Results and Discussion

Runoff Reduction Performance of PICP System

During the four-month period, a number of storm events were observed with rainfall intensity about 30, 60 and 120 mm/h. Infiltration of the underlying subgrade soil was the main factor for the rainfall runoff reduction through the PICP. Every rain event, hydrological properties for the inflow (INFLOW), PICP outflow (OUTFLOW) after infiltration the outflow runoff and peak flow reduction were calculated by using the volume reduction (VR, Equation (1) [26]), and the peak flow reduction (QR, Equation (2) [26]) as shown below.

$$VR = \frac{\sum_{i=1}^{n} V_{INFLOWi} - \sum_{i=1}^{n} V_{OUTFLOWi}}{\sum_{i=1}^{n} V_{INFLOWi}} \times 100 \qquad (1)$$

$$QR = \frac{Q_{PINFLOW} - Q_{POUTFLOW}}{Q_{PINFLOW}} \times 100 \qquad (2)$$

Equations (1) and (2) were used to analyze the runoff reduction performance of the PICP system during different time intervals. When the rainfall runoff falls on the PICP system, the water infiltrates the underlying soil and if the rainfall intensity is higher than the infiltration rate of the PICP, then runoff occurs on the surface. Different runoff volume reduction calculated on the basis of storm events and the analysis shows that the PICP system would have reduced 30–65% of the volume, which shows that it is a very effective technique to control flash flood issues in developed areas like Songpa, Seoul, Korea. Table 2 below shows that the permeable pavement is very effective in case of small rain events (of intensity about 40 (mm/h). It absorbed all rainwater and no discharge took place during these rain events. However, in case of bigger rain events, permeable pavement reduced rainfall runoff up to around 30–50% as shown in Table 2.

Table 2. Volume reduction of PICP system during different storm events.

Rainfall Events	Runoff Volume Reduction Performance
Rainfall intensity: 40 (mm/h)	100%
Rainfall intensity: 120 (mm/h)	30–50%
All storm events	Around 30–65%

During the field data analysis of the PICP system, the rain events hydrograph data give the details of the permeable interlocking concrete pavement response to the various storm event inputs. Hydrographs of various rain events were exhibited by the capture of rainwater runoff (Figures 7 and 8). The hydrographs responses indicate the variations in rainfall runoff outflow with various rain events. However, in a 40 mm/h small rain event, no discharge (rainfall runoff outflow) was found from the PICP system. It means that all the rainwater infiltrated through the PICP system. Another storm event on 3 July 2017 with rainfall intensity of 120 mm/h shows the larger rainfall runoff outflows of 3.2 L/s as shown in Figure 7 below. From the analysis during big storm events, the rainfall runoff outflow of PCIP system was found to about 3.5 L/s, which shows that PCIP system helps to control the surface runoff in urban areas. During the various storm events, the rainfall runoff outflows from PICP system varies from 1.0 L/s to 5.5 L/s. In this way, rainwater infiltrates to the ground surface and reduces the surface rainfall runoff. From the analysis, it was proven that PCIP can reduce the rainfall runoff about 35% to 65% of during small rain events. This is due to the ability of the PICP system to store large amount of rainwater in different layers.

In addition to reducing rainfall runoff, the PICP system notably decreased the peak flow on an average 10–25% as shown in Figures 7 and 8. This is because of the storage and infiltration of rainwater rainfall in the PICP system. Figure 8 shows the rainfall outflow from the PICP during different rain events of intensity 30–120 mm/h. This also shows that in case of small storm events of less than 40 mm/h rainfall intensity, the PICP system captures all the rainfall runoff and thus there was no runoff outflow found during these time periods.

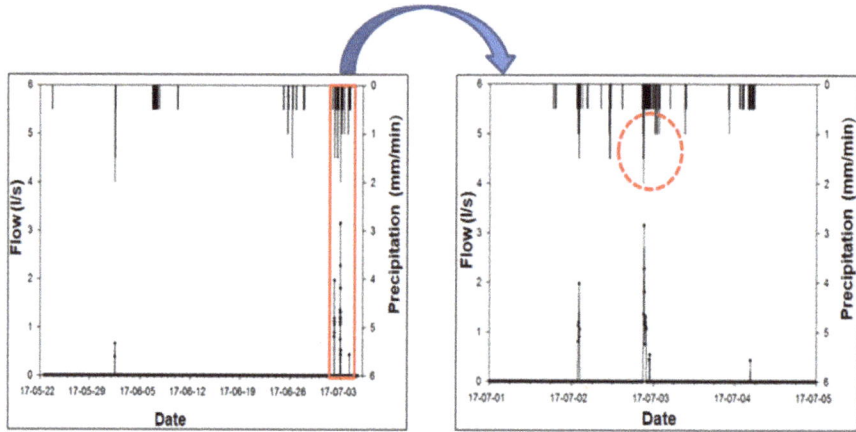

Figure 7. Indicates the rainfall runoff outflows (L/s) response during to the various storm events of 5 April–5 July 2017.

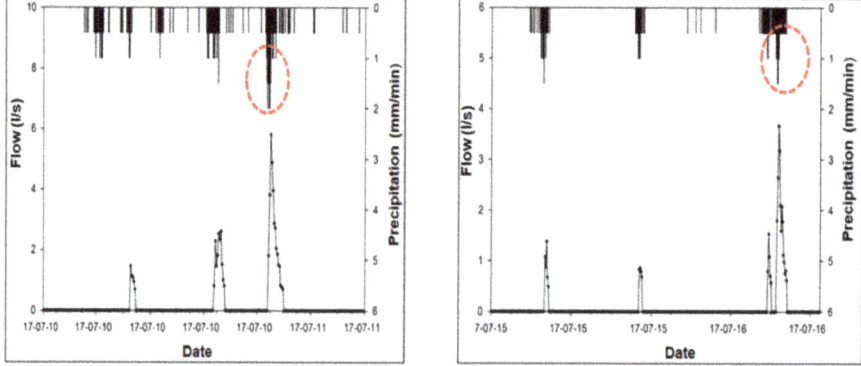

Figure 8. Shows the rainfall runoff outflows (L/s) response of the PICP system to the various rain events of 10–16 July 2017 with different rainfall intensities around 30–120 mm/h.

PICP system layers promote the infiltration of rainwater to the underground that helps to control the stormwater in urban areas. The runoff outflows were around 3–5 L/s during the big rain events of around 100–120 mm/h. This is because; during big storm events, the PICP system cannot detain and infiltrate all the rainwater in the underlying layers. From the analysis, it was found that the PICP system detain a significant amount of rainfall runoff. Therefore, a PICP technique is useful to control and infiltrate the rainfall runoff the ground surface and thus helps to enhance the ground water recharge in urban areas. If the depth of permeable pavement sub-base is greater, then it can capture and absorb more rainwater as compared to smaller depth sub-base. The results of rainfall runoff volume reduction showed that the PICP system is a very effective practice to retrieve the hydrological conditions of the urban area.

This field study manifested the capability of permeable pavements on rainfall runoff mitigation in an urban area. It gives useful information for stakeholders regarding the benefits of using the permeable pavements for controlling rainfall runoff in urban areas of Seoul, Korea. This research study mirrors the other field results [16] and indicates that the PICP system was very effective strategy in capturing a large amount of rainwater in a developed area. From the data analysis of field tests,

PCIP system performance in controlling rainfall runoff was found to be higher than other studies Drake [18] and Huang et al. [20]. Therefore, this PICP system is very effective for urban rainfall runoff management. However, in line with several research studies, there are certain limitations to the current field study. These research shortfalls the long term evaluation of the PICP system because long term field evaluation will also help to investigate the clogging of permeable pavement. As clogging is the factor which greatly affects the performance of permeable pavement over the time. Therefore, a long term evaluation of the PICP system should be carried out. There is also a need to select the best materials for the subgrade of the PICP system so that it can decrease the more rainfall runoff as well as to enhance the infiltrated water quality. In the future, to enhance the PCIP system in terms of rainfall runoff quantity and quality performance, it is necessary to combine the PICP system with other suited LID facilities [26] in such a way that it can provide multiple benefits (hydrological + water quality improvements) in the urban area.

4. Concluding Remarks

This research presented the hydrological performance of PICP in the urban catchment, in order to check their potential to mitigate the peak discharge in flood risk areas. The results have demonstrated that the retrofitted PICP system is very effective in handling the storm events to decrease chances of flash floods in the developed area of Seoul. The hydrology and stormwater quality performance provided by a retrofitted permeable interlocking concrete pavement was evaluated for about four consecutive months of (April–July) in 2017. Through this study, the following outcomes are drawn:

(1) Permeable interlocking concrete pavement showed the tremendous performance in controlling a large amount of rainfall runoff in urban areas. The overall runoff volume reduction was around 30–65% during various storm events. This not only decreases the rainfall runoff volume, but also the peak flow discharge which is very helpful in lowering flash flooding in urban areas.

(2) The hydrological performance of the PICP system was enormously influenced by the type of underlying native soil. This is because the native soil has higher infiltration rates and is thus capable of absorbing a large amount rainfall runoff. Because of this, it prevents the ponding on the surface of the pavement.

Above results showed that PICP system is an effective technique to reduce the rainfall runoff in a developed area of Seoul, Korea. However, to get multiple results, there is a need to connect the PICP system with other nearby LID facilities such as rain gardens, grass swales, etc. This will not only help to control more runoff volume but help in treating more pollutants in different ways.

Acknowledgments: The authors would like to express gratitude to the Korea Institute of Civil Engineering and Building Technology (KICT) for project funding. This research is supported by a grant (15 technology innovation C04) from the Advanced Water Management Research Program funded by the Ministry of Land, Infrastructure and Transport of the Korean government.

Author Contributions: The paper was guided by Reeho Kim. Muhammad Shafique and Kwon Kyung-Ho monitored the permeable pavement data at the site. The whole manuscript was analyzed, composed and written by Muhammad Shafique.

Conflicts of Interest: The authors declare no conflict of interest.

References

1. Franczyk, J.; Chang, H. The effects of climate change and urbanization on the runoff of the rock creek basin in the Portland Metropolitan Area, Oregon, USA. *Hydrol. Process.* **2009**, *23*, 805–815. [CrossRef]
2. Booth, D.; Hartley, D.; Jackson, R. Forest cover, impervious surface area, and the mitigation of storm-water impacts. *J. Am. Water Resour. Assoc.* **2002**, *38*, 835–845. [CrossRef]
3. Line, D.; White, N. Effects of development on runoff and pollutant export. *Water Environ. Res.* **2007**, *79*, 185–190. [CrossRef] [PubMed]
4. Moglen, G. Hydrology and impervious areas. *J. Hydrol. Eng.* **2009**, *14*, 303–304. [CrossRef]

5. Todeschini, S. Hydrologic and environmental impacts of imperviousness in an industrial catchment of Northern Italy. *J. Hydrol. Eng.* **2016**. [CrossRef]
6. Shafique, M.; Kim, R. Recent Progress in Low-Impact Development in South Korea: Water-Management Policies, Challenges and Opportunities. *Water* **2018**, *10*, 435. [CrossRef]
7. Pratt, C.J.; Wilson, S.; Cooper, P. Source control using constructed pervious pavements. In *Hydraulic, Structural and Water Quality Performance Issues*; CIRIA: London, UK, 2002.
8. Bean, E.Z.; Hunt, W.F.; Bidelspach, D.A. Evaluation of four permeable pavement sites in eastern North Carolina for runoff reduction and water quality impacts. *J. Irrig. Drain. Eng.* **2007**, *133*, 583–592. [CrossRef]
9. Brattebo, B.O.; Booth, D.B. Long-term stormwater quantity and quality performance of permeable pavement systems. *Water Res.* **2003**, *37*, 4369–4376. [CrossRef]
10. Scholz, M.; Grabowiecki, P. Review of permeable pavement systems. *Build. Environ.* **2007**, *42*, 3830–3836. [CrossRef]
11. Ahiablame, L.; Engel, B.; Chaubey, I. Effectiveness of low impact development practices: Literature review and suggestions for future research. *Water Air Soil Pollut.* **2012**, *223*, 4253–4273. [CrossRef]
12. Shafique, M.; Kim, R. Green stormwater infrastructure with low impact development concept: A review of current research. *Desalination Water Treat.* **2017**, *83*, 16–29. [CrossRef]
13. Dreelin, E.A.; Fowler, L.; Carroll, C.R. A test of porous pavement effectiveness on clay soils during natural storm events. *Water Res.* **2006**, *40*, 799–805. [CrossRef] [PubMed]
14. Park, S.B.; Tia, M. An experimental study on the water purification properties of porous concrete. *Cement Concr. Res.* **2004**, *34*, 177–184. [CrossRef]
15. Bernot, M.J.; Calkins, M.; Bernot, R.J.; Hunt, M. The influence of different urban pavements on water chemistry. *Road Mater. Pav. Des.* **2011**, *12*, 159–176. [CrossRef]
16. Collins, K.A.; Hunt, W.F.; Hathaway, J.M. Hydrologic comparison of four types of permeable pavement and standard asphalt in Eastern North Carolina. *J. Hydrol. Eng.* **2008**, *13*, 1146–1157. [CrossRef]
17. Palla, A.; Gnecco, I.; Carbone, M.; Garofalo, G.; Lanza, L.; Piro, P. Influence of stratigraphy and slope on the drainage capacity of permeable pavements: Laboratory results. *Urban Water J.* **2015**, *12*, 394–403. [CrossRef]
18. Drake, J. Performance and Operation of Partial Infiltration Permeable Pavement Systems in the Ontario Climate. Ph.D. Thesis, University of Guelph, Guelph, ON, Canada, 2013.
19. Gilbert, J.K.; Clausen, J.C. Stormwater runoff quality and quantity from asphalt, paver, and crushed stone driveways in Connecticut. *Water Res.* **2006**, *40*, 826–832. [CrossRef] [PubMed]
20. Huang, J.J.; Li, Y.; Niu, S.; Zhou, S.H. Assessing the performances of low impact development alternatives by long-term simulation for a semi-arid area in Tianjin, northern China. *Water Sci. Technol.* **2014**, *70*, 1740–1745. [CrossRef] [PubMed]
21. Xie, J.; Wu, C.; Li, H.; Chen, G. Study on storm-water management of grassed swales and permeable pavement based on SWMM. *Water* **2017**, *9*. [CrossRef]
22. Kumar, K.; Kozak, J.; Hundal, L.; Cox, A.; Zhang, H.; Granato, T. In-situ infiltration performance of different permeable pavements in a employee used parking lot—A four-year study. *J. Environ. Manag.* **2016**, *167*, 8–14. [CrossRef] [PubMed]
23. Gao, M.; Carmichael, G.R.; Saide, P.E.; Lu, Z.; Yu, M.; Streets, D.G.; Wang, Z. Response of winter fine particulate matter concentrations to emission and meteorology changes in North China. *Atmos. Chem. Phys.* **2016**, *16*, 11837–11851. [CrossRef]
24. Razzaghmanesh, M.; Beecham, S. A review of permeable pavement clogging investigations and recommended maintenance regimes. *Water* **2018**, *10*. [CrossRef]
25. Korea Meteorological Administration (KMA) 2018. Available online: http://web.kma.go.kr/eng/weather/climate/worldclimate.jsp (accessed on 19 January 2018).
26. Braswell, A.S.; Anderson, A.R.; Hunt, W.F. Hydrologic and water quality evaluation of a permeable pavement and biofiltration device in series. *Water* **2018**, *10*. [CrossRef]

© 2018 by the authors. Licensee MDPI, Basel, Switzerland. This article is an open access article distributed under the terms and conditions of the Creative Commons Attribution (CC BY) license (http://creativecommons.org/licenses/by/4.0/).

Article

Evaluating the Thermal Performance of Wet Swales Housing Ground Source Heat Pump Elements through Laboratory Modelling

Carlos Rey-Mahía [1], Luis A. Sañudo-Fontaneda [1,2,*], Valerio C. Andrés-Valeri [3,4], Felipe Pedro Álvarez-Rabanal [1], Stephen John Coupe [2] and Jorge Roces-García [5]

1. INDUROT Research Institute, GICONSIME Research Group, Department of Construction and Manufacturing Engineering, University of Oviedo, Campus of Mieres, Gonzalo Gutierrez Quiros s/n, 33600 Mieres, Spain; UO236881@uniovi.es (C.R.-M.); alvarezfelipe@uniovi.es (F.P.Á.-R.)
2. Centre for Agroecology, Water and Resilience, Coventry University, Ryton Gardens, Coventry CV8 3LG, UK; Stephen.coupe@coventry.ac.uk
3. Instituto de Obras Civiles, Facultad de Ciencias de la Ingeniería, Universidad Austral de Chile, General Lagos 2086, Campus de Miraflores, Valdivia 5090090, Chile; valerio.andres@uach.cl
4. GITECO Research Group, University of Cantabria, Avenida de los Castros 44, 39005 Santander, Spain
5. Department of Construction and Manufacturing Engineering, University of Oviedo, Campus of Gijón, Pedro Puig Adam s/n, EDO6, 33203 Gijón, Spain; rocesjorge@uniovi.es
* Correspondence: sanudoluis@uniovi.es or luis.sanudo-fontaneda@coventry.ac.uk; Tel.: +34-985-458-196

Received: 28 February 2019; Accepted: 28 May 2019; Published: 3 June 2019

Abstract: Land-use change due to rapid urbanization poses a threat to urban environments, which are in need of multifunctional green solutions to face complex future socio-ecological and climate scenarios. Urban regeneration strategies, bringing green infrastructure, are currently using sustainable urban drainage systems to exploit the provision of ecosystem services and their wider benefits. The link between food, energy and water depicts a technological knowledge gap, represented by previous attempts to investigate the combination between ground source heat pump and permeable pavement systems. This research aims to transfer these concepts into greener sustainable urban drainage systems like wet swales. A 1:2 scaled laboratory models were built and analysed under a range of ground source heat pump temperatures (20–50 °C). Behavioral models of vertical and inlet/outlet temperature difference within the system were developed, achieving high R^2, representing the first attempt to describe the thermal performance of wet swales in literature when designed alongside ground source heat pump elements. Statistical analyses showed the impact of ambient temperature and the heating source at different scales in all layers, as well as, the resilience to heating processes, recovering their initial thermal state within 16 h after the heating stage.

Keywords: ecosystem services; food-energy-water nexus; geothermal energy; LID; heating and cooling; stormwater BMP; SUDS; WSUD

1. Introduction

The built environment impacts the wider environments whilst threatening natural ecosystems in urban areas by reducing green spaces [1]. Rapid urbanization is at the core of the problem with its subsequent land-use change and being described as one of the most influential factors affecting flooding problems in urban environments [2]. In addition, Palazzo et al. [3] put a spotlight on stormwater management when referring to rainwater as a primary risk to urban resilience. Furthermore, Cai et al. [4] identified thermal changes in cities produced by urbanization processes. As a consequence, the concept of urban resilience has taken off in recent years, providing an insight into multidisciplinary contexts, such as socio-ecological systems and their sustainable management under highly complex and variable adaptive systems and climate change scenarios [5]. In this new urban context, Li et al. [1] suggested the implementation of multifunctional approaches through urban regeneration strategies, also highlighted by Peña et al. [6] under the concept of multifunctional landscapes. Widening this view, La Rosa et al. [7] pinpointed urban regeneration as the main way to achieve sustainable urban environments, especially when looking at health and wellbeing for citizens [8]. Therefore, transitioning towards a new paradigm of resilient cities through multifunctional green spaces, has been targeted under the concept of urban green infrastructure (UGI) [9].

Moving on towards the urban water paradigm shift, defined by Morison and Brown [10], authors, such as Perales-Momparler et al. [11] and Gonzales and Ajami [12], specified sustainable stormwater techniques as the main engineering and architectural route to water reuse and rainwater control with the aim to achieve secure water resources. Palazzo et al. [3] also brought into the picture the main philosophy behind this new approach to water management. It consists of outlining how the new concept of adaptive urban design works alongside rainwater rather than against it which has been historically the main way to deal with urban water. Following on from these new approaches, eco-hydrology has arisen as a new term for urban design and diagnosis which allows understanding of long-term patterns in urban climate and hydrology from an environmental appreciation [13]. 'Sponge cities' are perhaps the most easily identifiable eco-hydrological approach under the new urban water paradigm, bringing a wider comprehensive philosophy of urban development and water management [2]. The 'Sponge Cities' concept works with what it was defined as green corridors which allow landscape connectivity, supporting the overall ecosystem health and biodiversity conservation [14]. Continuing along these lines, Bortolini et al. [15] stressed the need to widen the ability of green spaces to ensure ecosystem services (ES) on the basis of multi-disciplinary approaches.

The United Nations (UN) released the 2030 Agenda for Sustainable Development as *'a plan of action for people, planet and prosperity. It also seeks to strengthen universal peace in larger freedom'* [16]. This document includes the millennium development goals [17] which relate directly to the previously defined urban resilient paradigm and the need to design and implement Nature-Based Solutions (NBS) at the very centre of the previously defined multifunctional approaches. The European Union (EU) has also stressed the importance of NBS and their urban implementation through green infrastructure (GI) strategies at all levels of society and the different stakeholders and sectors involved in the urban environment and its territorial planning [18]. Prior to this document, the EU defined holistically the role of GI in order to protect the ecosystem state and biodiversity, in promoting ES, societal health and wellbeing; the development of a green economy, and sustainable land and water management [19].

However, there are more questions that still are not fully answered, representing a knowledge gap to achieve the so-called food-energy-water (FEW) nexus [20]. Other authors, such as Zhang et al. [21] and Fan et al. [22], focused on the need to develop the FEW nexus in order to adopt the 2030 Agenda for Sustainable Development, incorporating other environmental, social and economic systems. Returning to the provision of ES and water management, Pappalardo et al. [23] stated how sustainable urban drainage systems (SUDS) have become the most utilized stormwater techniques to reach ES in GI-based urban plans. In this context, SUDS contribute to four main pillars as per defined by the UK CIRIA [24]: Water quantity, water quality, biodiversity and amenity. It is important to note that SUDS are often referred to as low impact development (LID) and stormwater best management practices (BMP) by other authors [25]. Just a few authors, such as Tota-Maharaj et al. [26] and del Castillo-García et al. [27], have explored the link between Energy and Water within the FEW nexus, combining ground source heat pump (GSHP) technology and permeable pavement systems (PPS). GSHP plays a key role in the production of clean energy as per stated by Gupta and Irving [28], who centred their efforts in helping dwellings to adapt to climate change by reducing carbon consumption. This point has been supported by literature, being represented by authors, such as Nathanail et al. [29], in order to achieve sustainability in the wider urban environment. This path has been also taken by Price et al. [30] through the creation of a new methodology for planning development using GSHP and SUDS as indicators, empowering the need for multifunctional purpose engineered elements in the city.

Charlesworth et al. [31] identified future prospects for GSHP and PPS, emphasising the application of horizontal heat pump technology in greener SUDS; an idea supported by Tota-Maharaj et al. [26] who identified paths towards the exploration of GSHP technology, previously used in PPS, in *'greener'* SUDS, such as wetlands. Andrés-Valeri et al. [32] pioneered the plan to housing GSHP elements in the structure of a wet swale, transferring the previously developed concepts for PPS into swales, highlighting the need to further develop research to fill this key gap in the current knowledge.

This research aims to further develop the use of GSHP combined with wet swales in order to lead the path towards the progress of the Energy-Water nexus. In addition, previous work by Abrahams et al. [33] depicted the potential of swales, designed under a new biological concept, including flood resilience, biomass production, sewage purification and biodiversity enhancement, to reach food production. This new scenario sets, in combination with the present work, the full FEW nexus, pioneering a new SUDS design. Specific objectives were also established, being condensed as follows:

- Overall description of how the structure of a 1:2 scaled laboratory model for a wet swale responds under a range of temperatures (20 up to 50 °C) and consequent performance of the GSHP system;
- Development of behavioral models for the vertical and inlet/outlet temperature difference within the wet swale structure.

With these specific objectives the hypotheses tested in this research relates to two main aspects: (a) The usual range of temperature of performance of GSHP devices might affect the overall thermal performance of a wet swale; (b) Green infrastructure, such as wet swales can be designed housing GSHP elements.

The main conclusion from this research is that wet swales presented good resilience to heating and cooling processes under standard performance temperature of GSHP. This research represents the first attempt to depict the thermal performance of wet swales when designed alongside GSHP elements based on the scientific literature consulted for this research.

2. Materials and Methods

2.1. Materials and Experimental Set-Up

The structure of the wet swale selected to be modelled at a laboratory scale in this research was designed after identifying the materials and geometries most commonly used in the literature. Fardel et al. [34] established four types of swales as follows: Standard, dry, wet and bioswales. In addition, wet swales were defined by Winston et al. [35] as swales functioning under conditions of ponded water or soil moisture at near saturation. Subsequently, Fardel et al. [34] carried an extensive literature review of the geometries and design specifications for 59 swales designed across the world, finding depths for the surface layer ranging between 15 mm up to 530 mm. Besides, media and other intermediate layers, as well as, the bottom layer when included in the profile, have been reported to reach deeper depths. As an example, Andrés-Valeri et al. [36] designed a field experiment using a total depth of 500 mm measured from the bottom part of the surface including the surface layer and the bottom layer. Supporting this work, the UK CIRIA SUDS Manual suggests depths between 500 mm and 2000 mm, recommending the use of a geomembrane liner at a minimum depth of 500 mm when infiltration to the ground needs to be prevented due to unfavorable groundwater conditions [24]. Thus, the layers and materials showed in Figure 1 constituted the structure of the laboratory models tailored made for this research.

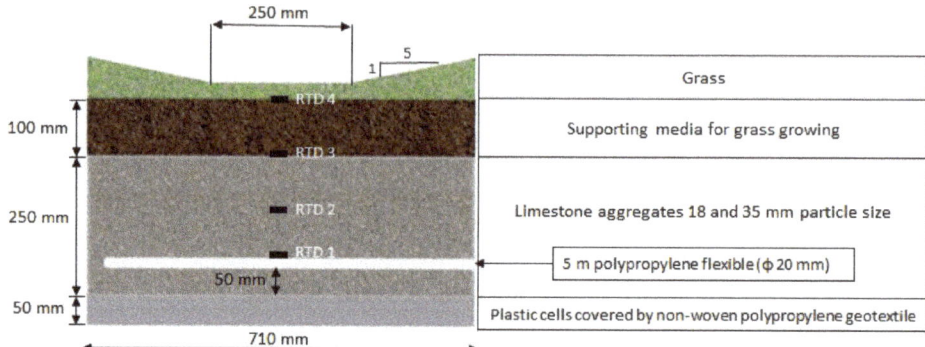

Figure 1. Cross-section of the laboratory wet swale model.

Then, three identical models at a 1:2 scale were built housed by 1110 mm × 710 mm × 610 mm high density polyethylene containers (HDPE) based upon the previous wet swale design (Figure 2). The temperature was monitored by using K thermocouple sensors placed at different heights measured from the bottom of the sub-base layer (100, 200, 300 and 400 mm) in order to identify different patterns of behaviour depending on the material and the depth of the profile, being respectively named as: RTD1, RTD2, RTD3 and RTD4. These sensors allowed the definition of the vertical temperature variation of the system. Furthermore, two K thermocouples were installed at the inlet and outlet points of the pipe in order to measure the horizontal variation of the circulating fluid within the simulated looping element during the whole experiment (Figure 2). In addition, the ambient temperature was registered all over the duration of the experiments.

Figure 2. Experimental set-up.

A 15 L insulated tank was utilized in order to function as a reservoir for the recirculation of the water through the GSHP simulated system. An electric resistor was introduced to the tank with the aim to heat the water up to the required temperature of the system. The tank was also connected to a 43 W hydraulic pump (Figure 2) which recirculated the water through the looping element.

Finally, a constant water height was maintained over the surface layer during the whole duration of the experiments, as it can be seen in Figure 2, in order to replicate the scenario of ponded water required by the literature to be considered as a wet swale [35].

2.2. Experimental Methodology

Constant water flow was circulated at a 1 L/min rate through the 5 m polypropylene flexible pipe which simulated the geothermal looping pipe for a GSHP device over the whole duration of the experiment. Temperatures were registered at 1 min intervals, using a computer connected to all sensors as shown in Figure 2, permitting data acquisition in real-time (Figure 2).

Water was then circulated through the system at 20, 30, 40 and 50 °C. These temperatures were selected as the usual operating temperature for most of the heat pumps utilized in GSHP systems which upper limit has been encountered to be around 50 °C based upon data from the Energy Saving Trust [37].

Three replicates were used for each test at all temperatures, supporting statistical soundness. Thus, statistical analyses were designed accordingly using MATLAB software. These analyses consisted on the development of regression models. The method of least squares (MLS) was utilized, as well as R^2 and co-linearity to reach the best fit. These later analyses also provide information about the quality of the models obtained as described by [38,39]. In addition, the accuracy of the models was quantified through adjusting the goodness of fit between the theoretical values and those from laboratory results. With this aim, relative and absolute errors, as well as the root mean square (RMS) values, were calculated in order to support the goodness of fit for the developed models following from the statistical methodology proposed by authors, such as Fernández-Martínez et al. [40]. This proposed method consisted of measuring uncertainty in civil engineering applications especially dedicated to structural materials.

Prior to that, pre-tests were conducted with the goal to identify the optimum duration for the experiment. For this reason, heating experiments were run between 73 and 95 h, reaching the temperature models obtained in Figure 3 a good level of fit after 8 h, based on the statistical models constructed. Quadratic models were the best fit for this stage. Based upon the previous finding, the heating stage of the experiment was fixed at 8 h duration, whilst the cooling stage was defined to last for 16 h. Therefore, each experiment was run over 24 h in total divided into the previously cited stages.

Furthermore, the heating stage consisted in heating the water in the tank (Figure 2) until it reaches the temperature of the experiment (20, 30, 40 or 50 °C depending on the experiment). Water is kept at that temperature for 8 h duration and, then the resistor is disconnected, commencing the cooling stage for the next 16 h. Thus, no heating is provided by the resistor during this later stage, allowing the system to cool down. This stage allows identifying whether the wet swale layers are more or less resilient to heating processes through the evaluation of the temperature variation within the system.

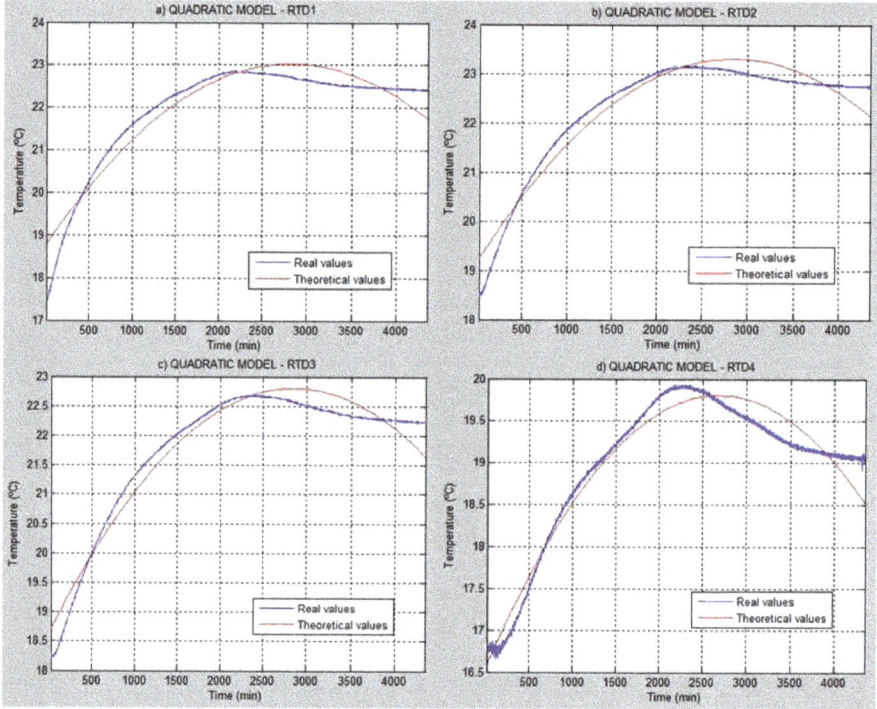

Figure 3. Statistical quadratic models for the temperature sensors RTD1, RTD2, RTD3 and RTD4 during the pre-tests.

3. Results and Discussions

The results from this research were divided into two main sections: Vertical and inlet/outlet temperature difference variabilities; including a general discussion at the end, presenting future research lines.

3.1. Vertical Temperature Variation

The temperature variation registered at each layer of the wet swale for the different temperatures of operation of the simulated GSHP system were represented through the development of behavioral models as can be seen in Table 1. All models obtained high values of R^2 and errors were also calculated in order to check the goodness of fit in all models.

These models depended on the temperature of operation and the duration time of the heating stage (Table 1).

Behavioral models obtained for 20 °C showed that the surface layer represented by the RTD4 sensor (the furthest from the heating source) had the lowest R^2 value, but presenting the lower error value (Table 1). Furthermore, higher temperatures of operation (30 °C, 40 °C and 50 °C) produced higher R^2 for the behavioral models in all depths within the wet swale profile as represented by the sensors (Table 1).

Table 1. Behavioral models of vertical thermal variation under the four temperatures of operation selected for the simulated ground source heat pump (GSHP) system in the experiment, depending on the time in minutes.

T (°C)	Sensor	Behavioral Model	R^2	Absolute Error (%)	Relative Error (%)	RMS Error (%)
20	RTD1	$T(°C) = -2.9079 \cdot 10^{-06} \cdot t^2 + 0.0038 \cdot t + 15.2110$	0.9845	15.0618	0.1967	0.6867
	RTD2	$T(°C) = -3.5333 \cdot 10^{-06} \cdot t^2 + 0.0038 \cdot t + 15.5638$	0.9842	13.9469	0.1775	0.6359
	RTD3	$T(°C) = -3.3094 \cdot 10^{-06} \cdot t^2 + 0.0038 \cdot t + 15.2652$	0.9919	9.6283	0.1252	0.4390
	RTD4	$T(°C) = 9.7249 \cdot 10^{-07} \cdot t^2 - 0.0002 \cdot t + 15.0351$	0.9025	7.1694	0.0989	0.3269
30	RTD1	$T(°C) = -6.3338 \cdot 10^{-06} \cdot t^2 + 0.0095 \cdot t + 15.9646$	0.9996	6.5708	0.0762	0.2996
	RTD2	$T(°C) = -5.0091 \cdot 10^{-06} \cdot t^2 + 0.0083 \cdot t + 16.8260$	0.9981	11.5799	0.1306	0.5280
	RTD3	$T(°C) = -2.8594 \cdot 10^{-06} \cdot t^2 + 0.0066 \cdot t + 16.3455$	0.9986	8.5591	0.0987	0.3902
	RTD4	$T(°C) = -2.8594 \cdot 10^{-06} \cdot t^2 + 0.0066 \cdot t + 16.3455$	0.9986	9.5416	0.1241	0.4350
40	RTD1	$T(°C) = -1.0367 \cdot 10^{-05} \cdot t^2 + 0.0152 \cdot t + 15.9599$	0.9954	29.8053	0.3278	1.3590
	RTD2	$T(°C) = -1.4382 \cdot 10^{-05} \cdot t^2 + 0.0177 \cdot t + 16.5443$	0.9969	29.6266	0.3093	1.3508
	RTD3	$T(°C) = -9.6683 \cdot 10^{-06} \cdot t^2 + 0.0143 \cdot t + 15.8760$	0.9991	14.2437	0.1580	0.6494
	RTD4	$T(°C) = 6.8632 \cdot 10^{-06} \cdot t^2 + 0.0143 \cdot t + 15.8760$	0.9991	25.3249	0.3296	1.1547
50	RTD1	$T(°C) = -1.7541 \cdot 10^{-05} \cdot t^2 + 0.0239 \cdot t + 15.4303$	0.9989	26.1399	0.2658	1.1918
	RTD2	$T(°C) = -1.9195 \cdot 10^{-05} \cdot t^2 + 0.0250 \cdot t + 16.5089$	0.9981	34.9741	0.3380	1.5947
	RTD3	$T(°C) = -1.3551 \cdot 10^{-05} \cdot t^2 + 0.0206 \cdot t + 16.5089$	0.9981	16.8782	0.1753	0.7695
	RTD4	$T(°C) = 8.4280 \cdot 10^{-06} \cdot t^2 + 0.0005 \cdot t + 15.1394$	0.9914	26.8387	0.3474	1.2237

Table 1 also shows that higher variation was found in those models for the simulated GSHP housed by the swale while operating under higher temperatures (40 °C and 50 °C) as per indicated by the values obtained for the absolute error. This error also highlighted that RTD1 and RTD2 sensors registered the highest values, indicating that lower areas within the wet swale cross-section (ranging between 100 and 200 mm) were more influenced by the heating source under high temperatures of operation in the system (Figure 4).

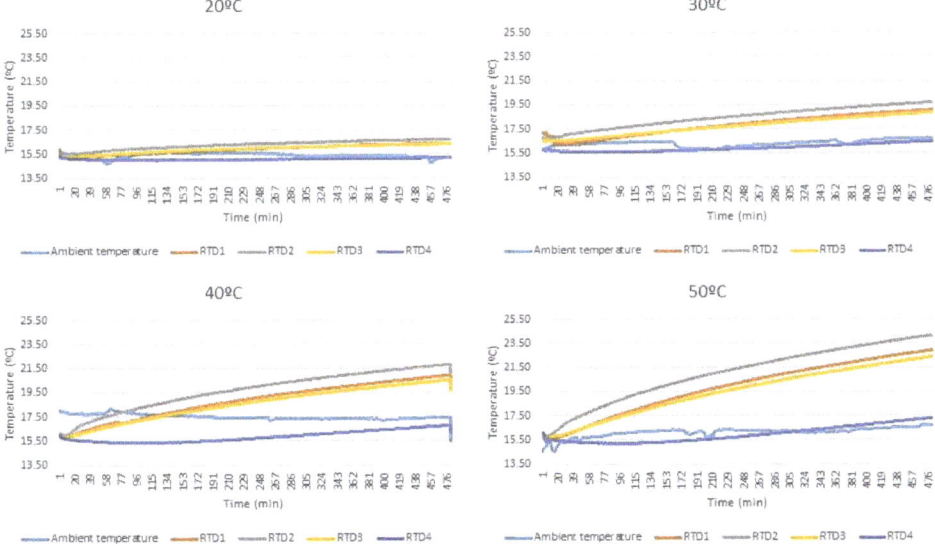

Figure 4. Vertical temperature variation during the cooling stage.

Furthermore, the ambient temperature was steady during all experiments as can be seen in Figure 4 with little variation registered (temperature range registered between 15.5 and 17.0 °C).

Ambient temperature has an influence on the surface water temperature as shown in Figure 4 for all temperatures of performance of the GSHP. The tendency for the temperature of the surface water registered by the sensor RTD4 is to converge at the same ambient temperature despite the temperature of performance of the GSHP, even in those cases with high temperatures (40 and 50 °C). The presence of water is key to cool the temperature down through evapotranspiration and heat transfer processes. Nevertheless, the remaining layers of the wet swale were affected by the temperature of performance of the system registering increases in temperature during the experiment ranging between 1.5 °C up to 8.0 °C for 20 and 50 °C respectively (Figure 4). This discussion provides a key point to consider when designing the wet swale from an ecological and biological view, especially when considering plant/vegetation growing as the temperature increase affects the supporting media for grass growing (Figure 1) as indicated by the temperature sensor RTD 3 in Figure 4. This increase in temperature might affect grass growth, and should be considered as future research looking at the best species to be used if the wet swale is designed in combination with GSHP. In addition, special attention should be taken when designing dry swales which performance is more variable from a hydrological perspective as they have a variable head of water and no presence of ponding water after infiltration. This difference of saturation would influence the heat transfer processes, modifying the temperatures within the profile of the dry swale.

A cooling stage was measured after the heating was disconnected, identifying the resilience of the system to recover the initial temperature of the wet swale layers before the experiment. Results showed that the temperature of operation of the GSHP impacted on the temperature range between the vertical sensors (RTD1, RTD2, RTD3 and RTD4) as can be seen in Figure 5.

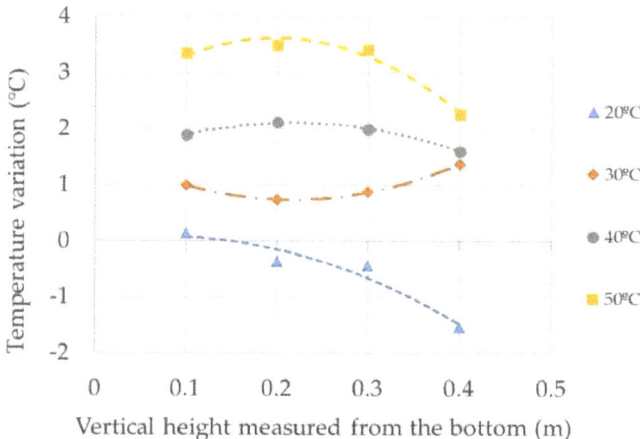

Figure 5. Vertical temperature variation during the cooling stage.

Temperatures registered by the upper sensor within the wet swale profile (RTD4) were usually inferior in comparison to those registered by intermediate sensors, such as RTD2 and RTD3 for temperatures of operation of the GSHP system, in the top temperature range (40 °C and 50 °C) as per indicated in Figure 4. On the contrary, for lower temperatures of performance (20 °C and 30 °C), sensors RTD1 and RTD4 showed higher variation. This outcome provides an interesting insight into how ambient temperature influences the surface layer of the system for temperatures of performance similar to those registered outside the wet swale. If the GSHP system is working under temperatures above the ambient ones, this later temperature contributes to blurring the temperature variation (Figure 5). As a consequence, the climate at the chosen location will influence the resilience of the wet swale to recover from the heating stage, influencing the future design of the system as a key parameter.

Trend lines for vertical temperature variation under the temperatures of performance of the GSHP elements were also developed to further depict these scenarios as can be seen in Table 2 and Figure 4. Height values are given in m from the bottom of the sub-base layer as indicated in Figure 1.

Table 2. Polynomic trend lines for vertical variation of the temperatures represented in Figure 4.

Temperature (°C)	Behavioral Model	R^2
20	$T(°C) = -3.250 \cdot z^2 + 12.933 \cdot z + 23.298$	0.9657
30	$T(°C) = -1.543 \cdot z^2 + 6.747 \cdot z + 13.720$	0.9968
40	$T(°C) = 1.889 \cdot z^2 - 8.169 \cdot z + 16.195$	0.9997
50	$T(°C) = -1.475 \cdot z^2 + 2.225 \cdot z + 0.025$	0.9308

3.2. Horizontal Thermal Variation

Inlet/outlet temperature difference variation between the inlet and outlet points of the system averaged between 2.15 °C in the experiments carried out under 20 °C of the operation performance of the GSHP up to 4.60 °C working under the top temperature of the range (50 °C) (Figure 6). Low variation was registered in those cases related to 30 °C and 40 °C whilst higher variation was found for the bottom and upper temperatures of operation (20 °C and 50 °C) (Figure 5).

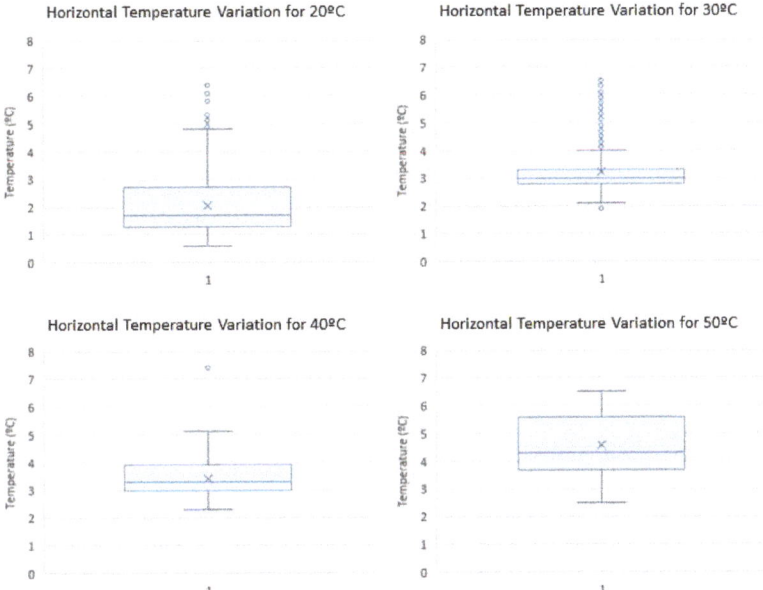

Figure 6. Box-plot showing the average values for the inlet/outlet temperature difference provided by the system under the different operational temperatures of the simulated GSHP system.

Results from Figure 6 are supplemented by those from the temperatures registered at the outlet point of the horizontal looping system (Figures 1 and 2) plotted versus time in Figure 7. Average values can be widely interpreted, presenting a steady behaviour during the heating stage. The water is heated inside the tank (Figure 2) and then recirculated through the simulated looping system during the 8 h of the duration of this stage. A steady temperature is reached between 10 and 20 min since the beginning of the experiment, maintaining temperatures slightly lower than those simulated for the GSHP elements (20, 30, 40 and 50 °C).

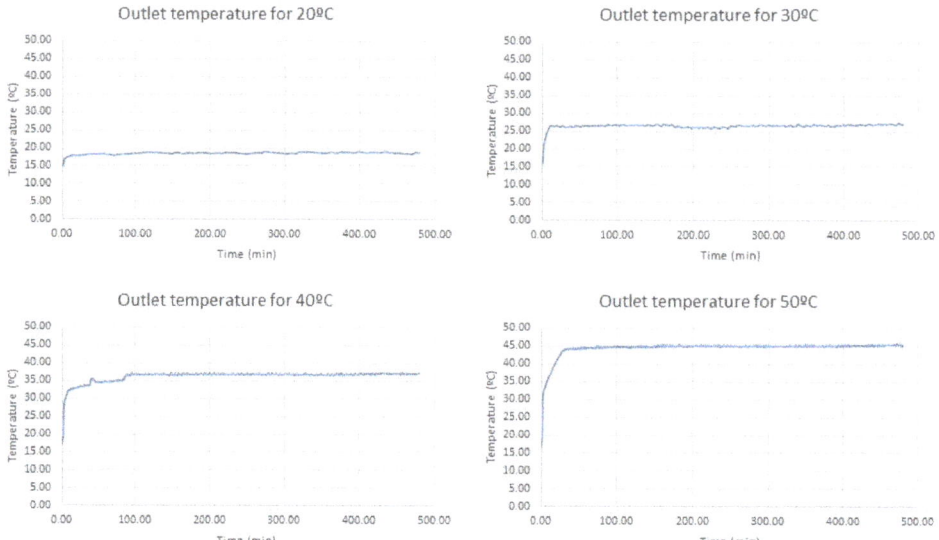

Figure 7. Average values for the outlet temperature difference under the different operational inlet temperatures of the simulated GSHP system.

3.3. General Discussions and Future Research Directions

Based upon the findings of this research, wet swales represent a good opportunity when compared with PPS combined with GSHP elements, considering the depth limitation of 500 mm for standard PPS design described by Charlesworth et al. [31]. Swale structures usually go beyond 500 mm depth, overcoming the limitation suggested by Charlesworth et al. [31] to the Coefficient of Performance (CoP). Land-take would be of a similar kind than the one necessary for the installation of a GSHP system within a PPS structure, and therefore, this solution could be used at a domestic scale, as well as commercial schemes. Swales are often used as the main drainage asset for roads and parks, representing a highly transferable technique in urban and rural environments.

Future research directions can be divided into two main steps. The first one being orientated towards laboratory experiments and modelling in order to better understand the heat transfer characteristics and processes within wet swale structures, prior to developing full-scale experiments which could validate first step findings. This addresses a key technological gap before possible commercialization of these systems. This study has provided relevant insights into how the different layers of a wet swale perform under different temperatures of operation by the GSHP system described by the Energy Saving Trust [37]. The outcomes obtained in this study have responded to the gaps identified by Andrés-Valeri et al., 2018 [32] and have further developed the understanding of the nexus between swales and GSHP, connecting to other studies carried out in other SUDS devices, such as Tota-Maharaj et al. [26].

Moreover, future research should follow on the analyses of heat transfer processes which allow further understanding of the thermal performance of the system whilst addressing key factors related to the CoP for these heat exchange systems. In this line, the use of a perimeter layer which isolates the system from horizontal losses is strongly recommended to improve the robustness of the experiment.

The next step of the laboratory experiments should also focus on numerical simulations in order to complement the laboratory experiments on the thermodynamic behaviour of the GSHP. A final step should look at characterising the CoP, allowing the determination of control strategy and balance of the plant when wet swales are exploited as a heat sink.

Further research into other types of swales, as described by Fardel et al. [34], could be conducted in order to apply the findings from this research to varying designs of swales. The next type of swale suggested for development would be the dry swale. Dry swales mainly differ from wet swales based on the lack of standing water over their surface layer at all time. This further research would supplement the findings from this research analysing the impact of standing water on temperature variation.

4. Conclusions

The hypotheses are confirmed as the application of temperatures within the usual range of performance of GSHP elements affected the overall thermal performance of the wet swale layers, presenting varying impacts. Furthermore, wet swales showed good resilience to heating processes in standard performance of GSHP elements, recovering the initial temperature in all their structural layers after 16 h. This represents an opportunity to use wet swales as multifunctional devices for stormwater management and energy saving.

Intermediate layers around 200 mm from the bottom part of the simulated laboratory model (represented by RTD2 sensor) were found to register higher variation as can be seen in the behavioral models for the vertical temperature trend of the system working under all temperatures of operation. The layer closest to the heating source was affected by the system under the heating stage, showing high resilience during the cooling stage for low temperatures (20 °C and 30 °C).

Surface layers were also affected by ambient temperature, as this effect is more noticeable for lower temperatures between 20 °C and 30 °C. However, higher temperatures of operation between 40 °C and 50 °C augmented the resilience of the surface layer, with this effect being blurred by the ambient temperature.

Inlet/outlet temperature difference was found to be lower when operating under temperatures between 30 °C and 40 °C, being higher for the extreme temperatures tested in these experiments (20 °C and 50 °C).

Development of behavioral models for the vertical and inlet/outlet temperature difference of a wet swale operating under a usual range of temperatures of the pump has been obtained, representing the first attempt in the literature consulted, to describe the thermal performance of this green infrastructure when designed alongside GSHP elements.

This research opens a new line to explore the Water and Energy nexus, contributing to new areas of development associated with 'greener' SUDS, such as swales. In addition, this research complements previous findings by Abrahams et al. [33] in meeting the FEW nexus, pioneering a new way of designing SUDS.

Author Contributions: Conceptualization, L.A.S.-F., S.J.C. and V.C.A.-V.; literature review J.R.-G and L.A.S.-F.; methodology, L.A.S.-F., S.J.C., V.C.A.-V. and C.R.-M.; formal analysis, C.R.-M.; investigation, V.C.A.-V., L.A.S.-F. and C.R.-M.; data curation, C.R.-M.; writing—original draft preparation, all authors; writing—review and editing, all authors; supervision, F.P.Á.-R., S.J.C., V.C.A.-V. and L.A.S.-F.; project administration, L.A.S.-F.; funding acquisition, L.A.S.-F., S.J.C. and F.P.Á.-R.

Funding: This research was funded by the following Institutions: Coventry University through the project "Investigation of green infrastructure as a combined technique for Bioretention, Flood Resilience and Renewable Energy"; the Gijon City Council and the IUTA through the projects SV-18-GIJON-1-23; the FICYT through the GRUPIN project Ref. IDI/2018/000221, co-financed with EU FEDER funds; and the University of Oviedo through the project PAPI-17-PEMERG-22.

Acknowledgments: The authors wish to acknowledge the CAWR, Coventry University, for the administrative support; and the GITECO Research Group, University of Cantabria, for housing the laboratory experiments.

Conflicts of Interest: The authors declare no conflict of interest. The funders had no role in the design of the study; in the collection, analyses, or interpretation of data; in the writing of the manuscript, or in the decision to publish the results.

References

1. Li, Q.; Yu, Y.; Jiang, X.; Guan, Y. Multifactor-based environmental risk assessment for sustainable land-use planning in Shenzhen, China. *Sci. Total Environ.* **2019**, *657*, 1051–1063. [CrossRef] [PubMed]
2. Chan, F.K.S.; Griffiths, J.A.; Higgitt, D.; Xu, S.; Zhu, F.; Tang, Y.-T.; Xu, Y.; Thorne, C.R. "Sponge City" in China—A breakthrough of planning and flood risk management in the urban context. *Land Use Policy* **2018**, *76*, 772–778. [CrossRef]
3. Palazzo, E. From water sensitive to floodable: Defining adaptive urban design for water resilient cities. *J. Urban Des.* **2019**, *24*, 137–157. [CrossRef]
4. Cai, D.; Fraedrich, K.; Guan, Y.; Guo, S.; Zhang, C. Urbanization and the thermal environment of Chinese and US-American cities. *Sci. Total Environ.* **2017**, *589*, 200–211. [CrossRef] [PubMed]
5. Meerow, S.; Newell, J.P.; Stults, M. Defining urban resilience: A review. *Landsc. Urban Plan.* **2016**, *147*, 38–49. [CrossRef]
6. Peña, L.; Onaindia, M.; Fernández de Manuel, B.; Ametzaga-Arregi, I.; Casado-Arzuaga, I. Analysing the Synergies and Trade-Offs between Ecosystem Services to Reorient Land Use Planning in Metropolitan Bilbao (Northern Spain). *Sustainability* **2018**, *10*, 4376. [CrossRef]
7. La Rosa, D.; Privitera, R.; Barbarossa, L.; La Greca, P. Assessing spatial benefits of urban regeneration programs in a highly vulnerable urban context: A case study in Catania, Italy. *Landsc. Urban Plan.* **2017**, *157*, 180–192. [CrossRef]
8. La Rosa, D. Accessibility to greenspaces: GIS based indicators for sustainable planning in a dense urban context. *Ecol. Indic.* **2014**, *42*, 122–134. [CrossRef]
9. Pauleit, S.; Ambrose-Oji, B.; Andersson, E.; Anton, B.; Buijs, A.; Haase, D.; Elands, B.; Hansen, R.; Kowarik, I.; Kronenberg, J.; et al. Advancing urban green infrastructure in Europe: Outcomes and reflections from the GREEN SURGE project. *Urban For. Urban Green.* **2018**, *40*, 4–16. [CrossRef]
10. Morison, P.J.; Brown, R.R. Understanding the nature of publics and local policy commitment to Water Sensitive Urban Design. *Landsc. Urban Plan.* **2011**, *99*, 83–92. [CrossRef]
11. Perales-Momparler, S.; Andrés-Doménech, I.; Hernández-Crespo, C.; Vallés-Morán, F.; Martín, M.; Escuder-Bueno, I.; Andreu, J. The role of monitoring sustainable drainage systems for promoting transition towards regenerative urban built environments: A case study in the Valencian region, Spain. *J. Clean. Prod.* **2017**, *163*, S113–S124. [CrossRef]
12. Gonzales, P.; Ajami, N.K. An integrative regional resilience framework for the changing urban water paradigm. *Sustain. Cities Soc.* **2017**, *30*, 128–138. [CrossRef]
13. Cai, D.; Fraedrich, K.; Guan, Y.; Guo, S.; Zhang, C.; Zhu, X. Urbanization and climate change: Insights from eco-hydrological diagnostics. *Sci. Total Environ.* **2019**, *647*, 29–36. [CrossRef] [PubMed]
14. Zhang, Z.; Meerow, S.; Newell, J.P.; Lindquist, M. Enhancing landscape connectivity through multifunctional green infrastructure corridor modeling and design. *Urban For. Urban Green.* **2019**, *38*, 305–317. [CrossRef]
15. Bortolini, L.; Semenzato, P.; Almási, B.; Csizmadia, D.; Kowalski, P.; Racoń-Leja, K.; Aarrevaara, E.; Scherzer, C. Multidisciplinary approaches for programming ecosystem services of urban green spaces. In Proceedings of the Acta Horticulturae, International Society for Horticultural Science (ISHS), Leuven, Belgium, 31 October 2018; pp. 411–414.
16. United Nations. *Transforming Our World: The 2030 Agenda for Sustainable Development*; United Nations: New York, NY, USA, 2015.
17. United Nations. *The Millennium Development Goals Report*; United Nations: New York, NY, USA, 2015.
18. European Commission. *Supporting the Implementation of Green Infrastructure*; European Commission: Brussels, Belgium, 2016.
19. European Commission. *The Multifunctionality of Green Infrastructure*; European Commission: Brussels, Belgium, 2012.
20. Huckleberry, J.K.; Potts, M.D. Constraints to implementing the food-energy-water nexus concept: Governance in the Lower Colorado River Basin. *Environ. Sci. Policy* **2019**, *92*, 289–298. [CrossRef]
21. Zhang, P.; Zhang, L.; Chang, Y.; Xu, M.; Hao, Y.; Liang, S.; Liu, G.; Yang, Z.; Wang, C. Food-energy-water (FEW) nexus for urban sustainability: A comprehensive review. *Resour. Conserv. Recycl.* **2019**, *142*, 215–224. [CrossRef]

22. Fan, J.-L.; Kong, L.-S.; Wang, H.; Zhang, X. A water-energy nexus review from the perspective of urban metabolism. *Ecol. Model.* **2019**, *392*, 128–136. [CrossRef]
23. Pappalardo, V.; La Rosa, D.; Campisano, A.; La Greca, P. The potential of green infrastructure application in urban runoff control for land use planning: A preliminary evaluation from a southern Italy case study. *Ecosyst. Serv.* **2017**, *26*, 345–354. [CrossRef]
24. Woods Ballard, B.; Wilson, S.; Udale-Clarke, H.; Illman, S.; Scott, T.; Ashley, R.; Kellagher, R. *The SuDS Manual*; CIRIA: London, UK, 2015; p. 968.
25. Fletcher, T.D.; Shuster, W.; Hunt, W.F.; Ashley, R.; Butler, D.; Arthur, S.; Trowsdale, S.; Barraud, S.; Semadeni-Davies, A.; Bertrand-Krajewski, J.-L.; et al. SUDS, LID, BMPs, WSUD and more—The evolution and application of terminology surrounding urban drainage. *Urban Water J.* **2015**, *12*, 525–542. [CrossRef]
26. Tota-Maharaj, K.; Scholz, M.; Coupe, S.J. Modelling Temperature and Energy Balances within Geothermal Paving Systems. *Road Mater. Pavement Des.* **2011**, *12*, 315–344. [CrossRef]
27. Del-Castillo-García, G.; Borinaga-Treviño, R.; Sañudo-Fontaneda, L.A.; Pascual-Muñoz, P. Influence of pervious pavement systems on heat dissipation from a horizontal geothermal system. *Eur. J. Environ. Civ. Eng.* **2013**, *17*, 956–967. [CrossRef]
28. Gupta, R.; Irving, R. Assessing the potential of ground source heat pumps to provide low-carbon heating and cooling in UK dwellings in a changing climate. In Proceedings of the Air Conditioning and the Low Carbon Cooling Challenge, London, UK, 27–29 July 2008; p. 14.
29. Nathanail, J.; Banks, V. Climate change: Implications for engineering geology practice. *Geol. Soc. Lond. Eng. Geol. Spec. Publ.* **2009**, *22*, 65–82. [CrossRef]
30. Price, S.J.; Terrington, R.L.; Busby, J.; Bricker, S.; Berry, T. 3D ground-use optimisation for sustainable urban development planning: A case-study from Earls Court, London, UK. *Tunn. Undergr. Sp. Technol.* **2018**, *81*, 144–164. [CrossRef]
31. Charlesworth, S.M.; Faraj-Llyod, A.S.; Coupe, S.J. Renewable energy combined with sustainable drainage: Ground source heat and pervious paving. *Renew. Sustain. Energy Rev.* **2017**, *68*, 912–919. [CrossRef]
32. Andrés-Valeri, V.C.; Sañudo-Fontaneda, L.A.; Rey-Mahía, C.; Coupe, S.J.; Álvarez-Rabanal, F.P. Thermal performance of wet swales designed as multifunctional green infrastructure systems for water management and energy saving. In Proceedings of the International Research Conference on Sustainable Energy, Engineering, Materials and Environment, Mieres, Spain, 25–27 September 2018; p. 18.
33. Abrahams, J.; Coupe, S.; Sañudo-Fontaneda, L.; Schmutz, U. The Brookside Farm Wetland Ecosystem Treatment (WET) System: A Low-Energy Methodology for Sewage Purification, Biomass Production (Yield), Flood Resilience and Biodiversity Enhancement. *Sustainability* **2017**, *9*, 147. [CrossRef]
34. Fardel, A.; Peyneau, P.-E.; Béchet, B.; Lakel, A.; Rodriguez, F. Analysis of swale factors implicated in pollutant removal efficiency using a swale database. *Environ. Sci. Pollut. Res.* **2019**, *26*, 1287–1302. [CrossRef]
35. Winston, R.; F Hunt, W.; Kennedy, S.; Wright, J. *Evaluation of Permeable Friction Course (PFC), Roadside Filter Strips, Dry Swales, and Wetland Swales for Treatment of Highway Stormwater Runoff*; North Carolina Department of Transportation: Raleigh, NC, USA, 2011.
36. Andrés-Valeri, V.C.; Castro-Fresno, D.; Sañudo-Fontaneda, L.A.; Rodriguez-Hernandez, J. Comparative analysis of the outflow water quality of two sustainable linear drainage systems. *Water Sci. Technol.* **2014**, *70*, 1341–1347. [CrossRef]
37. *Energy Saving Trust Domestic Ground Source Heat Pumps: Design and Installation of Closed-Loop Systems—A Guide for Specifiers, Their Advisors and Potential Users*; Energy Savin Trust™: London, UK, 2007; p. 24.
38. Steel, R.G.D.; Torrie, J.H.; Dickey, D.A. *Principles and Procedures of Statistics: A Biometrical Approach*, 3rd ed.; McGraw-Hill Series in Probability and Statistics; McGraw-Hill: New York, NY, USA, 1997; ISBN 9780070610286.
39. Novales, A. *Análisis de Regresión*; Universidad Complutense de Madrid: Madrid, Spain, 2010; p. 116.
40. Juan Luis, F.-M.; Zulima, F.-M.; Denys, B. The uncertainty analysis in linear and nonlinear regression revisited: Application to concrete strength estimation. *Inverse Probl. Sci. Eng.* **2018**, 1–25. [CrossRef]

© 2019 by the authors. Licensee MDPI, Basel, Switzerland. This article is an open access article distributed under the terms and conditions of the Creative Commons Attribution (CC BY) license (http://creativecommons.org/licenses/by/4.0/).

MDPI
St. Alban-Anlage 66
4052 Basel
Switzerland
Tel. +41 61 683 77 34
Fax +41 61 302 89 18
www.mdpi.com

Sustainability Editorial Office
E-mail: sustainability@mdpi.com
www.mdpi.com/journal/sustainability

www.ingramcontent.com/pod-product-compliance
Lightning Source LLC
LaVergne TN
LVHW070402100526
838202LV00014B/1370